Biocontrol of Major Grapevine Diseases: Leading Research

Biocontrol of Major Grapevine Diseases: Leading Research

Edited by

Stéphane Compant
AIT Austrian Institute of Technology, Tulln an der Donau, Austria

and

Florence Mathieu
Université de Toulouse, Castanet-Tolosan, France

www.cabi.org

CABI is a trading name of CAB International

CABI
Nosworthy Way
Wallingford
Oxfordshire OX10 8DE
UK

Tel: +44 (0)1491 832111
Fax: +44 (0)1491 833508
E-mail: info@cabi.org
Website: www.cabi.org

CABI
745 Atlantic Avenue
8th Floor
Boston, MA 02111
USA

Tel: +1 (617)682-9015
E-mail: cabi-nao@cabi.org

A catalogue record for this book is available from the British Library, London, UK.

Library of Congress Cataloging-in-Publication Data

Names: Compant, Stéphane, editor. | Mathieu, Florence, editor.
Title: Biocontrol of major grapevine diseases: leading research / edited by Stéphane Compant and Florence Mathieu.
Description: Boston, MA : CABI, [2016] | Includes bibliographical references and index.
Identifiers: LCCN 2015039032 | ISBN 9781780647128 (hbk : alk. paper)
Subjects: LCSH: Grapes--Diseases and pests--Biological control. | Grapes--Diseases and pests--Biological control--Research.
Classification: LCC SB608.G7 B56 2016 | DDC 634.8/22--dc23 LC record available at http://lccn.loc.gov/2015039032

ISBN-13: 978 1 78064 712 8

Commissioning editor: Rachael Russell
Editorial assistant: Emma McCann
Production editor: James Bishop

Typeset by SPi, Pondicherry, India
Printed and bound in the UK by CPI Group (UK) Ltd, Croydon, CRO 4YY

Contents

About the Book and the Editors ix
S. Compant and F. Mathieu

Preface xi
S. Compant and F. Mathieu

Acknowledgements xv

Contributors xvii

PART I GENERAL OVERVIEWS ON BIOCONTROL OF GRAPEVINE DISEASES

1 Major Biocontrol Studies and Measures Against Fungal and Oomycete Pathogens of Grapevine 1
A. Zanzotto and M. Morroni

2 Grapevine Endophytes and Plant Health: A Culture-Independent Approach 35
S. Yousaf, M. Anees and A. Campisano

PART II BIOCONTROL OF GREY MOULD

3 Ecology and Function of Grape-Associated Microorganisms with a Special Focus on Biocontrol of *Botrytis cinerea* 52
M. Cardinale and G. Berg

4 Beneficial Bacteria Prime Local and Systemic Immunity Against *Botrytis cinerea* in Grapevine 64
A. Aziz, B. Verhagen, S. Villaume, M. Höfte, F. Baillieul, C. Clément and P. Trotel-Aziz

5 Potential of Two *Bacillus* Antagonists for Biocontrol of Grey Mould 70
S. Ben-Maachia, R. Errakhi, F. Mathieu and A. Lebrihi

6 Isolation and Characterization of Antibiotics Produced by *Streptomyces* J-2 and their Role in Biocontrol of Plant Diseases, Especially Grey Mould 76
R. Errakhi, F. Bouteau, M. Barakate and A. Lebrihi

7 Use of *Saccharothrix algeriensis* NRRL B-24137 to Control *Botrytis cinerea*? 84
S. Muzammil, R. Saria, Z. Yu, C. Graillon, F. Mathieu, A. Lebrihi and S. Compant

8 DOPE-FISH, a Tool to Visualize Beneficial Bacteria Systemically Controlling *Botrytis cinerea*? 93
S. Compant, S. Muzammil and F. Mathieu

PART III BIOCONTROL OF DOWNY AND POWDERY MILDEWS

9 Endophytic Fungi in Asymptomatic *Vitis vinifera* L. and their Effects on *Plasmopara viticola* 98
S. Burruano, V. Mondello and G. Conigliaro

10 Enhancing the Aggressiveness of the Powdery Mildew Mycoparasite *Ampelomyces quisqualis* 113
D. Angeli, S. Micheli, M. Maurhofer and I. Pertot

11 Protection of Grapevines by *Pythium oligandrum* Strains Isolated from Bordeaux Vineyards Against Powdery Mildew 117
J. Gerbore, E. Bruez, J. Vallance, D. Grizard, C. Regnault-Roger and P. Rey

12 Bioactive Natural Products for Managing Downy Mildew Disease in Grapevine 125
M.T. Islam

PART IV BIOCONTROL OF GRAPEVINE TRUNK DISEASES

13 Biological Control of Grape Canker Pathogens 150
L. Costadone and W.D. Gubler

14 Esquive® WP Limits Development of Grapevine Trunk Diseases and Safeguards the Production Potential of Vineyards 160
E. Mounier, F. Boulisset, F. Cortes, M. Cadiou, P. Dubournet and E. Pajot

15 A Model for Biological Control Studies of Grapevine Trunk Diseases Under Laboratory Conditions 171
R.J.G. Pierron, J. Pouzoulet, A. Meziane, N. Mailhac and A. Jacques

PART V BIOCONTROL OF OTHER FUNGAL PATHOGENS AND FUNGAL PROBLEMS

16 Detection of Bacterial Endophytes in *Vitis vinifera* L. and Antibiotic Activity Against Grapevine Fungal Pathogens 182
S. Lo Piccolo, A. Alfonzo, S. Burruano and G. Moschetti

17 Impact of Grapevine Preharvest Treatments with Elicitor on the Occurrence and Toxigenesis of Ochratoxinogenic Fungi 191
C. Dachoupakan, C. Strub, V. Martinez, J.-C. Baccou and S. Schorr-Galindo

PART VI BIOCONTROL OF GRAPEVINE PHYTOPLASMAS

18 An Experimental Approach to Understanding Grapevine Yellows Associated with Phytoplasma Infections 210
G. Brader, S. Compant, H. Gangl and A. Sessitsch

19 Endophytic Bacteria Associated with Grapevine Plants: Putative Candidates for Phytoplasma Containment 215
D. Bulgari, P. Casati, F. Quaglino and P.A. Bianco

Conclusions 231
S. Compant and F. Mathieu

Index 233

About the Book and the Editors

For organic viticulture and integrated pest management (IPM) systems, research needs to be done on the development of new biocontrol agents as well as on the understanding of their use. The present volume provides a holistic overview of the advances in the use of biocontrol agents and plant resistance inducers to reduce various phytopathogenic infections of grapevine. The ecology, mechanisms involved in protection and application methods of these biocontrol agents and inducers are discussed in detail as applied to organic viticulture, IPM systems and different grapevine diseases, such as grey mould, trunk diseases, powdery and downy mildews and phytoplasmal diseases. A chapter is also included on the control of mycotoxin contamination of grapevines (chapter 17).

Some of the chapters are overviews that discuss in detail the reduction of infection by a particular phytopathogen using established biocontrol agents and plant resistance inducers. Other chapters describe new examples of biocontrol agents and the mechanisms that are involved in reducing the occurrence of several grapevine pathogen diseases. New models for understanding the mechanisms of action of biocontrol agents as well as reducing infection by pathogens are also presented. In addition, new technologies for understanding how microbes can protect plants are described, with discussion of their uses for organic farming and IPM systems as well as for future research in the field.

Key Features

1. The chapters are contributed by various experts, mainly scientists, from across the world. They provide diverse sets of knowledge on the use of new biocontrol strategies in relation to organic farming and IPM strategies, under both laboratory conditions and in the vineyard.
2. The application mechanisms involved are discussed in detail for specific biocontrol agents or plant resistance inducers to counteract infection caused by different phytopathogens.
3. Future directions in the biocontrol of grapevine diseases are discussed that can form the basis for further research.

The book is for students with advanced knowledge, industry and scientists.

The Editors
S. Compant
AIT Austrian Institute of Technology, Tulln an der Donau, Austria
and
F. Mathieu
Université de Toulouse, Castanet-Tolosan, France

Stéphane Compant is Associate Professor of microbiology, viticulture and quality of wines at the National Polytechnic Institute of Toulouse in France. He is currently working at the AIT Austrian Institute of Technology as Project Leader–Scientist on beneficial plant–microbe interactions and the biocontrol of plant diseases. Dr Compant is a member of two management committees of the European COST (European Cooperation in Science and Technology) actions, the first on beneficial microbes (endophytes, FA1103) and the second on sustainable control of grapevine trunk diseases (FA1303). His work is related to plant–microbe interactions as well as, in part, grapevine protection.

Florence Mathieu is a full Professor of microbiology at the National Polytechnic Institute of Toulouse in France. Dr Mathieu's work is related in part to the topic of beneficial bacteria and food safety.

Preface

Why produce a book about the biocontrol of major grapevine diseases? As the readers may know, more than 7.5 million ha of cultivated surfaces in the world are dedicated to grapevine, and 271 million hectolitres of wine are produced each year (FAOSTAT, 2011). Grapevine is, therefore, one of the most important economic crops. However, vineyards can suffer from various problems. In particular, grapevine plants can be infected by a large variety of pathogenic microorganisms, including fungi, oomycetes, viruses and bacteria (Gouadec *et al.*, 2007). These pathogens can have drastic effects on the host plants, on the berries and on wine quality (Compant *et al.*, 2013), as the organoleptic properties of the wine can be affected (Gouadec *et al.*, 2007); vine diseases can also result in important economic losses (van Helden, 2008).

Different methods have been used to counteract pathogenic infection. Pesticides, in particular, have been or are currently applied in the vineyard and, worldwide, around 35% of all pesticides are used in viticulture (Compant and Mathieu, 2011; Compant *et al.*, 2013). The use of these phytosanitary products means that outbreaks of vine pests or diseases can be avoided, as well as grape yields increased and the quality of wines ensured (Leroux, 2003; Pezet *et al.*, 2004).

However, pesticides have been overused, and the intensive use of phytosanitary products, such as chemical products, has been accompanied by an increasing awareness of the problems that they can cause. The intensive utilization of chemical products has led to their persistence in soils, the contamination of both soils and other parts of the environment, problems for human health, other non-target effects and also the appearance of pesticide-resistant pathogenic strains (Amaro and Mexia, 2003; Leroux, 2004).

To reduce the use of phytosanitary products, the use of genetically modified plants has been proposed (Ferreira *et al.*, 2004; Lemaire *et al.*, 2010), although this alternative strategy is still not widely accepted. Indeed, people in many regions of the world are generally not in favour of the cultivation of such crops (Marshall, 2009).

One other alternative strategy to reduce the use of chemical products in grapevine production is the use of biocontrol agents and of plant resistance inducers (Bent, 2006; van Helden, 2008). These approaches can aid in reducing the infection caused by different phytopathogens. Such strategies can be used in organic farming approaches as well as within an integrated pest management (IPM) strategy, and have minimal impact on the environment (van Helden, 2008). Some biocontrol agents and plant resistance inducers have already shown their potential when used in the field for the sustainable management of viticulture. Various products can now be found in the market. Thus, several new biocontrol agents can be used, and studied, to counteract diseases caused by several grapevine pathogens.

In this book, different authors from various countries describe which current – as well as new – biocontrol agents and plant resistance inducers can be used to control different grapevine pathogens. These accounts are presented with particular reference to the organic viticulture sector, but also in respect of IPM systems. The means by which biocontrol agents or plant resistance inducers can reduce grapevine diseases are also described in these chapters. Some chapters correspond to the use and understanding of the effects of these agents under laboratory conditions, whereas others demonstrate their field application. The reader will find a wealth of knowledge on the use of plant resistance inducers and biocontrol agents in the biocontrol of grey mould diseases, trunk diseases, downy mildews, powdery mildews, phytoplasmal diseases and mycotoxigenic fungi. The chapters include both reviews and original work, and some present new models for the biocontrol of grapevine diseases and provide important data for research and application.

There is now much more potential for reducing grapevine pathogens and diseases as well as other vineyard problems, as the reader will see in the different chapters.

The Editors
S. Compant
AIT Austrian Institute of Technology, Tulln an der Donau, Austria
and
F. Mathieu
Université de Toulouse, Castanet-Tolosan, France

References

Amaro, P. and Mexia, A. (2003) The pesticides very toxic to man, to natural enemies, to honey bees and to aquatic life must be prohibited or rigorously restricted for IPM in viticulture. *IOBC-WPRS Bulletin* 26(8), 277–282.

Bent, E. (2006) Induced systemic resistance mediated by plant growth-promoting rhizobacteria (PGPR) and fungi (PGPF). In: Tuzun, S. and Bent, E. (eds) *Multigenic and Induced Systemic Resistance in Plants*. Springer, New York, pp. 225–258.

Compant, S. and Mathieu, F. (2011) Bio-contrôle des maladies de la vigne avec des microbes bénéfiques. Une alternative à la chimie de synthèse en viticulture. *Revue des Œnologues* 141(Suppl.), 25.

Compant, S., Brader, G., Muzammil, S., Sessitsch, A., Lebrihi, A. and Mathieu, F. (2013) Use of beneficial bacteria and their secondary metabolites to control grapevine pathogen diseases. *BioControl* 58, 435–455.

FAOSTAT (2011) Data from the Statistics Division of the Food and Agriculture Organization of the United Nations, Rome. Available at: http://faostat.fao.org (accessed 14 September 2015).

Ferreira, R.B., Monteiro, S.S., Picarra-Pereira, M.A. and Teixeira, A.R. (2004) Engineering grapevine for increased resistance to fungal pathogens without compromising wine stability. *Trends in Biotechnology* 22, 168–173.

Gouadec, D., Blouin, J. and Groupe de Travail de la Vigne (eds) (2007) *Les Parasites de la Vigne: Stratégies de Protection Raisonnée*. Dunod, Paris.

Lemaire, O., Moneyron, A. and Masson, J.E. (2010) Interactive technology assessment and beyond: the field trial of genetically modified grapevines at INRA-Colmar. *PLoS Biology* 8(11): e1000551.

Leroux, P. (2003) Modes d'action des produits phytosanitaires sur les organismes pathogènes des plantes. *Comptes Rendus Biologies* 326, 9–21.

Leroux, P. (2004) Chemical control of *Botrytis* and its resistance to chemical fungicides. In: Elad, Y., Williamson, B., Tudzynski, P. and Delen, N. (eds) *Botrytis: Biology, Pathology and Control.* Kluwer, Dordrecht, The Netherlands, pp. 195–222.

Marshall, A. (2009) 13.3 million farmers cultivate GM crops. *Nature Biotechnology* 27, 221.

Pezet, R., Viret, O. and Gindro, K. (2004) Plant–microbe interaction: the *Botrytis* grey mould of grapes – biology, biochemistry, epidemiology and control management. In: Hemantaranjan, A. (ed.) *Advances in Plant Physiology*, Vol. 7. Scientific Publishers, Jodhpur, India, pp. 71–116.

van Helden, M. (2008) Protection intégrée: la protection intégrée vis-à-vis des ravageurs de la vigne. In: Kreiter, S. (ed.) *Ravageurs de la Vigne*, 2nd edn. Féret, Bordeaux, France, pp. 321–335.

Acknowledgements

We would like to acknowledge the contributions of all of the authors of the different book chapters. It would have not been possible to edit a book on the biocontrol of grapevine diseases without them. Special thanks, therefore, are due to them. Additional thanks are also given to Dr Robert W. Jackson of the University of Reading, UK, and to Dr Brian Murphy of Trinity College Dublin, Ireland, for checking some parts of this book, and to CABI. We also would like to thank people who have supported us, either personally or by providing us with funding for our research. Also, our thanks to the readers. We hope they will find the content of the book practically useful and that they will also be interested – or become even more interested – in the topic of the biocontrol of grapevine diseases.

Contributors

Antonio Alfonzo, Department SAF (Scienze Agrarie e Forestali), University of Palermo, Viale delle Scienze 4, 90128 Palermo, Italy. E-mail: antonio.alfonzo@unipa.it

Mariam Anees, Department of Biochemistry, Faculty of Biological Sciences, Quaid-i-Azam University, Islamabad, Pakistan. E-mail: mariam@qau.edu.pk and mariamanees@yahoo.com

Dario Angeli, Research and Innovation Centre, Fondazione Edmund Mach (FEM), Via E. Mach 1, 38010 S. Michele all'Adige (TN), Italy. E-mail: dario.angeli@fmach.it

Aziz Aziz, URVVC EA 4707, SDRP, University of Reims, F-51687 Reims Cedex 2, France. E-mail: aziz.aziz@univ-reims.fr

Jean-Claude Baccou, UMR QUALISUD 95 Joint Research Unit on Integrated Approach to Food Quality – Food Safety Team (CIRAD-UM, UAPV-SupAgro), cc023, Université de Montpellier, Place Eugène Bataillon, 34095 Montpellier Cedex 5, France. E-mail: jean-claude.baccou@umontpellier.fr

Fabienne Baillieul, URVVC EA 4707, SDRP, University of Reims, F-51687 Reims Cedex 2, France. E-mail: fabienne.baillieul@univ-reims.fr

Mustapha Barakate, Laboratoire de Microbiologie, Faculte des Sciences Semlalia, Université Cadi Ayyad, Marrakech, Morocco. E-mail: mbarakate.ucam.ac.ma

Sihem Ben-Maachia, LGC UMR 5503 (CNRS/INPT/UPS), Département Bioprocédés et Systèmes Microbiens, ENSAT-INP de Toulouse, Université de Toulouse, 1 Avenue de l'Agrobiopôle, BP 32607, F-31326 Castanet-Tolosan Cedex 1, France; and Centre Régional des Recherches en Agriculture Oasienne à Dégache, Tozeur, Tunisia. E-mail: sihem.benmaachia@gmail.com

Gabriele Berg, Institute of Environmental Biotechnology, Graz University of Technology, Petersgasse 12, 8010 Graz, Austria. E-mail: gabriele.berg@tugraz

Piero Attilio Bianco, Dipartimento di Scienze Agrarie e Ambientali, Università degli Studi di Milano, 20133 Milan, Italy. E-mail: piero.bianco@unimi.it

Florian Boulisset, Agrauxine, Groupe Lesaffre, 2 Rue Henri Becquerel, 49070 Beaucouze, France. E-mail: florian.boulisset@agrauxine.fr

François Bouteau, Laboratoire d'Électrophysiologies des Membranes, Institut de Biologie des Plantes, Université Paris Diderot, Sorbonne Paris Cité, 91405 Orsay, France. E-mail: fbouteau@gmail.com

Günter Brader, Bioresources Unit, Health and Environment Department, AIT Austrian Institute of Technology GmbH, 3430 Tulln an der Donau, Austria. E-mail: guenter.brader@ait.ac.at

Emilie Bruez, INRA, ISVV, UMR1065 Santé et Agroécologie du Vignoble (SAVE), F-33140 Villenave d'Ornon, France; and ISVV, UMR1065 SAVE, Bordeaux Sciences Agro, Université de Bordeaux, F-33140 Villenave d'Ornon, France. E-mail: emilie.bruez@bordeaux.inra.fr

Daniella Bulgari, Dipartimento di Scienze Agrarie e Ambientali, Università degli Studi di Milano, 20133 Milan, Italy. E-mail: daniela.bulgari@unimi.it

Santella Burruano, Department SAF (Scienze Agrarie e Forestali), University of Palermo, Viale delle Scienze 4, 90128 Palermo, Italy. E-mail: santella.burruano@unipa.it

Mickael Cadiou, Agrauxine, Groupe Lesaffre, 2 Rue Henri Becquerel, 49070 Beaucouze, France. E-mail: mickael.cadiou@agrauxine.fr

Andrea Campisano, Sustainable Agro-ecosystems and Bioresources Department, IASMA Research and Innovation Centre, Fondazione Edmund Mach (FEM), Via E. Mach 1, 38010 San Michele all'Adige (TN), Italy. E-mail: andrea.campisano@fmach.it

Massimiliano Cardinale, Institute of Applied Microbiology, Justus-Liebig-University, Heinrich-Buff-Ring 26-32, 35392 Giessen, Germany. E-mail: massimiliano.cardinale@umwelt.uni-giessen.de

Paola Casati, Dipartimento di Scienze Agrarie e Ambientali, Università degli Studi di Milano, 20133 Milan, Italy. E-mail: paola.casati@unimi.it

Christophe Clément, URVVC EA 4707, SDRP, University of Reims, F-51687 Reims Cedex 2, France. E-mail: christophe.clement@univ-reims.fr

Stéphane Compant, LGC UMR 5503 (CNRS/INPT/UPS), Département Bioprocédés et Systèmes Microbiens, ENSAT-INP de Toulouse, Université de Toulouse, 1 Avenue de l'Agrobiopôle, BP 32607, F-31326 Castanet-Tolosan Cedex 1, France. Present address: Bioresources Unit, Health and Environment Department, AIT Austrian Institute of Technology GmbH, 3430 Tulln an der Donau, Austria. E-mail: stephane.compant@ait.ac.at

Gaetano Conigliaro, Department SAF (Scienze Agrarie e Forestali), University of Palermo, Viale delle Scienze 4, 90128 Palermo, Italy. E-mail: gaetano.conigliaro@unipa.it

Felipe Cortes, Agrauxine, Groupe Lesaffre, 2 Rue Henri Becquerel, 49070 Beaucouze, France. E-mail: unanochedeinvierno@gmail.com

Laura Costadone, Department of Plant Pathology, University of California, Davis, CA 95616, USA. E-mail: costadone@ucdavis.edu

Cheewanun Dachoupakan, Department of Microbiology, Faculty of Science, Chulalongkorn University, Bangkok, 10330 Thailand. E-mail: cheewanun@hotmail.com

Patrice Dubournet, Bayer SAS, 16 rue Jean-Marie Leclair, CP310, 69337 Lyon Cedex 09, France. E-mail: patrice.dubournet@bayer.com

Rafik Errakhi, Plateforme de Biotechnologie, AGRONUTRITION, 3 Av. de l'Orchidée, 31390 Carbonne, France; LGC UMR 5503 (CNRS/INPT/UPS), Département Bioprocédés et Systèmes Microbiens, ENSAT-INP de Toulouse, Université de Toulouse, 1 Avenue de l'Agrobiopôle, BP 32607, F-31326 Castanet-Tolosan Cedex 1, France; and Laboratoire d'Électrophysiologie des Membranes, Institut de Biologie des Plantes, Université Paris Diderot, 91405 Orsay, France. Present address: Faculté des Sciences Meknès, Université Moulay Ismail, Meknès, Morocco. E-mail: r.errakhi@umi.ac.ma

Helmut Gangl, Bundesamt für Weinbau, Gölbeszeile 1, 7000 Eisenstadt, Austria. E-mail: office@ganglwines.at

Jonathan Gerbore, INRA, ISVV, UMR1065 Santé et Agroécologie du Vignoble (SAVE), F-33140 Villenave d'Ornon, France; BIOVITIS, 15400 Saint Etienne de Chomeil, France; and UMR CNRS 5254/IPREM-EEM, IBEAS, Université de Pau et des Pays de l'Adour, 64013 Pau, France. E-mail: jonathan.gerbore@sabiovitis.fr

Clotilde Graillon, LGC UMR 5503 (CNRS/INPT/UPS), Département Bioprocédés et Systèmes Microbiens, ENSAT-INP de Toulouse, Université de Toulouse, 1 Avenue de l'Agrobiopôle, BP 32607, F-31326 Castanet-Tolosan Cedex 1, France. E-mail: clotilde.graillon@hotmail.fr

Damien Grizard, BIOVITIS, 15400 Saint Etienne de Chomeil, France. E-mail: damien.grizard@sabiovitis.fr

Walter Douglas Gubler, Department of Plant Pathology, University of California, Davis, CA 95616, USA. E-mail: wdgubler@ucdavis.edu

Monica Höfte, Laboratory of Phytopathology, Faculty of Bioscience, University of Ghent, B-9000 Ghent, Belgium. E-mail: monica.hofte@ugent.be

Md. Tofazzal Islam, Department of Biotechnology, Bangabandhu Sheikh Mujibur Rahman Agricultural University, Gazipur-1706, Bangladesh. E-mail: tofazzalislam@yahoo.com

Alban Jacques, Equipe Agrophysiologie et Agromolécules, Département des Sciences Agronomiques et Agroalimentaires, INP-EI Purpan, Université de Toulouse, 75 Voie du T.O.E.C., BP 57611, 31076 Toulouse Cedex 3, France. E-mail: alban.jacques@purpan.fr

Ahmed Lebrihi, LGC UMR 5503 (CNRS/INPT/UPS), Département Bioprocédés et Systèmes Microbiens, ENSAT-INP de Toulouse, Université de Toulouse, 1 Avenue de l'Agrobiopôle, BP 32607, F-31326 Castanet-Tolosan Cedex 1, France. E-mail: lebrihi@ensat.fr or ahmed.lebrihi@ensat.fr

Sandra Lo Piccolo, Department SAF (Scienze Agrarie e Forestali), University of Palermo, Viale delle Scienze 4, 90128 Palermo, Italy. R-mail: sandra.lopiccolo@email.it

Nathalie Mailhac, Equipe Agrophysiologie et Agromolécules, Département des Sciences Agronomiques et Agroalimentaires, INP-EI Purpan, Université de Toulouse, 75 Voie du T.O.E.C., BP 57611, 31076 Toulouse Cedex 3, France. E-mail: nathalie.mailhac@yahoo.fr

Veronique Martinez, UMR QUALISUD 95 Joint Research Unit on Integrated Approach to Food Quality – Food Safety Team (CIRAD-UM, UAPV-SupAgro), cc023, Université de Montpellier, Place Eugène Bataillon 34095 Montpellier Cedex 5, France. E-mail: veronique.martinez@umontpellier.fr

Florence Mathieu, LGC UMR 5503 (CNRS/INPT/UPS), Département Bioprocédés et Systèmes Microbiens, ENSAT-INP de Toulouse, Université de Toulouse, 1 Avenue de l'Agrobiopôle, BP 32607, F-31326 Castanet-Tolosan Cedex 1, France. E-mail: mathieu@ensat.fr or florence.mathieu@ensat.fr

Monika Maurhofer, Institute of Integrative Biology, Swiss Federal Institute of Technology (ETH), 8092 Zurich, Switzerland. E-mail: monika.maurhofer@usys.ethz.ch

Adel Meziane, Equipe Agrophysiologie et Agromolécules, Département des Sciences Agronomiques et Agroalimentaires, INP-EI Purpan, Université de Toulouse, 75 Voie du T.O.E.C., BP 57611, 31076 Toulouse Cedex 3, France. E-mail: adel.mez@gmail.com

Susanna Micheli, Research and Innovation Centre, Fondazione Edmund Mach (FEM), Via E. Mach 1, 38010 S. Michele all'Adige (TN), Italy. E-mail: susanna.micheli@fmach.it

Vincenzo Mondello, Department SAF (Scienze Agrarie e Forestali), University of Palermo, Viale delle Scienze 4, 90128 Palermo, Italy. E-mail: vincenzo.mondello@unipa.it

Marco Morroni, Consiglio per la ricerca in agricoltura e l'analisi dell'economia agraria, Centro di ricerca per la Viticoltura, Viale XXVIII Aprile, 26, 31015 Conegliano (TV), Italy. E-mail: marco.morroni@gmail.com or morroni.marco@libero.it

Giancarlo Moschetti, Department SAF (Scienze Agrarie e Forestali), University of Palermo, Viale delle Scienze 4, 90128 Palermo, Italy. E-mail: giancarlo.moschetti@unipa.it

Emmanuelle Mounier, Agrauxine, Groupe Lesaffre, 2 Rue Henri Becquerel, 49070 Beaucouze, France. E-mail: emmanuelle.mounier@agrauxine.fr

Saima Muzammil, LGC UMR 5503 (CNRS/INPT/UPS), Département Bioprocédés et Systèmes Microbiens, ENSAT-INP de Toulouse, Université de Toulouse, 1 Avenue de l'Agrobiopôle, BP 32607, F-31326 Castanet-Tolosan Cedex 1, France. E-mail: saimamuzammil83@yahoo.com

Emmanuel Pajot, Agrauxine, Groupe Lesaffre, 2 Rue Henri Becquerel, 49070 Beaucouze, France. E-mail: emmanuel.pajot@agrauxine.fr

Ilaria Pertot, Research and Innovation Centre, Fondazione Edmund Mach (FEM), Via E. Mach 1, 38010 S. Michele all'Adige (TN), Italy. E-mail: ilaria.pertot@fmach.it

Romain J G. Pierron, Equipe Agrophysiologie et Agromolécules, Département des Sciences Agronomiques et Agroalimentaires, INP-EI Purpan, Université de Toulouse, 75 Voie du T.O.E.C., BP 57611, 31076 Toulouse Cedex 3, France. E-mail: rjg.pierron@gmail.com

Jérôme Pouzoulet, Equipe Agrophysiologie et Agromolécules, Département des Sciences Agronomiques et Agroalimentaires, INP-EI Purpan, Université de Toulouse, 75 Voie du T.O.E.C., BP 57611, 31076 Toulouse Cedex 3, France. E-mail: jerome.pouzoulet@ucr.edu

Fabio Quaglino, Dipartimento di Scienze Agrarie e Ambientali, Università degli Studi di Milano, 20133 Milan, Italy. E-mail: fabio.quaglino@unimi.it

Catherine Regnault-Roger, UMR CNRS 5254/IPREM-EEM, IBEAS, Université de Pau et des Pays de l'Adour, 64013 Pau, France. E-mail: catherine.regnault-roger@univ-pau.fr

Patrice Rey, INRA, ISVV, UMR1065 Santé et Agroécologie du Vignoble (SAVE), F-33140 Villenave d'Ornon, France; and ISVV, UMR1065 SAVE, Bordeaux Sciences Agro, Université de Bordeaux, F-33140 Villenave d'Ornon, France. E-mail: patrice.rey@agro-bordeaux.fr

Rayenne Saria, LGC UMR 5503 (CNRS/INPT/UPS), Département Bioprocédés et Systèmes Microbiens, ENSAT-INP de Toulouse, Université de Toulouse, 1 Avenue de l'Agrobiopôle, BP 32607, F-31326 Castanet-Tolosan Cedex 1, France. E-mail: s-rayenne@hotmail.com

Sabine Schorr-Galindo, UMR QUALISUD 95 Joint Research Unit on Integrated Approach to Food Quality – Food Safety Team (CIRAD-UM, UAPV-SupAgro), cc023, Université de Montpellier, Place Eugène Bataillon 34095 Montpellier Cedex 5, France. E-mail: sabrine.galindo@umontpellier.fr

Angela Sessitsch, Bioresources Unit, Health and Environment Department, AIT Austrian Institute of Technology GmbH, 3430 Tulln an der Donau, Austria. E-mail: angela.sessitsch@ait.ac.at

Caroline Strub, UMR QUALISUD 95 Joint Research Unit on Integrated Approach to Food Quality – Food Safety Team (CIRAD-UM, UAPV-SupAgro), cc023, Université de Montpellier, Place Eugène Bataillon 34095 Montpellier Cedex 5, France. E-mail: caroline.strub@umontpellier.fr

Patricia Trotel-Aziz, URVVC EA 4707, SDRP, University of Reims, F-51687 Reims Cedex 2, France. E-mail: patricia.trotel-aziz@univ-reims.fr

Jessica Vallance, INRA, ISVV, UMR1065 Santé et Agroécologie du Vignoble (SAVE), F-33140 Villenave d'Ornon, France; and ISVV, UMR1065 SAVE, Bordeaux Sciences Agro, Université de Bordeaux, F-33140 Villenave d'Ornon, France. E-mail: jessica.vallance@agro-bordeaux.fr

Bas Verhagen, URVVC EA 4707, SDRP, University of Reims, F-51687 Reims Cedex 2, France. E-mail: bas.verhagen@gmail.com

Sandra Villaume, URVVC EA 4707, SDRP, University of Reims, F-51687 Reims Cedex 2, France. E-mail: sandra.villaume@univ-reims.fr

Sohail Yousaf, Department of Environmental Sciences, Faculty of Biological Sciences, Quaid-i-Azam University, Islamabad, Pakistan. E-mail: syousaf@qau.edu.pk and rasy_1023@yahoo.com

Zhuoran Yu, LGC UMR 5503 (CNRS/INPT/UPS), Département Bioprocédés et Systèmes Microbiens, ENSAT-INP de Toulouse, Université de Toulouse, 1 Avenue de l'Agrobiopôle, BP 32607, F-31326 Castanet-Tolosan Cedex 1, France. E-mail: amaocat@gmail.com

Alessandro Zanzotto, Consiglio per la ricerca in agricoltura e l'analisi dell'economia agraria, Centro di ricerca per la Viticoltura, Viale XXVIII Aprile, 26, 31015 Conegliano (TV), Italy. E-mail: alessandro.zanzotto@entecra.it

1 Major Biocontrol Studies and Measures Against Fungal and Oomycete Pathogens of Grapevine

A. Zanzotto* and M. Morroni

Consiglio per la ricerca in agricoltura e l'analisi dell'economia agraria, Centro di ricerca per la Viticoltura, Conegliano, Italy

Introduction

Grapevine and major fungal and oomycetal diseases

The productivity of grapevine cultivation around the world is challenged by the harmful action of many plant pathogens, including fungi, oomycetes, prokaryotes (bacteria, phytoplasmas), arthropods (insects, mites), nematodes and viruses.

In this review, we report on and discuss information from the available literature about the biological control of some major grapevine (*Vitis vinifera* L.) diseases caused by fungi or oomycetes:

- downy mildew, caused by the oomycete *Plasmopara viticola* (Berk. & M.A. Curtis) Berl. and De Toni;
- powdery mildew, caused by the fungus *Erysiphe* (syn. *Uncinula*) *necator* (Schwein.) Burrill;
- grey mould, caused by the fungus *Botrytis cinerea* Pers.;
- diseases of grapevine woody tissue caused by the fungi *Eutypa lata* (Pers.) Tul. and C. Tul., Botryosphaeriaceae spp., *Phaeomoniella chlamydospora* (W. Gams, Crous, M.J. Wingf. and Mugnai) Crous and W. Gams, *Phaeoacremonium* spp. and *Fomitiporia mediterranea* M. Fisch, including esca disease of mature grapevines, which is associated with the latter three species.

Since the end of the Second World War, disease management has mostly been based on chemical pesticides (fungicides, insecticides, etc.), but the continuous use of these chemicals has raised widespread concern about their possible risks towards consumers, winegrowers and, also, bystanders. The collateral, negative effects of chemical pesticides on the environment were debated as well and have been the subject of recent European Union (EU) legislation (Directive 2009/128/EC, Regulation (EC) No. 1107/2009). The induction of resistance to the action of chemical pesticides

*Corresponding author: alessandro.zanzotto@entecra.it

in the pathogens, through repeated use of the same active ingredient, is another factor that has stimulated the use of alternative, non-chemical, control agents.

To support passage from a chemical-based disease control strategy towards sustainable agricultural practices, several microorganisms were tested for their ability to affect pathogen development in plant tissues. In addition, the mechanisms of some plant–pathogen interactions have been investigated in order to clarify the capacity of various chemical compounds to induce resistance reactions in the host plants.

Induced resistance

Resistance to pathogen infection can be limited to a few cells, as in localized acquired resistance (LAR), or it can be extended to other cells that are not directly affected by the pathogen, as in systemic acquired resistance (SAR) (Kessmann *et al.*, 1994; Sticher *et al.*, 1997). SAR is mediated by salicylic acid (SA) (or its chemical analogues) acting against a wide range of organisms and it is characterized by the expression of pathogenesis-related (PR) proteins (Durrant and Dong, 2004). In induced systemic resistance (ISR), the enhancement of plant defences is mediated by jasmonic acid (JA) and can be expressed after applications of non-pathogenic rhizobacteria, which are called plant growth-promoting rhizobacteria (PGPR) (Van Loon *et al.*, 1998; Pieterse *et al.*, 2001; Bakker *et al.*, 2007) or other biocontrol agents (BCAs).

ISR plant defences are primed but not really activated until the infection occurs, and then higher levels of defence gene products, reactive oxygen species (ROS) and phenolic compounds are produced (reviewed in Gessler *et al.*, 2011). Among these, resveratrol has an important role, both in plant defence and as a nutraceutical element. This compound has both pharmacological and antioxidant properties (Bavaresco *et al.*, 2012). Comprehensive descriptions of the complex mechanisms that form the basis of the various plant defence systems have recently been given in reviews by Fu and Dong (2013) and Gozzo and Faoro (2013) for SAR, by Pieterse *et al.* (2014) for ISR and the defence priming system, and by Conrath (2011) for molecular aspects of the defence priming system.

Many studies have highlighted the role in plant disease resistance mechanisms of several substances and microorganisms, and they were eventually considered for exploitation in crop protection. The possibilities for the practical application of induced resistance in disease management are discussed by Walters *et al.* (2013).

In the review presented in this chapter, the various substances that have been reported as showing effects against plant pathogens and that could be considered as putative alternatives to chemical fungicides, have been divided into two main groups: inorganic compounds; organic chemical inducers and natural extracts – the latter being further subdivided into plant, fungal and compost extracts. The last part of the chapter looks specifically at biocontrol agents – fungi, yeast-like fungi, bacteria and oomycetes.

Inorganic Compounds

Copper is a common element used against downy mildew in viticulture. Additionally to its direct effect on the pathogen, a Bordeaux mixture, which is based on copper sulfate – and to a lesser extent a copper hydroxide formulation – increased the level of plant defence indicators such as peroxidases (POXs), phenols, resveratrol and anthocyanins (Coulomb *et al.*, 1999). Aziz *et al.* (2006) demonstrated that copper sulfate can induce the synthesis of the phytoalexins *cis*- and *trans*-resveratrol, *cis*- and *trans*-ε-viniferine and *cis*- and *trans*-piceid in leaves of the Chardonnay grapevine cultivar, effects that were increased by joint treatment with chitosan oligomers; chitosan oligomers alone reduced the diameter of *B. cinerea* lesions and the amount of leaf surface infected by *P. viticola*, and these effects were increased by the joint administration of copper sulfate. Together with copper, Sulfur is a substance permitted in organic viticulture where it is used to control powdery mildew. **Aluminium chloride**, used in combination with seaweed extract in the commercial product Synermix®, has been shown to enhance resveratrol production in detached grapevine leaves and increase the efficiency of anti-*Botrytis* treatments in a field trial (Jeandet *et al.*, 2000).

Potassium silicate (Bowen *et al.*, 1992) and **calcium**, acting on the cell wall structure, have a role in increasing plant protection. Reynolds *et al.* (1996) reported that potassium silicate (K_2SiO_3) treatments of grapevine cv. Bacchus, reduced the incidence of powdery mildew in 2 out of 3 years of trials. Calcium chelates the cell wall pectic components of the host, reducing their degradation by *B. cinerea*, but without significant effect on the infection rate (Chardonnet *et al.*, 1997). Chardonnet *et al.* (1999) also observed that calcium can have a direct effect on the pathogen's cell walls. Miceli *et al.* (1999) reported that calcium application effectively controlled grey mould in bunches of vines grown outdoors, as well as in the cold room.

Nigro *et al.* (2006) obtained positive results in the control of *B. cinerea* on table grapes, using calcium chloride, **sodium bicarbonate** and other carbonate salts. Sodium bicarbonate was also tested by Karabulut *et al.* (2003) and found to effectively control postharvest diseases of grapes. Zerbetto *et al.* (2002, 2004), reporting results obtained in some vineyards treated with sodium bicarbonate and a mixture of sodium bicarbonate with paraffinic oil, observed that the treatment with sodium bicarbonate was ineffective against *P. viticola* and *B. cinerea*, while it positively protected the vineyard against *E. necator* in low-risk situations. With high disease pressure, better results were obtained with a mixture of sodium bicarbonate and paraffinic oil. Limited phytotoxic effects were observed on the leaves (0.25 to 2%), but interveinal necroses occurred at higher concentrations (6 and 9%).

The effective use of **potassium bicarbonate** against *E. necator* and *P. viticola* has also been reported by Sawant and Sawant (2008) and Dagostin *et al.* (2011), respectively, with some phytotoxic effect shown in the latter case. Positive results from trials using potassium bicarbonate against *B. cinerea* on grapevine have been presented by Di Martino *et al.* (2014) from field trials and Youssef and Ruffo Roberto (2014a), also of table grapes; in the latter case also with the use of potassium sorbate. A possible induction of resistance was hypothesized after the postharvest incidence of grey mould was reduced by potassium sorbate in 2 out of 3 years of trials (Feliziani *et al.*, 2013). Effective control of grey mould on table grapes was also observed by Youssef and Ruffo Roberto (2014b) with three other salts (sodium silicate, iron sulfate and ammonium bicarbonate), also after preharvest and postharvest treatments.

Organic Chemical Inducers and Natural Extracts

Details of the use of these agents in the control of grapevine diseases are summarized in Table 1.1.

Table 1.1. The use of organic chemical inducers and natural extracts in the control of grapevine diseases.

Inducer/extract	Notes	Field tested[a]	Examples of common product names	Reference(s)
Grey mould				
Chitosan	Concentration dependent; decrease of efficacy under high disease pressure	Yes	Armour-Zen®, Chitogel®, Elexa®, Chito Plant®	Ait Barka *et al.* (2004), Amborabé *et al.* (2004), Aziz *et al.* (2006), Trotel-Aziz *et al.* (2006), Reglinski *et al.* (2010), Romanazzi *et al.* (2012), Feliziani *et al.* (2013)
Laminarin	Reduction in lesion size	No	–	Aziz *et al.* (2003)
Oligogalacturonides (OGAs)	Dose dependent	No	–	Aziz *et al.* (2004), De Miccolis Angelini *et al.* (2009)
Rhamnolipids	Good stimulator of plant defence	No	–	Varnier *et al.* (2009)
Compost extracts	Growth chamber	No	–	Elad and Shtienberg (1994)
Reynoutria sachalinensis[b]	Efficacy similar to sulfur or copper	Yes	Milsana®	Elmer and Reglinski (2006)
Saccharomyces[c]	–	Yes	Romeo™ (in progress)	Pujos *et al.* (2014)
Seaweed[b] + $AlCl_3$	Enhances efficacy of iprodione	Yes	Synermix®	Jeandet *et al.* (2000)
Viola odorata[b]	Not yet tested on grapevine	No	–	Hammami *et al.* (2011a)
Downy mildew				
Beta-aminobutyric acid (BABA)	Able to reduce sporulation	Yes	–	Cohen *et al.* (1999), Reuveni *et al.* (2001), Hamiduzzaman *et al.* (2005), Dubreuil-Maurizi *et al.* (2010)
Chitosan	Possibility to use different molecules/oligomers	Yes	Chito Plant®	Aziz *et al.* (2006), Dagostin *et al.* (2011), Romanazzi *et al.* (2014)
Laminarin	Reduction in lesion size, effects on stomatal closure	Yes	Frontiere®	Aziz *et al.* (2003), Trouvelot *et al.* (2008), Allègre *et al.* (2009), Romanazzi *et al.* (2014)

Rhamnolipids	Provided control	Yes	ZONIX™	Dagostin *et al.* (2011)
Salicylic acid	Disease reduction; concn > 2 mM can be phytotoxic	Yes	–	Kast (2000), Elmer and Reglinski (2006), Tamm *et al.* (2011)
Aqueous/hydroalcoholic solutions of extracts of various plant species	Good effect in combination with copper	Yes (pots)	–	Arnault *et al.* (2013)
Ascophyllum nodosum[b]	Tested on potted vines	No	–	Lizzi *et al.* (1998)
Compost extract	Tested on detached leaves/no effect	–	–	Sackenheim *et al.* (1994)/Thuerig *et al.* (2011)
Frangula alnus[b]	Fungitoxic and fungistatic	No	–	Godard *et al.* (2009)
Inula viscosa[b]	Efficacy variable	Yes	–	Cohen *et al.* (2006), Dagostin *et al.* (2011)
Melaleuca alternifolia[b]	Various degrees of efficacy	Yes	Timorex®, BM-608	Dagostin *et al.* (2011), La Torre *et al.* (2014)
Penicillium chrysogenum[c]	Different results	Yes	–	Thuerig *et al.* (2006), Harm *et al.* (2011)
Rheum palmatum[b]	Fungitoxic and fungistatic	No	–	Godard *et al.* (2009)
Saccharomyces[c]	–	Yes	K&A Oomisine®, Romeo™[,d]	Pujos *et al.* (2014)
Salvia officinalis[b]	Efficacy reduced by rainfall	Yes	–	Dagostin *et al.* (2010)
Solidago canadensis[b]	Tested on potted vines	No	–	Harm *et al.* (2011)
Yucca schidigera[b]	Efficacy variable	Yes	–	Dagostin *et al.* (2011)
Powdery mildew				
Chitosan	Possibility to use different molecules/oligomers	Yes	Kendal-COPS®	Iriti *et al.* (2011), Van Aubel *et al.* (2014)
Jasmonates	Enhanced tolerance	Yes	–	Belhadj *et al.* (2006)
P. chrysogenum[c]	As effective as sulfur and copper	Yes	–	Thuerig *et al.* (2006)
Reynoutria sachalinensis[b]	As effective as sulfur and copper	Yes	Milsana®, Sakalia®[,d]	Elmer and Reglinski (2006), Konstantinidou-Doltsinis *et al.* (2007), Ortugno *et al.* (2014)
Saccharomyces[c]	–	Yes	Romeo™[,d]	Pujos *et al.* (2014)
Trunk diseases				
Ascophyllum nodosum[b]	–	Yes	Marvita®	Di Marco (2010)

[a]In at least one of the reviewed papers; [b]Indicates plant extract; [c]Indicates fungal extract; [d]In registration.

Beta-aminobutyric acid (BABA) is an isomer of aminobutyric acid and a non-protein amino acid that occurs only rarely in plants; it acts as a resistance inducer in BABA-induced resistance (BABA-IR) mechanism (Jakab *et al.*, 2001). Cohen *et al.* (1999) reported good activity of BABA against downy mildew on grapevine leaf discs, with sporulation reduced by 85–94%, and with higher concentrations able to protect potted plants. Field evaluation has been conducted on grapevine leaves of cvs Chardonnay and Cabernet Sauvignon; these cultivars were effectively protected against downy mildew when BABA or a mixture of BABA and different fungicides were used (Reuveni *et al.*, 2001). Hamiduzzaman *et al.* (2005) indicated that protection against downy mildew induced by BABA was induced by the deposition of callose and the potentiation of *PR* (pathogenesis-related) gene expression. Dubreuil-Maurizi *et al.* (2010) observed that in grapevine leaves treated with BABA, there was a stronger production of ROS after *P. viticola* infection.

Chitosans are natural non-toxic polymers of β-1,4-linked D-glucosamine and *N*-acetyl-D-glucosamine produced by treating (deacetylating) chitin from crustacean shells and fungal cell walls with sodium hydroxide. Their biological activity is related to the size of the oligomer, degree of deacetylation and number of amino groups (Kauss *et al.*, 1989; Hadwiger *et al.*, 1994). Repka (2001) observed an accumulation of PR proteins after the treatment of grapevine cell suspension cultures with chitosan. New insights into the structure–activity relationships of chitosan have been recently presented by Sahariah *et al.* (2014). The *in vitro* activity of chitosan against *B. cinerea* on excised Chardonnay cultivar leaves was reported by Trotel-Aziz *et al.* (2006), and by Aziz *et al.* (2006), who also reported the activity of chitosan against *P. viticola.* In the first study, treatments of the leaves induced the activity of lipoxygenase, phenylalanine ammonia-lyase (PAL) and chitinase. The best protection was achieved at the dose of 75–100 mg l^{-1}. In the second study, the authors compared the effects of treating the leaves with chitosan oligomers either alone or in combination with copper sulfate. Chitosan oligomers induced the stimulation of chitinase and glucanase activities, in addition to an accumulation of stilbene phytoalexins. When chitosan oligomers were used in combination with copper sulfate, higher production of phytoalexins and increased resistance were observed. The results also seemed to indicate that the action of chitosan is related to the induction of resistance rather than a direct effect on the pathogens.

Results obtained by Reglinski *et al.* (2010) supported the work of Trotel-Aziz *et al.* (2006) in indicating that chitosan acted in a concentration-dependent manner, but also reporting a direct antifungal activity. The same authors tested chitosan in combination with *Ulocladium oudemansii* for the control of grey mould in the vineyard, and demonstrated effective control under conditions of low disease pressure. Interestingly, the application of chitosan was correlated with an increased activity of POX, an enzyme catalysing reactions that strengthen plant cell walls. The application of a chitosan solution to vines showed both preventive and curative control of infection by *B. cinerea* (Amborabé *et al.*, 2004). Iriti *et al.* (2009) observed a partial stomatal closure in bean leaves induced by chitosan, thus making it an antitranspirant compound. Alternative methods to control postharvest grey mould on table grapes have been reviewed by Romanazzi *et al.* (2012), and further research by Feliziani *et al.* (2013) described the effects of preharvest chitosan treatments using different commercial formulations of chitosan on the quality and postharvest decay of table grapes.

In field trials, chitosan effectively reduced the severity of infection by *P. viticola* (Romanazzi *et al*., 2014), but showed phytotoxicity at high doses (Dagostin *et al*., 2011). Iriti *et al*. (2011) reported an increase of total polyphenol content and antioxidant power in the berry tissues of a *V. vinifera* cultivar treated with Kendal-COPS® (a formulation based on chitosan, Cu and Mn), together with a reduction in powdery mildew severity on leaves and bunches. Van Aubel *et al*. (2014), using a complex molecule formed by the combination of oligopectin and oligochitosan, reported positive results in protection against powdery mildew.

A product named Chitogel® enhanced the development of cv. Chardonnay plantlets and inhibited the growth of *B. cinerea* (Ait Barka *et al*., 2004). Use of the products Elexa®, Chito Plant® and Armour-Zen® have been cited in reviews by Elmer and Reglinski (2006) and Romanazzi *et al*. (2012).

Jasmonic acid is a plant hormone, and this substance and its derivative, **methyl jasmonate**, are involved in the regulation of several plant processes (reviewed in Avanci *et al*., 2010). They are concerned with the activation of plant defence responses, such as the elicitation of several defence-related genes, and with the accumulation of PR proteins, callose deposition, hypersensitive response-like cell death and stilbene phytoalexin biosynthesis (Repka *et al*., 2004, 2013; Vezzulli *et al*., 2007; Faurie *et al*., 2009). Belhadj *et al*. (2006), after treating cv. Cabernet Sauvignon with methyl jasmonate, obtained an enhanced tolerance against powdery mildew.

Laminarin is a glucan made up of glucose units linked by glycosidic bonds, and is extracted from the brown seaweed *Laminaria digitata* (Vera *et al*., 2011). It has been reported to induce the activation of defence-related genes in grape cells, and to elicit resistance to *B. cinerea* and *P. viticola* (Aziz *et al*., 2003). Sulfated laminarin induces priming mechanisms for resistance (Trouvelot *et al*., 2008). Allègre *et al*. (2009) studied the effects of laminarin in the *P. viticola*–*V. vinifera* pathosystem and observed that the elicitor induced the closure of stomata but that this effect alone was not sufficient to protect the leaves from downy mildew. Romanazzi *et al*. (2014) tested the effectiveness of laminarin against downy mildew in field trials, and observed a lack of protection when it was used alone but better results when it was used in combination with copper salts at lower doses and *Saccharomyces* spp. extracts.

Oligogalacturonides (OGAs) are released from plant cell walls and are homopolymers of D-galacturonic acid (Reymond *et al*., 1995). They have the ability to induce resistance to *B. cinerea* in detached grapevine leaves in a dose-dependent relationship (Aziz *et al*., 2004), triggering many defence-related genes as well as increasing chitinase and glucanase activities. The mode of action of OGAs also involves oxidative burst and protein phosphorylation mechanisms. De Miccolis Angelini *et al*. (2009), in a short review, cited promising results obtained with plant-derived elicitors, such as oligogalacturonides and cellodextrins.

Rhamnolipids are glycolipid biosurfactants produced by bacteria (reviewed in Vatsa *et al*., 2010). Their protective effects against *B. cinerea* have been evaluated on cell suspension cultures and *in vitro* plantlets by Varnier *et al*. (2009). Ca^{2+} influx, mitogen-activated protein kinase (MAPK) activation, ROS production and the induction of defence gene expression and the hypersensitive response (HR) are involved in the resistance mechanism. They act as MAMPs (microbe-associated molecular patterns), in that they are recognized by plants and induce a defence response involving different signalling sectors according to the type of pathogen (Sanchez *et al*., 2012). A direct effect on the fungal pathogen concerned has also been identified on spore germination and mycelium growth (Varnier *et al.*, 2009). Dagostin *et al.* (2011) observed good control of the severity of *P. viticola* infection in field trials with the commercial product ZONIX™.

Salicylic acid (SA) is a plant hormone implicated in the coordination of plant disease resistance, and induces PR proteins in grapevine (reviewed in Elmer and Reglinski, 2006). Tamm *et al.* (2011) reported that, under controlled conditions, SA reduced downy mildew incidence on leaves. A small reduction in downy mildew severity (max. 50%) was observed in field trials using 0.2% SA (Kast, 2000) but, as reported by Elmer and Reglinski (2006), a concentration higher than 2 mM may cause phytotoxicity, thus raising problems for its practical use.

Plant extracts

Reynoutria sachalinensis (giant knotweed) extracts induce defence reactions in various crops (Schmitt, 2006). Elmer and Reglinski (2006) reported positive results with a formulation of *R. sachalinensis* extract (Milsana®), against grapevine powdery mildew and grey mould, obtained by some researchers in Germany. Milsana® was also evaluated by Konstantinidou-Doltsinis *et al.* (2007) in vineyard trials in which a significant reduction of powdery mildew disease severity was observed; the product was more effective when applied at the early stages of disease development. The effectiveness of a new formulation of *R. sachalinensis* extract against *E. necator* in vineyard trials was reported by Ortugno *et al.* (2014). *R. sachalinensis* extract has recently been added under code P5 (host plant defence inducers – plant extracts) to the Fungicide Resistance Action Committee (FRAC) list (FRAC, 2015).

Lizzi *et al.* (1998) reported a positive effect of extracts of the brown seaweed ***Ascophyllum nodosum*** against *P. viticola* in laboratory and greenhouse experiments, in which it acted as a resistance inducer on grapevine leaves. Di Marco (2010) gave a preliminary report on the activity of the commercial product Marvita® (which is based on *A. nodosum* extract) against esca disease. As mentioned above, a formulation of **seaweed extract** and **aluminium chloride** induced resveratrol accumulation in detached grapevine leaves. In vineyard trials, the efficacy of iprodione against *B. cinerea* was strengthened by the use of Synermix® (Jeandet *et al.*, 2000).

The antifungal activity of ***Melaleuca alternifolia*** (tea tree) components was investigated by Hammer *et al.* (2003). The effectiveness of some tea tree extracts against downy mildew under controlled conditions was reduced in vineyard applications (Dagostin *et al.*, 2011). La Torre *et al.* (2014) observed a positive effect of a product containing 23.8% tea tree oil against downy mildew, but with lower effectiveness than that of a cupric reference product in a field trial test. *M. alternifolia* extract is also reported in the FRAC (2015) list under the code F7 (lipid synthesis and membrane integrity – proposed cell membrane disruption).

The essential oil from ***Viola odorata*** flowers has been tested *in vitro* against *B. cinerea* (Hammami *et al.*, 2011a), who obtained a strong inhibition of the pathogen, even at low concentrations. Harm *et al.* (2011) tested the efficacy of ***Solidago canadensis*** extract as an inducer of a defence reaction against *P. viticola* in potted vines grown outdoors and obtained more than 80% protection.

Arnault *et al.* (2013) reported the preliminary results of tests with hydroalcoholic solutions (***Salix alba*, *Equisetum arvense*, *Artemisia vulgaris***) or aqueous solutions (***Frangula alnus*, *Rheum palmatum***) of extracts of various plant species against *P. viticola,* in field conditions. Three products (from *F. alnus* and *E. arvense*, fructose solutions), combined with copper at 100g ha^{-1} dose, were as effective as copper alone at 600 g ha^{-1}; no effects were reported with the *A. vulgaris* solution. Godard *et al.* (2009) evaluated the effectiveness of ***Rheum palmatum*** roots and ***Frangula alnus*** bark extracts, reporting fungitoxic and defence reaction effects against *P. viticola* on leaves and increased stilbenic phytoalexins and POX activity.

The antifungal activity of a *Salvia officinalis* (sage) alcoholic extract against *P. viticola,* on potted grapevine plants, was investigated by Dagostin *et al.* (2010). The extract gave effective control on the potted plants but in field trials its efficacy was lower, as its persistence was reduced by rainfall. In a large-scale study, Dagostin *et al.* (2011) also investigated the use of ***Yucca schidigera*** extract against downy mildew, and considered it to be worth further development. Another extract tested was that of ***Inula viscosa***; this provided good control over *P. viticola* in one out of two field trials, but caused widespread phytotoxicity. Cohen *et al.* (2006), after growth chamber tests, reported that an oily paste extract of *I. viscosa* applied to grapevine leaves was effective in controlling downy mildew infections.

Fungal extracts

An extract of ***Penicillium chrysogenum*** was reported to control downy and powdery mildew as copper and sulfur fungicides or resistance inducers, under greenhouse and vineyard conditions (Thuerig *et al.*, 2006). In certain conditions phytotoxic side-effects were observed. Since the fungal extract had no direct fungicidal effects, its protecting properties are likely to depend on the activation of defence mechanisms. Also Harm *et al.* (2011) tested the effectiveness of *P. crysogenum* extracts against downy mildew on grapevine, observing a minimal protection but the induction of a wide range of resistance-related metabolites.

A commercial formulation containing ***Saccharomyces cerevisiae*** extracts is available with the name of K&A Oomisine® to stimulate the natural control of *P. viticola* on grapevine. A new product containing cerevisane, an inert fraction from a selected strain of *S. cerevisiae,* gave interesting results against downy mildew, powdery mildew and grey mould after vineyard trials carried out in Italy (Pujos *et al.*, 2014).

Compost extracts

The effect of water extracts of fermented composts from animal and plant sources (cattle manure, chicken–cattle manure and grape marc compost) was tested by Elad and Shtienberg (1994), who observed the reduction of *B. cinerea* infections on grape berries in growth chamber experiments. Sackenheim *et al.* (1994) observed a positive result in the control of *P. viticola* on grapevine cuttings in a growth chamber after the application of aqueous extracts of composted microbiologically active substrates.

The use of grape marc extract to elicit plant defence responses has been recently investigated by Goupil *et al.* (2012). Treatments of tobacco leaves with marc extracts of red *V. vinifera* cultivars induced HR-like effects with local and systemic up-regulation of PR1 and PR2 encoding genes. On the other hand Thuerig *et al.* (2011) studied the possibility of a site-specific resistance modulation by the application of organic amendments, but they did not observe any difference in leaf susceptibility to *P. viticola.*

Biocontrol Agents (BCAs)

Plants interact with many microbial organisms living inside the plants, as endophytes, or on the external surface, as epiphytes. Special attention has been given to fungi and bacteria, acting as biocontrol agents, plant growth-promoting agents and resistance inducers.

Fungi

Details of the use of fungi in the biocontrol of grapevine diseases are summarized in Table 1.2.

***Acremonium* spp.:** *A. byssoides* is an endophytic fungus, isolated from grapevine leaves by Burruano *et al.* (2008), who evaluated its activity against *P. viticola*. The biocontrol agent (BCA) parasitized the pathogen, with invasion of the mycelium and deformation of the asexual structures. Culture filtrates and extracts from *A. byssoides* completely inhibited the development of *P. viticola* sporangia. *A. cephalosporium* has been tested as a BCA against *Botrytis*, *Aspergillus* and *Rhizopus* rots by Zahavi *et al.* (2000).

***Alternaria* spp.:** Fungi belonging to this genus are ubiquitous. *A. alternata* has been observed to inhibit the sporulation of *P. viticola* on grapevine leaves (Musetti *et al.*, 2006) and the production of secondary metabolites, such as diketopiperazines, has also been reported (Musetti *et al.*, 2007).

***Ampelomyces quisqualis*:** This species includes many strains that are natural antagonists of powdery mildews, such as *E. necator* (Kiss, 1998), although it should be noted that it has been concluded after a series of molecular studies that *A. quisqualis* should not be considered as a single species, but as a species complex (Kiss and Nakasone, 1998). More recently, Angeli *et al.* (2012) also observed the existence of different physiological forms within the *A. quisqualis* species. The parasitization of *E. necator* cleistothecia and mycelium by *A. quisqualis* has been reported by Falk *et al.* (1995). One strain – M10 – has been commercialized as AQ10® and has been tested in vineyards, with good results under low–medium disease pressure and in alternation with systemic fungicides. When used in alternation with sulfur under high disease pressure, the efficacy was significantly lower (Monchiero *et al.*, 1996; Zanzotto *et al.*, 2005). In some experiments, late treatments with *A. quisqualis* caused a reduction in the production of *E. necator* chleistothecia (Zanzotto *et al.*, 2005; Caffi *et al.*, 2010; Legler *et al.*, 2011).

***Aphanocladium album*:** *A. album* is a BCA that both expresses a mycoparasitic activity and produces chitinolytic enzymes (Kuntz *et al.*, 1992). Ciccarese *et al.* (2006) observed good activity when it was applied to grapes preharvest against rot diseases caused by *B. cinerea* and other microbial agents.

***Chaetomium* spp.:** The *in vitro* activity of this ascomycete was preliminarily reported by Spagnolo *et al.* (2012) in the biocontrol of grapevine trunk diseases. The

Table 1.2. Fungal biocontrol agents (BCAs) for the treatment of grapevine diseases.

BCA	Notes	Field tested[a]	Examples of common product names	Reference(s)
Grey mould				
Acremonium cephalosporium	Variable effectiveness on bunch rots	Yes	–	Zahavi *et al.* (2000)
Aphanocladium album	Good activity when applied preharvest	Yes	–	Ciccarese *et al.* (2006)
Epicoccum nigrum	Different results	Yes	–	Fowler *et al.* (1999), Elmer and Reglinski (2006)
Gliocladium spp.	Different results	Yes	–	Elmer and Reglinski (2006)
Trichoderma spp.	Various degrees of efficacy	Yes	Trichodex®	Elmer and Reglinski (2006), Harman *et al.* (1996), O'Neil *et al.* (1996), Woo *et al.* (2014)
Ulocladium spp.	High disease pressure could reduce its efficacy	Yes	Botry-Zen®	Fowler *et al.* (1999), Elmer *et al.* (2005), Elmer and Reglinski (2006), Calvo-Garrido *et al.* (2014a)
Downy mildew				
Acremonium byssoides	Hyperparasite	No	–	Burruano *et al.* (2008)
Alternaria spp.	Inhibit sporulation	No	–	Musetti *et al.* (2006)
E. nigrum	Questionable	No	–	Kortekamp (1997)
Fusarium proliferatum	More effective in reducing severity than incidence	Yes	–	Falk *et al.* (1996), Bakshi *et al.* (2001)
Trichoderma spp.	–	Yes	Trichodex®	Perazzolli *et al.* (2008), Dagostin *et al.* (2011), Palmieri *et al.* (2012), Banani *et al.* (2014), Woo *et al.* (2014)
Powdery mildew				
Ampelomyces quisqualis	Reduced efficacy with high disease pressure	Yes	AQ10®	Monchiero *et al.* (1996), Zanzotto *et al.* (2005), Caffi *et al.* (2010), Legler *et al.* (2011)
Verticillium lecanii	Hyperparasite	No	–	Heintz and Blaich (1990)

Continued

Table 1.2. Continued.

BCA	Notes	Field tested[a]	Examples of common product names	Reference(s)
Trunk diseases				
Chaetomium spp.	Active against some strains of *Neofusicoccum parvum*, *Diplodia seriata*, *Phaeomoniella chlamydospora*, *Phaeoacremonium aleophilium*	No	–	Spagnolo *et al.* (2012)
F. lateritium	Wound protectant	Yes	–	John *et al.* (2005)
Trichoderma spp.	Active against wood diseases and as wound protectant	Yes	Trichoseal®, Biotricho®, Eco 77®, Esquive® WP, Patriot Dry®, Rootshield®, Remedier®, Vinevax®	John *et al.* (2005), Di Marco and Osti (2007), Pertot *et al.* (2009), Halleen *et al.* (2010), Kotze *et al.* (2011), Mutawila *et al.* (2011, 2013), Pajot *et al.* (2012), Reggiori *et al.* (2014), Woo *et al.* (2014)

[a]In at least one of the reviewed papers.

strain Cha1, isolated from asymptomatic trunk wood, was able to overgrow colonies of two *Neofusicoccum parvum* and *Diplodia seriata* strains and one strain each of both *Phaeomoniella chlamydospora* and *Phaeoacremonium aleophilum*.

Epicoccum nigrum: The effectiveness of this BCA against downy mildew was investigated by Kortekamp (1997), who eventually questioned the success of biological control by this species. *E. nigrum* gave good performances against *B. cinerea* (reviewed in Elmer and Reglinski, 2006). Suppression of infection and sporulation on detached rachii was observed, but field tests on the capacity of this BCA to reduce *B. cinerea* overwintering inoculum gave weak and inconsistent results (Fowler *et al.*, 1999). The saprophytic activity of *E. nigrum* is considered to be its primary mode of action, but an effect due to the production of antimicrobial metabolites has also been hypothesized (reviewed in Elmer and Reglinski, 2006).

Fusarium proliferatum: *F. proliferatum* has been reported as an integrative BCA against grapevine downy mildew (Falk *et al.*, 1996); multi-year vineyard treatments on two susceptible *Vitis* interspecific hybrids significantly reduced infection severity in some years. *F. proliferatum* was more effective in reducing disease severity than disease incidence. The authors considered that it might be used together with other products, such as sulfur or copper, on poorly susceptible varieties or in areas where the disease is not severe. A cold-tolerant strain of *F. proliferatum* (isolate 1505) produced through UV mutagenesis (Bakshi *et al.*, 2001) was demonstrated to be capable of controlling *P. viticola* on detached grapevine leaves. Pruning wounds treated with *Fusarium lateritium* 1–14 days before artificial infection by *Eutypa lata* reduced recovery of the pathogen (John *et al.*, 2005).

Gliocladium roseum **(now reclassified as** ***Clonostachys rosea*****):** Sutton *et al.* (1997) reviewed the activity of this BCA against *B. cinerea* on various crops. Elmer and Reglinski (2006) reviewed two studies that evaluated the activity of members of *Gliocladium* spp. against *B. cinerea* on grapes. Both were published in 1998, but reported contrasting results. The mode of action of *Gliocladium* spp. as a BCA has been studied on different crops, and the results indicated antibiosis and mycoparasitism as possible mechanisms of action.

Trichoderma **spp.:** *Trichoderma* spp. are soil inhabiting ascomycota that are distributed worldwide, and colonize various substrates as well as the plant rhizosphere, acting as antagonists of several plant pathogens (reviewed in Verma *et al.*, 2007; Sawant, 2014). Their biocontrol mechanism is based on the production of lytic enzymes and other secondary metabolites, and has been reviewed in Reino *et al.* (2008). *Trichoderma* spp. have positive effects on their host, enhancing growth and inducing plant defence responses (Harman, 2006; Vinale *et al.*, 2008); Woo *et al.* (2006) reported that enzymes produced by the species cause the release of low molecular weight oligosaccharides that induce resistance.

In recent decades, a wide variety of *Trichoderma* spp. has been proposed as biocontrol agents in agriculture, and some strains have given promising results against *B. cinerea*, which have resulted in the formulation of commercial products, e.g. Trichodex®, based on *T. harzianum* (T39) (O'Neil *et al.*, 1996; Elad, 2000). Strain T39 competes with the pathogen for nutrients, interfering with *B. cinerea* pectolytic enzymes and inducing resistance in the plant (reviewed in Elmer and Reglinski, 2006). *T. harzianum* (strains 1295-22 and P1) has proved to be effective against *B. cinerea* on grapes in multi-year field trials (Harman *et al.*, 1996).

Trichoderma spp. have also been evaluated as BCAs against the grapevine trunk pathogens *E. lata, Phaeomoniella chlamydospora, Phomopsis viticola* and species of the Botryosphaeriaceae (Kotze *et al.*, 2011). In particular, *T. atroviride* strain USPP-T1 was very effective in reducing infection by these pathogens in artificially inoculated pruning wounds. The utilization of *T. harzianum* as a BCA against *E. lata* had previously been evaluated by John *et al.* (2005) and Halleen *et al.* (2010), showing a good performance in the protection of pruning wounds. Pertot *et al.* (2009) gave a preliminary report of good control by *T. atroviride* (strain SC1) of esca disease agents on potted grapevines. Mutawila *et al.* (2011) applied *Trichoderma* spp. to grapevine pruning wounds as BCAs against trunk pathogens, and observed significant treatment × cultivar interactions in the incidence of *Trichoderma* in table and wine grapevines. The best time for application was 6 h after pruning, for both early and late vine pruning (Mutawila *et al.*, 2013). Pajot *et al.* (2012) reported promising results obtained from testing *T. atroviride* strain I-1237 (EsquiveWP®) against canker diseases of grapevine. Reggiori *et al.* (2014) have recently described good results with the utilization of Remedier® (*T. asperellum* and *T. gamsii*) against esca disease in 4 years of experimentation in six Italian regions. Rootshield®, a commercial formulation based on the *T. harzianum* strain T22, gave good results in nurseries only if applied at the rooting stage (Di Marco and Osti, 2007).

The activity of *T. harzianum* T39 against downy mildew has been studied by Perazzolli *et al.* (2008), whose studies were subsequently integrated by the work of Palmieri *et al.* (2012), who investigated the relationships among *T. harzianum* T39, *V. vinifera* and *P. viticola*. They provided evidence of systemic resistance induced by the BCA and showed how a large number of proteins related to energy metabolism, redox signalling and stress are overexpressed in plants treated with T39 and infected by *P. viticola*. The effect of T39 on the expression of defence-related genes involved in contrasting *P. viticola* infections on different cultivars has been reported by Banani *et al.* (2014), suggesting an important role of the host genetic background in the level of biocontrol. After indoor and field trials, Dagostin *et al.* (2011) considered *T. harzianum* a suitable candidate for further development. Activity by the *T. atroviride* strain SC1 was observed by Lenzi *et al.* (2013) against *P. viticola* sporangia *in vitro* – authors who had also previously observed that *T. harzianum* T39 induced resistance against downy mildew in grapevine.

A general review on the SAR and ISR mechanisms triggered by *Trichoderma* spp. in plants has been recently published by Nawrocka and Małolepsza (2013), while an overview of all of the *Trichoderma*-based commercial products and their targets is given by Woo *et al.* (2014).

***Ulocladium* spp.:** This genus contains mostly plant pathogens, but some strains and species, predominantly saprophytes, are used as BCAs. A strain of *U. oudemansii* (HRU3) has been reported to be an effective reducer of *Botrytis* bunch rot in the vineyard (reviewed by Elmer and Reglinski, 2006) and is commercialized as Botry-Zen® for the control of *B. cinerea*. Its activity has been briefly reviewed by Elmer *et al.* (2005), who reported a high degree of efficacy in field trials on cv. Chardonnay. In particular, Calvo-Garrido *et al.* (2014a) showed how effective *U. oudemansii* is in reducing *B. cinerea* incidence on aborted flowers and calyptras. Another strain of *Ulocladium* that is widely studied is *U. atrum* U385, which has been field tested in various vineyards worldwide, and shown to provide a reduction of *Botrytis* bunch rot (reviewed by Elmer and Reglinski, 2006). *In vitro* studies conducted on strains U13 and U16 by Ronseaux *et al.* (2013) suggested an activity of *Ulocladium* spp. in promoting plant growth and a direct (antagonism) and indirect (resistance) mechanisms of action against *B. cinerea*.

***Verticillium lecanii*:** *V. lecanii* is a well-known entomopathogenic fungus, which also acts as an antagonist of plant moulds, rusts and mildews (Benhamou, 2004). The capacity of this fungus to penetrate and hyperparasitize *E. necator* on *V. vinifera* leaves was observed by Heintz and Blaich (1990).

Yeast-like fungi

Details of the use of yeast-like fungi in the control of grapevine diseases are summarized in Table 1.3.

***Aureobasidium pullulans*:** *A. pullulans* is a cosmopolitan yeast-like fungus that has been reported as being very effective on table grapes against *B. cinerea* (Lima *et al.*, 1997) and also against *A. niger* and *Rhizopus stolonifer* (Schena *et al.*, 1999; Castoria *et al.*, 2001). The strain L47 was tested under field conditions on table grapes, and showed a significant reduction of incidence and severity of grey mould (Schena *et al.*, 1999). Schena *et al.* (2003) tested another strain (named 547) against postharvest decays on table grape, with better results than those from the reference

Table 1.3. Yeast-like fungal biocontrol agents (BCAs) for the treatment of grapevine diseases.

BCA	Notes	Field tested[a]	Examples of common product names	Reference(s)
Grey mould				
Aureobasidium pullulans	Well tested on table grapes	Yes	Botector®	Lima *et al.* (1997), Schena *et al.* (1999, 2003), Castoria *et al.* (2001), Raspor *et al.* (2010), Achleitner and Kunz (2013), Parafati *et al.* (2015)
Bulleromyces albus	Efficacy tends to decrease	No	–	Raspor *et al.* (2010)
Candida	Effective	Yes	–	Zahavi *et al.* (2000), Raspor *et al.* (2010), Cañamás *et al.* (2011), Calvo-Garrido *et al.* (2013, 2014b)
Cryptococcus	Different results	No	–	Lima *et al.* (1998), Raspor *et al.* (2010)
Hanseniaspora	Different results	No	–	Rabosto *et al.* (2006), Raspor *et al.* (2010), Qin *et al.* (2015)
Metschnikowia	Various degrees of efficacy	Yes	Shemer®	Nigro *et al.* (1999), Karabulut *et al.* (2003), Tasin *et al.* (2009), Raspor *et al.* (2010), Parafati *et al.* (2015)
Pichia	Very promising efficacy *in vitro*	No	–	Masih *et al.* (2000), Masih and Paul (2002), Santos and Marquina (2004), Raspor *et al.* (2010)
Saccharomyces	Various degrees of efficacy	–	Saccharopulvin 25PU®	Salmon (2009), Raspor *et al.* (2010), Nally *et al.* (2012, 2013)

Continued

Table 1.3. Continued.

BCA	Notes	Field tested[a]	Examples of common product names	Reference(s)
Rhodotorula	Poorly effective	No	–	Raspor *et al.* (2010)
Sporidiobolus	Poorly effective	No	–	Raspor *et al.* (2010)
Sporobolomyces	Poorly effective	No	–	Raspor *et al.* (2010)
Schizosaccharomyces pombe	Good performance	No	–	Nally *et al.* (2012)
Wickerhamomyces anomalus	Good performance	No	–	Parafati *et al.* (2015)
Downy mildew				
A. pullulans	Minimally effective	No	–	Harm *et al.* (2011)
Powdery mildew				
Pseudozyma flocculosa	Declining efficacy with high disease pressure	Yes	Sporodex®	Konstantinidou-Doltsinis *et al.* (2007)

[a]In at least one of the reviewed papers.

strain L47. Further studies confirmed a high *in vitro* antagonism towards *B. cinerea* (Raspor *et al.*, 2010).

A. pullulans has been tested also in potted vines against *P. viticola,* by Harm *et al.* (2011), but provided incomplete protection and no resistance induction response. The application of an autoclaved *A. pullulans* filtrate to grape calli elicited an increase in gene expression and enzymatic activity (PAL and stilbene synthase), thus demonstrating the possibility of a resistance induction mechanism (Rühmann *et al.*, 2013). During *in vitro* assays, a wide range of lytic enzymatic activities has been observed (Parafati *et al.*, 2015). As reported by Elmer and Reglinski (2006), a BCA product containing *A. pullulans* has been registered and commercialized against *Erwinia amylovora* (Blossom-Protect®), but its efficacy against *B. cinerea* in grapes had not yet been assessed. A formulation containing two strains of *A. pullulans* (DSM 14940 and DSM 14941) is commercially available against *B. cinerea* on grapevine (Botector®). Achleitner and Kunz (2013) gave preliminary reports of positive effects of this biotechnological botryticide in field trials.

***Candida* spp.:** *In vivo* studies (Cañamás *et al.*, 2011) have illustrated the antagonism between *Candida sake* CPA-1 and *B. cinerea* on detached berries of table grape. Subsequent field tests on cv. Cabernet Sauvignon (Cañamás *et al.*, 2011), on cultivar Macabeu (Calvo-Garrido *et al.*, 2013) and on cv. Merlot (Calvo-Garrido *et al.*, 2014b) confirmed this activity. Both the incidence and severity of bunch rot were much reduced in plants pretreated with the BCA, an effect that was further improved when an adjuvant (Fungicover®) was added. The activity of *C. sake* CPA-1 and the adjuvant has also been tested in organic viticulture; under conditions of high disease pressure, four applications during the most critical periods were sufficient for an effective control of grey mould control on grape, without affecting wine quality parameters (Calvo-Garrido *et al.*, 2013). Nally *et al.* (2012) reported that a new commercial product named Candifruit®, based on *C. sake* CPA-1, was going to be released to the market.

Other *Candida* spp. tested for their biocontrol activities include: *C. oleophila* being used as the basis for a new biological fungicide named NEXY (Nally *et al.*, 2012); *C. guilliermondii* (A42) reported as active in the field against *B. cinerea* and *A. niger* (Zahavi *et al.*, 2000) on table and wine grapes; *C. glabrata*, *C. norvegica*, *C. vini* and *C. diversa*, all tested *in vitro* and *in semi-vivo* by Raspor *et al.* (2010), and giving higher protection than *C. oleophila*, but with activity mostly decreasing with time; and *C. zemplinina*, which showed just a moderate effect as a BCA (Raspor *et al.*, 2010).

***Cryptococcus* spp.:** *C. laurentii* (LS-28) was shown to be effective in reducing bunch decay on table grapes (Lima *et al.*, 1998), although *Cryptococcus* spp. showed negligible activity against *B. cinerea* when tested *in vitro* by Raspor *et al.* (2010).

***Metschnikowia* spp.:** The yeast *M. fructicola* was evaluated against *B. cinerea* and *A. niger* by Karabulut *et al.* (2003) with promising results. In grapevine, a preliminary report by Tasin *et al.* (2009) indicated how *M. fructicola,* applied before artificial inoculation of the pathogen, significantly reduced grey mould infections. A biofungicide based on this BCA had been marketed (Shemer®) against postharvest fruit rots.

M. reukaufii has been compared with other yeasts and with the commercially used yeast *Candida oleophila* by Raspor *et al.* (2010) against *B. cinerea*, and gave *in vitro* biocontrol of around 30%. Better results have been reported in the same work for a strain of *M. pulcherrima*, which gave biocontrol of 39.1% (*C. oleophila*, used as reference, gave 24.1% control). *M. pulcherrima* (anamorph: *Candida pulcherrima*), isolate 320, has also been shown to be an effective BCA against *Botrytis* causing storage decay on table grapes (Nigro *et al.*, 1999). Parafati et al. (2015) observed that the *in vivo* effective antagonistic action of this yeast against postharvest bunch rot was correlated with iron availability.

***Pichia* spp.:** *P. anomala* (FY-102) has been described as a strong antagonist of *B. cinerea* by Masih *et al.* (2000). The contact between the yeast and the pathogen brought about the coagulation of the protoplast and the hyphae and emptied the latter. Subsequent experiments on plant growth *in vitro* after co-inoculation of the two fungi confirmed complete protection against the pathogen (Masih *et al.*, 2000).

P. membranifaciens has been studied for its potential use against *B. cinerea*. Strain FY-101, an isolate from grape skin, expressed a strong β-1,3-glucanase activity (Masih and Paul, 2002). *In vitro* analyses showed the inhibition of conidial production by *B. cinerea* in the hyphae nearest to the yeast. Microscopic examination of the pathogen mycelium revealed complete rupture and fragmentation of the hyphal filaments; grapevines inoculated *in vitro* were fully developed and vigorous (Masih and Paul, 2002). Santos and Marquina (2004) identified another strain of *P. membranifaciens* (CYC 1106) with a particularly strong zymocidal activity. Subsequent tests against *B. cinerea* identified a killer protein as partly responsible for the fungicidal effect (Santos and Marquina, 2004). *In vitro* tests showed 34% biocontrol activity 7 days after inoculation, but this decreased later (Raspor *et al.*, 2010). Other strains tested against *B. cinerea* by the same authors were: *P. guilliermondii* (ZIM624), which showed very promising antagonistic activity, and *P. kluyveri* (Raspor *et al.*, 2010).

***Pseudozyma flocculosa*:** *P. flocculosa* is a yeast identified as a natural antagonist of powdery mildews (Hammami *et al.*, 2011b). The efficacy of a commercial formulation of one strain of this species (Sporodex®) in two field trials was reported by Konstantinidou-Doltsinis *et al.* (2007), who observed that it was effective on grape bunches but had decreasing efficacy under conditions of high disease pressure.

***Saccharomyces* spp.:** As reviewed in Elmer and Reglinski (2006), an *S. chevalieri* isolate has been used in the commercial formulation Saccharopulvin 25 PU® and field tested, when it gave an average protection of grapevine against *B. cinerea* of 91%. A few reports have indicated the possible utilization of strains of *S. cerevisiae* as BCAs against bunch rot: Salmon (2009) highlighted the antagonism of industrial isolates against *B. cinerea* and *Aspergillus carbonarius* on grapes, and *S. cerevisiae* was further investigated by Nally *et al.* (2012) for the control of *B. cinerea* on the table grape cultivar Red Globe. In this latter work, the authors tested several strains of *S. cerevisiae*, both *in vitro* and *in semi-vivo*, at different temperatures. Not all of the strains that showed good activity *in vitro* have been confirmed as effective when tested directly on grapes, but the *S. cerevisiae* isolate BSc68 was able to effectively inhibit *B. cinerea*. A further report detailed the effectiveness *in vivo* of other *Saccharomyces* and non-*Saccharomyces* isolates in inhibiting the growth of fungi involved in the 'sour rot' disease of grapes, which is caused by a complex of microorganisms in which filamentous fungi play a key role (Nally *et al.*, 2013). Raspor *et al.* (2010) highlighted how the *S. cerevisiae* isolate ZIM2180 exhibited an extremely high activity against *B. cinerea in vitro*, on media with high concentration of glucose. The production of volatile organic compounds was observed during *in vitro* trials against *B. cinerea*, with a significant growth reduction under acidic conditions (Parafati *et al.*, 2015).

Other yeast-like fungi: The work carried out by Raspor *et al.* (2010) also compared other yeast species for their activity against *B. cinerea*. ***Bulleromyces albus*** showed an *in vitro* control activity of 25.7%, but this decreased with time; members of the genera ***Hanseniaspora***, ***Rhodotorula***, ***Sporidiobolus*** and ***Sporobolomyces*** were all classified as poorly effective. ***H. uvarum*** has been reported to effectively reduce *Botrytis* on grape (Rabosto *et al.*, 2006; reviewed in Romanazzi *et al.*, 2012), with a remarkable improvement in control when used in combination with SA or sodium bicarbonate (Qin *et al.*, 2015). ***Schizosaccharomyces pombe*** gave a good performance both *in vitro* and *in semi-vivo* on detached grapes (Nally *et al.*, 2012). Parafati *et al.* (2015) observed significant control of postharvest *Botrytis* bunch rot by ***Wickerhamomyces anomalus***, which was related to the production of volatile organic compounds (VOCs), glucanase activity and biofilm formation.

Bacteria

Details of the use of bacteria in the control of grapevine diseases are summarized in Table 1.4.

The utilization of bacteria to control grapevine pathogens has been recently reviewed by Compant *et al.* (2013), and a comprehensive description of the role of plant growth-promoting rhizobacteria in improving crop productivity under critical conditions can be found in Nadeem *et al.* (2014).

***Acinetobacter lwoffii*:** Two strains of *A. lwoffii* (PTA-113 and PTA-152) have been widely studied for their activity in the protection of grapevine against *B. cinerea*. Strains induced systemic resistance and enhanced chitinase and glucanase activities in leaves and berries (Magnin-Robert *et al.*, 2007), as well as lipoxygenase (LOX), PAL (Trotel-Aziz *et al.*, 2008) oxidative bursts and phytoalexin production (Verhagen *et al.*, 2011).

Table 1.4. Bacterial biocontrol agents (BCAs) for the treatment of grapevine diseases.

BCA	Notes	Field tested[a]	Examples of common product names	Reference(s)
Grey mould				
Acinetobacter lwoffii	Good level of resistance, especially when used as a soil drench	Yes	–	Magnin-Robert *et al.* (2007, 2013), Trotel-Aziz *et al.* (2008), Verhagen *et al.* (2011)
Bacillus	Some defence-related responses	Yes	Serenade®, AMYLO-X®	Paul *et al.* (1997), Magnin-Robert *et al.* (2007, 2013), Trotel-Aziz *et al.* (2008), Verhagen *et al.* (2011)
Burkholderia phytofirmans	PGPR (plant growth-promoting rhizobacteria)	No	–	Ait Barka *et al.* (2002, 2006)
Cupriavidus campinensis	Degrades oxalic acid	No	–	Schoonbeek *et al.* (2007)
Micromonospora	Tested *in vitro* in a greenhouse	No	–	Loqman *et al.* (2009)
Pantoea agglomerans	Mostly well performing	Yes	–	Magnin-Robert *et al.* (2007, 2013), Trotel-Aziz *et al.* (2008), Verhagen *et al.* (2011)
Pseudomonas	Induction of resistance	Yes	–	Verhagen *et al.* (2010, 2011), Magnin-Robert *et al.* (2007, 2013), Trotel-Aziz *et al.* (2008)
Serratia liquefaciens	Reduces pathogen inoculum	No	–	Whiteman and Stewart (1998)
Streptomyces	Tested *in vitro* in a greenhouse	No	–	Loqman *et al.* (2009), Lebrihi *et al.* (2009)
Downy mildew				
Bacillus	Control on bunches	Yes	Serenade®	Dagostin *et al.* (2011)
Lysobacter capsici	Preliminary indications	No	–	Puopolo *et al.* (2014)
Pseudomonas	Induces systemic resistance (ISR)	No	–	Archana *et al.* (2011)
Powdery mildew				
Bacillus	Good efficacy under low–moderate disease pressure	Yes	Milastin K®	Compant *et al.* (2013), Sawant *et al.* (2011)

Continued

Table 1.4. Continued.

BCA	Notes	Field tested[a]	Examples of common product names	Reference(s)
Pseudomonas	Able to significantly reduce disease incidence	No	–	Sendhilvel *et al.* (2007)
Trunk diseases				
Bacillus	Reduction of infection by several trunk pathogens	Yes		Ferreira *et al.* (1991), Schmidt *et al.* (2001), Alfonzo *et al.* (2009), Kotze *et al.* (2011)
P. agglomerans	Tested on autoclaved grape wood	No	–	Schmidt *et al.* (2001)

[a]In at least one of the reviewed papers.

These strains induced a good level of resistance when applied as a soil drench, highlighting their role as PGPRs (Magnin-Robert *et al.*, 2007). Subsequent studies underlined the differential effect on *B. cinerea* protection given by *A. lwoffii* in combination with other beneficial bacteria (*Bacillus subtilis* PTA-271, *Pantoea agglomerans* PTA-AF1 and PTA-AF2, *Pseudomonas fluorescens* PTA-268 and PTA-CT2) (Magnin-Robert *et al.*, 2013).

***Bacillus* spp. strains** have been widely investigated for their activity as BCAs. Ferreira *et al.* (1991) observed a positive effect of these bacteria in reducing the *in vitro* growth of *E. lata*; a significant reduction of fungal infection was visible if pruning wounds were sprayed with a suspension of a *B. subtilis* isolate before inoculation with *E. lata*. The efficacy of *B. subtilis* in the biocontrol of pruning wound infections was assessed in field trials, and it was shown to significantly reduce experimental infection by *E. lata* and other trunk disease pathogens as assessed 8 months after inoculation (Kotze *et al.*, 2011). Crude metabolites produced by *B. subtilis* strain AG1 (later reclassified as *B. amyloliquefaciens*) that had previously been found to reduce the growth of esca fungi *in vitro* have been extracted and applied in laboratory assays, which confirmed their antagonistic action against *Phaeoacremonium aleophilum* and *Phaeomoniella chlamydospora* (Alfonzo *et al.*, 2009). Alfonzo *et al.* (2012) observed effective *in vitro* activity of antifungal peptides produced by the *B. amyloliquefaciens* strain AG1 against many grapevine fungal pathogens. A commercial product of *B. amyloliquefaciens* subsp. *plantarum*, strain D747, is currently available in Italy for use against grey mould (AMYLO-X®).

Some *Bacillus* strains have also been reported for their activity against *E. necator*, such as *B. pumilus* (B-30087), which is able to protect cv. Chardonnay plantlets as well as myclobutanil, and *Bacillus* strains ATCC 55608 and 55609 (reviewed in Compant *et al.*, 2013). A multi-year field test conducted by Sawant *et al.* (2011) with the commercial *Bacillus*-based product Milastin K® highlighted its effectiveness under conditions of low–moderate disease pressure.

A formulation of the *B. subtilis* strain QST 713 is commercialized (Serenade® Max) and has been registered for the protection of grapevine from *B. cinerea*. The activity of *B. subtilis* strain PTA-271 against *B. cinerea* has been also tested, both alone and in combination with other beneficial bacteria, with variable results (Magnin-Robert *et al.*, 2007, 2013; Trotel-Aziz *et al.*, 2008; Verhagen *et al.*, 2011). Dagostin *et al.* (2011) reported positive results with Serenade® after field trials against downy mildew. Another *Bacillus* species, *B. circulans*, was able to suppress *B. cinerea* infections in *in vitro* plantlets (Paul *et al.*, 1997).

Burkholderia phytofirmans (strain PsJN) is an endophyte which also colonizes grapevine tissues and xylem vessels (Compant *et al.*, 2008). *B. phytofirmans* PsJN acts as a PGPR, enhancing protection against *B. cinerea* in grapevine, and also protecting plants against low-temperature stress (Ait Barka *et al.*, 2002, 2006). Bordiec *et al.* (2011) showed that *B. phytofirmans* PsJN caused a transient extracellular alkalinization of grapevine cells together with an increment in SA production, but neither accumulation of ROS nor cell death.

***Lysobacter capsici*:** Puopolo *et al.* (2014) reported the potential use of *L. capsici* strain AZ78 in the control of *P. viticola*. The strain survived on grapevine leaves and it reduced the presence of downy mildew. Environmental stresses and copper did not have any negative influence on its development.

Pantoea agglomerans **(syn. *Erwinia herbicola*):** Strains PTA-AF1 and PTA-AF2 have been widely studied for their BCA activities against *B. cinerea* in several studies, in combination with other beneficial bacteria (Magnin-Robert *et al.*, 2007, 2013; Trotel-Aziz *et al.*, 2008). *P. agglomerans* generated chitinase and glucanase activity in co-inoculated plants (Magnin-Robert *et al.*, 2013), with an increase in oxidative burst and resveratrol production (Verhagen *et al.*, 2011). Mutated strains of *P. agglomerans* showed 100% efficiency against *E. lata* on autoclaved grape wood (Schmidt *et al.*, 2001).

Pseudomonas **spp.:** Sendhilvel *et al.* (2007) tested the efficacy of the *P. fluorescens* strain Pf1 against powdery mildew; under glasshouse conditions, foliar application at 2% significantly reduced the incidence of the disease and induced defence gene products (chitinase, POX and polyphenol oxidase – PPO). Verhagen *et al.* (2010, 2011) demonstrated the ability of a *P. fluorescens* and a *P. aeruginosa* strain to induce resistance against *B. cinerea* in grapevine. The two strains triggered an oxidative burst and phytoalexin accumulation, and primed leaves for a resistance response to *B. cinerea* attacks *in vitro*. Trotel-Aziz *et al.* (2008) observed an apparent antagonistic effect of the *P. fluorescens* strain PTA-CT2. The combination of *P. fluorescens* and other beneficial bacteria as BCAs against *Botrytis* has been mentioned above; mixtures with *A. lwoffii* resulted in slight induction of resistance in berries. Finally, the induction of systemic resistance to *P. viticola* triggered by *P. fluorescens* was investigated by Archana *et al.* (2011).

Other bacteria indicated as possible BCAs include *Serratia liquefaciens* (Whiteman and Stewart, 1998), which is able to reduce the sporulation of *B. cinerea*; *Cupriavidus campinensis* (Schoonbeek *et al.*, 2007), which degrades oxalic acid (a virulence factor of *B. cinerea*); and members of the Actinomycetales, such as *Streptomyces* spp. and *Micromonospora* spp. (Loqman *et al.*, 2009), which give protection against grey rot *in vitro* and in greenhouse plants (Lebrihi *et al.*, 2009). The mechanism by which disease incidence was reduced by *Streptomyces* spp. was not elucidated, but it was hypothesized that these actinomycetes directly inhibited the hyphal growth and structure of fungal pathogens (Loqman *et al.*, 2009).

Oomycetes

***Pythium* spp.:** This genus of the oomycetes consists of saprophytic and parasitic species (Benhamou *et al.*, 2012). Some are mycoparasites of *B. cinerea* and have been tested as BCAs, such as *P. radiosum* and *P. periplocum* (reviewed in Elmer and Reglinski, 2006) and *P. paroecandrum* (Abdelghani *et al.*, 2004). Paul (2004) identified a new *Pythium* species, named *P. citrinum* (S-12), with a high level of antagonism. *P. oligandrum* is also known for its biocontrol properties, through the direct control of plant pathogens or via the induction of resistance (reviewed in Gerbore *et al.*, 2014). Gerbore *et al.* (2011) reported the effectiveness of some *P. oligandrum* strains against *Phaeomoniella chlamydospora*. A study by Bala and Paul (2012) clarified that the biocontrol of *B. cinerea* by *Pythium* spp. could be related to an increment of phenolic compounds in grapevine leaves.

Conclusions

A long series of studies on the biological control of plant pathogens has yielded a great amount of information on this topic. In light of the reduction in pesticide use that is presently being urged, it is now possible that this could be supported by the use of alternative products such as biocontrol agents (BCAs) and inducers of natural resistance.

The elicitation of plant resistance by natural compounds or living microorganisms could be highly beneficial as it could ensure the low environmental impact of vineyard management and contribute to improving the total antioxidant content of grapevines and grapes (Iriti *et al.*, 2011; Flamini *et al.*, 2013). Apart from an exclusive use of natural products, as in organic viticulture, biocontrol measures could be considered as usefully supporting integrated pest management (IPM) and fungicide resistance management (De Miccolis Angelini *et al.*, 2009; Gozzo and Faoro, 2013).

However, as reported in this review, natural compounds alone are mostly incapable of ensuring sufficient protection under high levels of disease pressure. Furthermore, many BCAs with good protection capacities under laboratory conditions did not have their performances confirmed in field tests (Gessler *et al.*, 2011). The variability of results could be related to the climatic conditions, plant–soil interactions and abiotic or biotic stresses, which can interfere with the exploitation of the antagonistic capacities of BCAs (Pertot *et al.*, 2013; Roatti *et al.*, 2013). Also, the specific microbial strain or the type of formulation can strongly affect the level of efficacy, while the effectiveness of the same BCA can vary on different grape cultivars (Banani *et al.*, 2014). Moreover, the industry plays a decisive role in the implementation of the best formulations to be commercialized.

The partial efficacy of many biological products, which are usually weaker in their actions than traditional fungicides, should stimulate the development of specific control strategies in vineyards. In the open field, BCAs and plant resistance activators should be applied with particular care and consideration of their preventive activities; the frequency of their application can have particular relevance as well (Walters *et al.*, 2013). The type of crop management (organic or IPM) should also be considered, as Campisano *et al.* (2014) have highlighted the differences in the respective bacterial grapevine-associated communities under these two regimes. An important supportive role could be played by forecast models for the risk of infections, and the use of BCAs preferred under conditions of low disease pressure.

In conclusion, future studies should focus on the efficacy of treatments in the vineyard and on the combined strategies that could be adopted, while also considering the role played, for example, by agronomic practices in the induction/reduction of plant pathogen attacks.

Note

Commercial product names are given only as an example and the current authorizations for their application should be verified by users.

References

Abdelghani, E.Y., Bala, K. and Paul, B. (2004) Characterisation of *Pythium paroecandrum* and its antagonism towards *Botrytis cinerea*, the causative agent of grey mould disease of grape. *FEMS Microbiology Letters* 230, 177–183.

Achleitner, D. and Kunz, S. (2013) Biotechnological protection against *Botrytis* bunch rot on grapes. *Future IPM in Europe, 19–21 March 2013, PalaCongressi, Riva del Garda, Italy. Book of Abstracts: Biocontrol*. Fondazione Edmund Mach, Michele all'Adige and Land- und Forstwirtschaftliches Versuchszentrum Laimburg, Pfatten, Italy with PURE project (Innovative Crop Protection for Sustainable Agriculture), European Commission, p. 33. Available at: http://www.pure-ipm.eu/sites/default/files/content/files/BOOK%20OF%20ABSTRACTS_BIOCONTROL_0.pdf (accessed 16 September 2015).

Ait Barka, E., Gognies, S., Nowak, J., Audran, J.C. and Belarbi, A. (2002) Inhibitory effect of endophyte bacteria on *Botrytis cinerea* and its influence to promote the grapevine growth. *Biological Control* 24, 135–142.

Ait Barka, E., Eullaffroy, P., Clément, C. and Vernet, G. (2004) Chitosan improves development, and protects *Vitis vinifera* L. against *Botrytis cinerea*. *Plant Cell Reports* 22, 608–614.

Ait Barka, E., Nowak, J. and Clément, C. (2006) Enhancement of chilling resistance of inoculated grapevine plantlets with a plant growth-promoting rhizobacterium, *Burkholderia phytofirmans* strain PsJN. *Applied and Environmental Microbiology* 72, 7246–7252.

Alfonzo, A., Conigliaro, G., Torta, L., Burruano, S. and Moschetti, G. (2009) Antagonism of *Bacillus subtilis* strain AG1 against vine wood fungal pathogens. *Phytopathologia Mediterranea* 48, 155–158.

Alfonzo, A., Lo Piccolo, S., Conigliaro, G., Ventorino, V., Burruano, S. and Moschetti, G. (2012) Antifungal peptides produced by *Bacillus amyloliquefaciens* AG1 active against grapevine fungal pathogens. *Annals of Microbiology* 62, 1593–1599.

Allègre, M., Héloir, M.C., Trouvelot, S., Daire, X., Pugin, A., Wendehenne, D. and Adrian, M. (2009) Are grapevine stomata involved in the elicitor-induced protection against downy mildew? *Molecular Plant-Microbe Interactions* 22, 977–986.

Amborabé, E., Aziz, A., Trotel-Aziz, P., Quantinet, D., Dhuicq, L. and Vernet, G. (2004) Stimulation des défenses naturelles de la vigne. Essais d'emploi du chitosan contre *Botrytis cinerea* [Chitosan against *Botrytis cinerea* in the vineyard]. *Phytoma* 571, 26–29. [English abstract.]

Angeli, D., Maurhofer, M., Gessler, C. and Pertot, I. (2012) Existence of different physiological forms within genetically diverse strains of *Ampelomyces quisqualis*. *Phytoparasitica* 40, 37–51.

Archana, S., Prabakar, K., Raguchander, T., Hubballi, M., Valarmathi, P. and Prakasam, V. (2011) Defense response of grapevine to *Plasmopara viticola* induced by azoxystrobin and *Pseudomonas fluorescens*. *International Journal of Sustainable Agriculture* 3, 30–38.
Arnault, I., Chovelon, M. and Derridj, S. (2013) Preliminary tests in field conditions of alternatives [sic] substances against grape downy mildew in organic farming. *IOBC-WPRS Bulletin* 86, 257–262.
Avanci, N.C., Luche, D.D., Goldman, G.H. and Goldman, M.H. (2010) Jasmonates are phytohormones with multiple functions, including plant defense and reproduction. *Genetics and Molecular Research* 9, 484–505.
Aziz, A., Poinssot, B., Daire, X., Adrian, M., Bézier, A., Lambert, B., Joubert, J.M. and Pugin, A. (2003) Laminarin elicits defense responses in grapevine and induces protection against *Botrytis cinerea* and *Plasmopara viticola*. *Molecular Plant-Microbe Interactions* 16, 1118–1128.
Aziz, A., Heyraud, A. and Lambert, B. (2004) Oligogalacturonide signal transduction, induction of defense-related responses and protection of grapevine against *Botrytis cinerea*. *Planta* 218, 767–774.
Aziz, A., Trotel-Aziz, P., Dhuicq, L., Jeandet, P., Couderchet, M. and Vernet, G. (2006) Chitosan oligomers and copper sulfate induce grapevine defense reactions and resistance to gray mold and downy mildew. *Phytopathology* 96, 1188–1194.
Bakker, P.A.H.M., Pieterse, C.M.J. and Van Loon, L.C. (2007) Induced systemic resistance by fluorescent *Pseudomonas* spp. *Phytopathology* 97, 239–243.
Bakshi, S., Sztejnberg, A. and Yarden, O. (2001) Isolation and characterization of a cold-tolerant strain of *Fusarium proliferatum*, a biocontrol agent of grape downy mildew. *Phytopathology* 91, 1062–1068.
Bala, K. and Paul, B. (2012) *Pythium* induced phenolic compounds in the resistance of *Vitis vinifera* to *Botrytis cinerea*. *Plant Pathology and Quarantine* 2, 16–23.
Banani, H., Roatti, B., Ezzahi, B., Giovannini, O., Gessler, G., Pertot, I. and Perazzolli, M. (2014) Characterization of resistance mechanisms activated by *Trichoderma harzianum* T39 and benzothiadiazole to downy mildew in different grapevine cultivars. *Plant Pathology* 63, 334–343.
Bavaresco, L., Mattivi, F., De Rosso, M. and Flamini, R. (2012) Effects of elicitors, viticultural factors, and enological practices on resveratrol and stilbenes in grapevine and wine. *Mini-Reviews in Medicinal Chemistry* 12, 1366–1381.
Belhadj, A., Saigne, C., Telef, N., Cluzet, S., Bouscaut, J., Corio-Costet, M.F. and Merillon, J.M. (2006) Methyl jasmonate induces defense responses in grapevine and triggers protection against *Erysiphe necator*. *Journal of Agricultural and Food Chemistry* 54, 9119–9125.
Benhamou, N. (2004) Potential of the mycoparasite, *Verticillium lecanii*, to protect citrus fruit against *Penicillium digitatum*, the causal agent of green mold: a comparison with the effect of chitosan. *Phytopathology* 94, 693–795.
Benhamou, N., Le Floch, G., Vallance, J., Gerbore, J., Grizard, D. and Rey, P. (2012) *Pythium oligandrum*: an example of opportunistic success. *Microbiology* 158, 2679–2694.
Bordiec, S., Paquis, S., Lacroix, H., Dhondt, S., Ait Barka, E., Kauffmann, S., Jeandet, P., Mazeyrat-Gourbeyre, F., Clément, C., Baillieul, F. and Dorey, S. (2011) Comparative analysis of defence responses induced by the endophytic plant growth-promoting rhizobacterium *Burkholderia phytofirmans* strain PsJN and the non-host bacterium *Pseudomonas syringae* pv. *pisi* in grapevine cell suspensions. *Journal of Experimental Botany* 62, 595–603.
Bowen, P., Menzies, J., Ehret, D., Samuels, L. and Glass, A.D.M. (1992) Soluble silicon sprays inhibit powdery mildew development on grape leaves. *Journal of the American Society for Horticultural Science* 117, 906–912.
Burruano, S., Alfonzo, A., Lo Piccolo, S., Conigliaro, C., Mondello, V., Torta, L., Moretti, M. and Assante, G. (2008) Interaction between *Acremonium byssoides* and *Plasmopara viticola*. *Phytopathologia Mediterranea* 47, 122–131.

Caffi, T., Legler, S., Rossi, V., Kiss, L. and Pintye, A. (2010) New biocontrol strategy against the sexual fruiting bodies of grapevine powdery mildew. *Phytopathology* 100(6S), S19.
Calvo-Garrido, C., Elmer, P.A.G., Viñas, I., Usall, J., Bartra, E. and Teixidó, N. (2013) Biological control of botrytis bunch rot in organic wine grapes with the yeast antagonist *Candida sake* CPA-1. *Plant Pathology* 62, 510–519.
Calvo-Garrido, C., Viñas, I., Elmer, P.A.G., Usall, J. and Teixidó, N. (2014a) Suppression of *Botrytis cinerea* on necrotic grapevine tissues by early-season applications of natural products and biological control agents. *Pest Management Science* 70, 595–602.
Calvo-Garrido, C., Teixidó, N., Roudet, J., Viñas, I., Usall, J. and Fermaud, M. (2014b) Biological control of *Botrytis* bunch rot in Atlantic climate vineyards with *Candida sake* CPA-1 and its survival under limiting conditions of temperature and humidity. *Biological Control* 79, 24–35.
Campisano, A., Antonielli, L., Pancher, M., Yousaf, S., Pindo, M. and Pertot, I. (2014) Bacterial endophytic communities in the grapevine depend on pest management. *PloS One* 9(11): e112763.
Cañamás, T.P., Viñas, I., Torres, R., Usall, J., Solsona, C. and Teixidó, N. (2011) Field applications of improved formulations of *Candida sake* CPA-1 for control of *Botrytis cinerea* in grapes. *Biological Control* 56, 150–158.
Castoria, R., De Curtis, F., Lima, G., Caputo, L., Pacifico, S. and De Cicco, V. (2001) *Aureobasidium pullulans* (LS30) an antagonist of postharvest pathogens of fruits: study on its modes of action. *Postharvest Biology and Technology* 22, 7–17.
Chardonnet, C., L'Hyvernay, A.L. and Doneche, B. (1997) Effect of calcium treatment prior to *Botrytis cinerea* infection on the changes in pectic composition of grape berry. *Physiological and Molecular Plant Pathology* 50, 213–218.
Chardonnet, C.O., Sams, C.E. and Conway, W.S. (1999) Calcium effect on the mycelial cell walls of *Botrytis cinerea*. *Phytochemistry* 52, 967–973.
Ciccarese, F., Longo, O., Ambrico, A., Schiavone, D. and Ziadi, T. (2006) Impiego di *Aphanocladium album* (isolato MX-95) nella lotta contro gli agenti dei marciumi dell'uva da tavola in pre e post-raccolta [Use of *Aphanocladium album* (isolate MX-95) in the control of grape rot diseases in pre and post-harvest. *Atti Giornate Fitopatologiche* 2, 443–448. [English abstract.]
Cohen, Y., Reuveni, M. and Baider, A. (1999) Local and systemic activity of BABA (DL-3-aminobutyric acid) against *Plasmopara viticola* in grapevines. *European Journal of Plant Pathology* 105, 351–361.
Cohen, Y., Wang, W., Ben-Daniel, B. and Ben-Daniel, Y. (2006) Extracts of *Inula viscosa* control downy mildew of grapes caused by *Plasmopara viticola*. *Phytopathology* 96, 417–424.
Compant, S., Kaplan, H., Sessitsch, A., Nowak, J., Ait Barka, E. and Clément, C. (2008) Endophytic colonization of *Vitis vinifera* L. by *Burkholderia phytofirmans* strain PsJN: from the rhizosphere to inflorescence tissues. *FEMS Microbiology Ecology* 63, 84–93.
Compant, S., Brader, G., Muzammil, S., Sessitsch, A., Lebrihi, A. and Mathieu, F. (2013) Use of beneficial bacteria and their secondary metabolites to control grapevine pathogen diseases. *BioControl* 58, 435–455.
Conrath, U. (2011) Molecular aspects of defence priming. *Trends in Plant Science* 16, 524–531.
Coulomb, C., Lizzi, Y., Coulomb, P.J., Roggero, J.P., Coulomb, P.O. and Agulhon, O. (1999) Le cuivre a-t-il un effect eliciteur? [Can copper have an elicitor effect?]. *Phytoma* 512, 41–46. [English abstract.]
Dagostin, S., Formolo, T., Giovannini, O., Pertot, I. and Schmitt, A. (2010) *Salvia officinalis* extract can protect grapevine against *Plasmopara viticola*. *Plant Disease* 94, 575–580.
Dagostin, S., Schärer, H., Pertot, I. and Tamm, L. (2011) Are there alternatives to copper for controlling grapevine downy mildew in organic viticulture? *Crop Protection* 30, 776–788.
De Miccolis Angelini, R.M., Pollastro, S. and Faretra, F. (2009) Induction of plant resistance in crop protection against fungal diseases. *Petria* 19, 57–59.
Di Marco, S. (2010) Effetti di trattamenti con biostimolanti sul mal dell'esca. In: Bucciantini, V., Braccini, P., Surico, G. and Mugnai, L. (eds) *Il Mal dell'Esca della Vite. Interventi di Ricerca e Sperimentazione per il Contenimento della Malattia. Progetto MesVit. Rete Interregionale*

per la Ricerca Agraria, Forestale, Acquacoltura e Pesca. Agenzia Regionale per lo Sviluppo e l'Innovazione nel Settore Agricolo-forestale (ARSIA), Florence, Italy, pp. 479–482. Available at: http://www.agrinnovazione.regione.sicilia.it/reti/Viticoltura/progetti_di_ricerca/contenimento_mal_esca_della_vite%20.pdf (accessed 16 September 2015).

Di Marco, S. and Osti, F. (2007) Applications of *Trichoderma* to prevent *Phaeomoniella chlamydospora* infections in organic nurseries. *Phytopathologia Mediterranea* 46, 73–83.

Di Martino, M.A., Amadei, M., Medico, E. and Arbizzani, A. (2014) Bicarbonato di potassio (Karma® e Armicarb®): nuovo fungicida per il controllo di oidio e botrite [Karma® and Armicarb®: new fungicides based on potassium bicarbonate to control powdery mildew and gray mould]. *Atti Giornate Fitopatologiche* 2, 31–40.

Dubreuil-Maurizi, C., Trouvelot, S., Frettinger, P., Pugin, A., Wendehenne, D. and Poinssot, B. (2010) β-Aminobutyric acid primes an NADPH oxidase-dependent reactive oxygen species production during grapevine-triggered immunity. *Molecular Plant-Microbe Interactions* 23, 1012–1021.

Durrant, W.E. and Dong, X. (2004) Systemic acquired resistance. *Annual Review of Phytopathology* 42, 185–209.

Elad, Y. (2000) *Trichoderma harzianum* T39 preparation for biocontrol of plant diseases – control of *Botrytis cinerea*, *Sclerotinia sclerotiorum* and *Cladosporium fulvum*. *Biocontrol Science and Technology* 10, 499–507.

Elad, Y. and Shtienberg, D. (1994) Effect of compost water extracts on grey mould (*Botrytis cinerea*). *Crop Protection* 13, 109–114.

Elmer, P.A.G. and Reglinski, T. (2006) Biosuppression of *Botrytis cinerea* in grapes. *Plant Pathology* 55, 155–177.

Elmer, P.A.G., Hoyte, S.M., Vanneste, J.L., Reglinski, T., Wood, P.N. and Parry, F.J. (2005) Biological control of fruit pathogens. *New Zealand Plant Protection* 58, 47–54.

Falk, S.P., Gadoury, D.M., Cortesi, P., Pearson, R.C. and Seem, R.C. (1995) Parasitism of *Uncinula necator* cleistothecia by the mycoparasite *Ampelomyces quisqualis*. *Phytopathology* 85, 794–800.

Falk, S.P., Pearson, R.C., Gadoury, D.M., Seem, R.C. and Sztejnberg, A. (1996) *Fusarium proliferatum* as a biocontrol agent against grape downy mildew. *Phytopathology* 86, 1010–1017.

Faurie, B., Cluzet, S. and Mérillon, J.M. (2009) Implication of signaling pathways involving calcium, phosphorylation and active oxygen species in methyl jasmonate-induced defense responses in grapevine cell cultures. *Journal of Plant Physiology* 166, 1863–1877.

Feliziani, E., Smilanick, J.L., Margosan, D.A., Mansour, M.F., Romanazzi, G., Gu, S., Gohil, H.L. and Rubio Ames, Z. (2013) Preharvest fungicide, potassium sorbate, or chitosan use on quality and storage decay of table grapes. *Plant Disease* 97, 307–314.

Ferreira, J.H.S., Matthee, F.N. and Thomas, A.C. (1991) Biological control of *Eutypa lata* on grapevine by an antagonistic strain of *Bacillus subtilis*. *Phytopathology* 81, 283–287.

Flamini, R., Mattivi, F., De Rosso, M., Arapitsas, P. and Bavaresco, L. (2013) Advanced knowledge of three important classes of grape phenolics: anthocyanins, stilbenes and flavonols. *International Journal of Molecular Sciences* 14, 19651–19669.

Fowler, S.R., Jaspers, M.V., Walter, M. and Stewart, A. (1999) Suppression of overwintering *Botrytis cinerea* inoculum on grape rachii using antagonistic fungi. *Proceedings of the New Zealand Plant Protection Conference* 52, 141–147.

FRAC (2015) FRAC Code List© 2015: Fungicides Sorted by Mode of Action (including FRAC Code Numbering). Fungicide Resistance Action Committee, CropLife International, Brussels. Available at: http://www.frac.info/docs/default-source/publications/frac-code-list/frac-code-list-2015-finalC2AD7AA36764.pdf?sfvrsn=4 (accessed 15 September 2015).

Fu, Z.Q. and Dong, X. (2013) Systemic acquired resistance: turning local infection into global defence. *Annual Review of Plant Biology* 64, 839–863.

Gerbore, J., Bruez, E., Vallance, J., Grizard, D., Regnault-Roger, C. and Rey, P. (2011) Protection against a vine trunk attack by *Phaeomoniella chlamydospora* is concomitant with root colonization by the oomycete *Pythium oligandrum*. *IOBC-WPRS Bulletin* 71, 31–35.

Gerbore, J., Benhamou, N., Vallance, J., Le Floch, G., Grizard, D., Regnault-Roger, C. and Rey, P. (2014) Biological control of plant pathogens: advantages and limitations seen through the case study of *Pythium oligandrum*. *Environmental Science and Pollution Research* 21, 4847–4860.

Gessler, C., Pertot, I. and Perazzolli, M. (2011) *Plasmopara viticola*: a review of knowledge on downy mildew of grapevine and effective disease management. *Phytopathologia Mediterranea* 50, 3–44.

Godard, S., Slacanin, I., Viret, O. and Gindro, K. (2009) Induction of defence mechanisms in grapevine leaves by emodin- and anthraquinone-rich plant extracts and their conferred resistance to downy mildew. *Plant Physiology and Biochemistry* 47, 827–837.

Goupil, P., Benouaret, R., Charrier, O., Ter Halle, A., Richard, C., Eyheraguibel, B., Thiery, D. and Ledoigt, G. (2012) Grape marc extract acts as elicitor of plant defence responses. *Ecotoxicology* 21, 1541–1549.

Gozzo, F. and Faoro, F. (2013) Systemic acquired resistance (50 years after discovery): moving from the lab to the field. *Journal of Agricultural and Food Chemistry* 61, 12473–12491.

Hadwiger, L.A., Ogawa, T. and Kuyama, H. (1994) Chitosan polymer sizes effective in inducing phytoalexin accumulation and fungal suppression are verified with synthesized oligomers. *Molecular Plant-Microbe Interactions* 7, 531–533.

Halleen, F., Fourie, P.H. and Lombard, P.J. (2010) Protection of grapevine pruning wounds against *Eutypa lata* by biological and chemical methods. *South African Journal for Enology and Viticulture* 31, 125–132.

Hamiduzzaman, M.M., Jakab, G., Barnavon, L., Neuhaus, J.M. and Mauch-Mani, B. (2005) β-Aminobutyric acid induced resistance against downy mildew in grapevine acts through the potentiation of callose formation and jasmonic acid signaling. *Molecular Plant-Microbe Interactions* 18, 819–829.

Hammami, I., Kamoun, N. and Rebay, A. (2011a) Biocontrol of *Botrytis cinerea* with essential oil and methanol extract of *Viola odorata* L. flowers. *Archives of Applied Science Research* 3, 44–51.

Hammami, W., Quiroga Castro, C., Rémus-Borel, W., Labbé, C. and Bélanger, R.R. (2011b) Ecological basis of the interaction between *Pseudozyma flocculosa* and powdery mildew fungi. *Applied and Environmental Microbiology* 77, 926–933.

Hammer, K.A., Carson, C.F. and Riley, T.V. (2003) Antifungal activity of the components of *Melaleuca alternifolia* (tea tree) oil. *Journal of Applied Microbiology* 95, 853–860.

Harm, A., Kassemeyer, H.H., Seibicke, T. and Regner, F. (2011) Evaluation of chemical and natural resistance inducers against downy mildew (*Plasmopara viticola*) in grapevine. *American Journal of Enology and Viticulture* 62, 184–192.

Harman, G.E. (2006) Overview of mechanisms and uses of *Trichoderma* spp. *Phytopathology* 96, 190–194.

Harman, G.E., Latorre, B., Agosin, E., San Martin, R., Riegel, D.G., Nielsen, P.A., Tronsmo, A. and Pearson, R.C. (1996) Biological and integrated control of *Botrytis* bunch rot of grape using *Trichoderma* spp. *Biological Control* 7, 259–266.

Heintz, C. and Blaich, R. (1990) *Verticillium lecanii* als Hyperparasite des Rebmehltaus (*Uncinula necator*). *Vitis* 29, 229–232.

Iriti, M., Picchi, V., Rossoni, M., Gomarasca, S., Ludwig, N., Gargano, M. and Faoro, F. (2009) Chitosan antitranspirant activity is due to abscisic acid-dependent stomatal closure. *Environmental and Experimental Botany* 66, 493–500.

Iriti, M., Vitalini, S., Di Tommaso, G., D'Amico, S., Borgo, M. and Faoro, F. (2011) New chitosan formulation prevents powdery mildew infection and improves polyphenols content and free radical scavenging activity of grape and wine. *Australian Journal of Grape and Wine Research* 17, 263–269.

Jakab, G., Cottier, V., Toquin, V., Rigoli, G., Zimmerli, L., Métraux, J.P. and Mauch-Mani, B. (2001) β-Aminobutyric acid-induced resistance in plants. *Journal of Plant Pathology* 107, 29–37.

Jeandet, P., Adrian, M., Breuil, A.C., Sbaghi, M., Debord, S., Bessis, R., Weston, L.A. and Harmon, R. (2000) Chemical induction of phytoalexin synthesis in grapevines: application to the control of grey mould in the vineyard. *Acta Horticulturae* 528, 591–596.

John, S., Wicks, T.J., Hunt, J.S., Lorimer, M.F., Oakey, H. and Scott, E.S. (2005) Protection of grapevine pruning wounds from infection by *Eutypa lata* using *Trichoderma harzianum* and *Fusarium lateritium*. *Australasian Plant Pathology* 34, 569–575.

Karabulut, O.A., Smilanick, J.L., Mlikota Gabler, F., Mansour, M. and Droby, S. (2003) Near harvest applications of *Metschnikowia fructicola*, ethanol, and sodium bicarbonate to control postharvest diseases of grape in central California. *Plant Disease* 87, 1384–1389.

Kast, W.K. (2000) Salicylic acid and phosphorous acid – possible alternative of copper? In: Willer, H. and Meier, U. (eds) *Proceedings of the 6th International Congress on Organic Viticulture, 25–26 August, Convention Center Basel*. Forschungsinstitut für Biologischen Landbau (FiBL), Frick, Switzerland and Stiftung Ökologie & Landbau (SÖL), Bad Dürkheim, Germany, pp. 177–179. Available at: http://orgprints.org/548/1/willer-meier-2000-winecongress.pdf (accessed 16 September 2015).

Kauss, H., Jeblick, W. and Domard, A. (1989) The degrees of polymerization and N-acetylation of chitosan determine its ability to elicit callose formation in suspension cells and protoplasts of *Catharanthus roseus*. *Planta* 178, 385–392.

Kessmann, H., Staub, T., Hofmann, C., Maetzke, T. and Herzog, J. (1994) Induction of systemic acquired disease resistance in plants by chemicals. *Annual Review of Phytopathology* 32, 439–459.

Kiss, L. (1998) Natural occurrence of *Ampelomyces quisqualis* intracellular mycoparasites in mycelia of powdery mildew. *New Phytologist* 140, 709–714.

Kiss, L. and Nakasone, K.K. (1998) Ribosomal DNA internal transcribed spacer sequences do not support the species status of *Ampelomyces quisqualis*, a hyperparasite of powdery mildew fungi. *Current Genetics* 33, 362–367.

Konstantinidou-Doltsinis, S., Markellou, E., Kasselaki, A.M., Siranidou, E., Kalamarakis, A., Tzembelikou, K., Schmitt, A., Koumakis, C. and Malathrakis, N. (2007) Control of powdery mildew of grape in Greece using Sporodex® L and Milsana®. *Journal of Plant Diseases and Protection* 114, 256–262.

Kortekamp, A. (1997) *Epicoccum nigrum* link – a biological control agent of *Plasmopara viticola* (Berk. et Curt.) Berl. et De-Toni? *Vitis* 36, 215–216.

Kotze, C., Van Niekerk, J., Mostert, L., Halleen, F. and Fourie, P. (2011) Evaluation of biocontrol agents for grapevine pruning wound protection against trunk pathogen infection. *Phytopathologia Mediterranea* 50, S247–S263.

Kuntz, C., Sellam, O. and Bertheau, Y. (1992) Purification and characterization of a chitinase from *Aphanocladium album*. *Physiological and Molecular Plant Pathology* 40, 117–131.

La Torre, A., Mandalà, C., Pezza, L., Caradonia, F. and Battaglia, V. (2014) Evaluation of essential plant oils for the control of *Plasmopara viticola*. *Journal of Essential Oil Research* 26, 282–291.

Lebrihi, A., Errakhi, R. and Barakate, M. (2009) New *Streptomyces barakatei* strain, culture filtrate, derived active compounds and use thereof in the treatment of plants. Patent WO/2009/156687–156688. Available at: http://www.google.com/patents/WO2009156687A2?cl=en (accessed 16 September 2015).

Legler, S.E., Caffi, T., Benuzzi, M., Ladurner, E. and Rossi, V. (2011) New perspectives for the use of *Ampelomyces*-based biofungicides for effective control of powdery mildew on grapevine. In: *4ème Conférence Internationale sur les Méthodes Alternatives en Protection des Cultures. Evolution des Cadres Réglementaires Européen et Français. Nouveaux Moyens et Stratégies Innovantes, Nouveau Siècle, Lille, France, 8–10 Mars 2011*. Association Française de Protection des Plantes, Alforteville, France, pp. 546–551.

Lenzi, L., Palmieri, M.C., Caruso, C., Pertot, I. and Perazzolli, M. (2013) Characterization of efficient resistance inducers for control of crop disease. *IOBC-WPRS Bulletin* 89, 137–138.

Lima, G., Ippolito, A., Nigro, F. and Salerno, M. (1997) Lotta biologica contro la muffa grigia dell'uva da tavola in conservazione mediante trattamenti pre-raccolta con *Aureobasidium pullulans* e *Candida oleophila* [Biological control of grey mould of stored table grapes by pre-harvest applications of *Aureobasidium pullulans* and *Candida oleophila*]. *Dell'Historia delle Piante* 20, 21–28. [English abstract.]

Lima, G., De Curtis, F., Castoria, R. and De Cicco, V. (1998) Activity of the yeasts *Cryptococcus laurentii* and *Rhodotorula glutinis* against post-harvest rots in different fruits. *Biocontrol Science and Technology* 8, 257–267.

Lizzi, Y., Coulomb, C., Polian, C., Coulomb, P.J. and Coulomb, P.O. (1998) L'algue face au mildiou: quel avenir? Des résultats de laboratoire très encourageants [Seaweed and mildew: what does the future hold?]. *Phytoma* 508, 29–30. [English abstract.]

Loqman, S., Barka, E.A., Clément, C. and Ouhdouch, Y. (2009) Antagonistic actinomycetes from Moroccan soil to control the grapevine gray mold. *World Journal of Microbiology and Biotechnology* 25, 81–91.

Magnin-Robert, M., Trotel-Aziz, P., Quantinet, D., Biagianti, S. and Aziz, A. (2007) Biological control of *Botrytis cinerea* by selected grapevine-associated bacteria and stimulation of chitinase and β-1,3-glucanase activities under field conditions. *European Journal of Plant Pathology* 118, 43–57.

Magnin-Robert, M., Quantinet, D., Couderchet, M., Aziz, A. and Trotel-Aziz, P. (2013) Differential induction of grapevine resistance and defense reactions against *Botrytis cinerea* by bacterial mixtures in vineyards. *BioControl* 58, 117–131.

Masih, E.I. and Paul, B. (2002) Secretion of beta-1,3-glucanases by the yeast *Pichia membranifaciens* and its possible role in the biocontrol of *Botrytis cinerea* causing grey mould disease of the grapevine. *Current Microbiology* 44, 391–395.

Masih, E.I., Alie, I. and Paul, B. (2000) Can the grey mould disease of the grapevine be controlled by yeast? *FEMS Microbiology Letters* 189, 233–237.

Miceli, A., Ippolito, A., Linsalata, V. and Nigro, F. (1999) Effect of preharvest calcium treatments on decay and biochemical changes in table grape during storage. *Phytopathologia Mediterranea* 38, 47–53.

Monchiero, M., Piano, S. and Gullino, M.L. (1996) Valutazione dell'attività di *Ampelomyces quisqualis* nei confronti di *Uncinula necator* nell'Italia centro-settentrionale [Evaluation of the efficacy of *Ampelomyces quisqualis* against grape powdery mildew in North-Central Italy]. *Informatore Fitopatologico* 9, 19–23. [English abstract.]

Musetti, R., Vecchione, A., Stringher, L., Borselli, S., Zulini, L., Marzani, C., D'Ambrosio, M., Sanità di Toppi, L. and Pertot, I. (2006) Inhibition of sporulation and ultrastructural alterations of grapevine downy mildew by the endophytic fungus *Alternaria alternata*. *Phytopathology* 96, 689–698.

Musetti, R., Polizzotto, R., Vecchione, A., Borselli, S., Zulini, L., D'Ambrosio, M., Sanità di Toppi, L. and Pertot, I. (2007) Antifungal activity of diketopiperazines extracted from *Alternaria alternata* against *Plasmopara viticola:* an ultrastructural study. *Micron* 38, 643–650.

Mutawila, C., Fourie, P.H., Halleen, F. and Mostert, L. (2011) Grapevine cultivar variation to pruning wound protection by *Trichoderma* species against trunk pathogens. *Phytopathologia Mediterranea* 50, S264–S276.

Mutawila, C., Halleen, F. and Mostert, L. (2013) Optimisation of time of application of *Trichoderma* bio-control agents for grapevine pruning wound protection. *IOBC-WPRS Bulletin* 86, 175–176.

Nadeem, S.M., Ahmad, M., Zahir, Z.A., Javaid, A. and Ashraf, M. (2014) The role of mycorrhizae and plant growth promoting rhizobacteria (PGPR) in improving crop productivity under stressful environments. *Biotechnology Advances* 32, 429–448.

Nally, M.C., Pesce, V.M., Maturano, Y.P., Muñoz, C.J., Combina, M., Toro, M.E., Castellanos de Figueroa, L.I. and Vazquez, F. (2012) Biocontrol of *Botrytis cinerea* in table grapes by non-pathogenic indigenous *Saccharomyces cerevisiae* yeasts isolated from viticultural environments in Argentina. *Postharvest Biology and Technology* 64, 40–48.

Nally, M.C., Pesce, V.M., Maturano, Y.P., Toro, M.E., Combina, M., Castellanos de Figueroa, L.I. and Vazquez, F. (2013) Biocontrol of fungi isolated from sour rot infected table grapes by *Saccharomyces* and other yeast species. *Postharvest Biology and Technology* 86, 456–462.

Nawrocka, J. and Małolepsza, U. (2013) Diversity in plant systemic resistance induced by *Trichoderma*. *Biological Control* 67, 149–156.

Nigro, F., Finetti Sialer, M.M. and Gallitelli, D. (1999) Transformation of *Metschnikowia pulcherrima* 320, biocontrol agent of storage rot, with the green fluorescent protein gene. *Journal of Plant Pathology* 81, 205–208.

Nigro, F., Schena, L., Ligorio, A., Pentimone, I., Ippolito, A. and Salerno, M.G. (2006) Control of table grape storage rots by pre-harvest applications of salts. *Postharvest Biology and Technology* 42, 142–149.

O'Neil, T.M., Elad, Y., Shtienberg, D. and Cohen, A. (1996) Control of grapevine grey mould with *Trichoderma harzianum* T39. *Biocontrol Science and Technology* 6, 139–146.

Ortugno, C., Galeazzi, M., Rubboli, V., Valente, M., Busin, B. and Gualco, A. (2014) Sakalia®: risultati di prove sperimentali contro oidio su orticole e vite [Sakalia®: field trials against powdery mildew on vegetables and grapes]. *Atti Giornate Fitopatologiche* 2, 261–270. [English abstract.]

Pajot, E., Mounier, E., Blal, B., Cortes, F. and Coutant, A. (2012) Selection steps of the *Trichoderma atroviride* strain i-1237, and some practical results of the development of the fungi within the cankers diseases of grapevine and soil borne diseases of vegetables. In: Barker, E.A. and Clément, C. (Organizers) *IOBC, Reims, 24–27 June 2012, Working Group, Biological Control of Plant Pathogens: Biocontrol of Plant Pathogens in Sustainable Agriculture. Book of Abstracts*. Université de Reims, Reims, Champagne-Ardenne, France, p. 40. Available at: http://www.envirochange.eu/english/2012_Reims/20120624_IOBC_Reims_program_final.pdf (accessed 15 September 2015).

Palmieri, M.C., Perazzolli, M., Matafora, V., Moretto, M., Bachi, A. and Pertot, I. (2012) Proteomic analysis of grapevine resistance induced by *Trichoderma harzianum* T39 reveals specific defence pathways activated against downy mildew. *Journal of Experimental Botany* 63, 6237–6251.

Parafati, L., Vitale, A., Restuccia, C. and Cirvilleri, G. (2015) Biocontrol ability and action mechanism of food-isolated yeast strains against *Botrytis cinerea* causing post-harvest bunch rot of table grape. *Food Microbiology* 47, 85–92.

Paul, B. (2004) A new species of *Pythium* isolated from Burgundian vineyards and its antagonism towards *Botrytis cinerea,* the causative agent of the grey mould disease. *FEMS Microbiology Letters* 234, 269–274.

Paul, B., Girard, I., Bhatnagar, T. and Bouchet, P. (1997) Suppression of *Botrytis cinerea* causing grey mould disease of grapevine (*Vitis vinifera*) and its pectinolytic activities by a soil bacterium. *Microbiological Research* 152, 413–420.

Perazzolli, M., Dagostin, S., Ferrari, A., Elad, Y. and Pertot, I. (2008) Induction of systemic resistance against *Plasmopara viticola* in grapevine by *Trichoderma harzianum* T39 and benzothiadiazole. *Biological Control* 47, 228–234.

Pertot, I., Oliveira Longa, C.M., Savazzini, F., Michelon, L., Prodorutti, D., Pellegrini, C. and Ferrari, A. (2009) *Trichoderma atroviride* SC1 for biocontrol of fungal diseases in plants *IOBC-WPRS Bulletin* 43, 185–188.

Pertot, I., Puopolo, G., Hosni, T., Pedrotti, L., Jourdan, E. and Ongena, M. (2013) Limited impact of abiotic stress on surfactin production in planta and on disease resistance induced by *Bacillus amyloliquefaciens* S499 in tomato and bean. *FEMS Microbiology Letters* 86, 505–519.

Pieterse, C.M.J., Van Pelt, J.A., Van Wees, S.C.M., Ton, J., Léon-Kloosterziel, K.M., Keurentjes, J.J.B., Verhagen, B.W.M., Knoester, M., Van der Sluis, I., Bakker, P.A.H.M. and Van Loon, L.C.

(2001) Rhizobacteria-mediated induced systemic resistance: triggering, signaling and expression. *European Journal of Plant Pathology* 107, 51–61.
Pieterse, C.M.J., Zamioudis, C., Berendsen, R.L., Weller, D.M., Van Wees, S.C.M. and Bakker, P.A.H.M. (2014) Induced systemic resistance by beneficial microbes. *Annual Review of Phytopathology* 52, 347–375.
Pujos, P., Martin, A., Farabullini, F. and Pizzi, M. (2014) Risultati dell'attività sperimentale condotta in Italia su vite con il biofungicida Romeo™, a base di cerevisane [Experimental results of Romeo™, cerevisane-based biofungicide on grape in Italy]. *Atti Giornate Fitopatologiche* 2, 255–260. [English abstract.]
Puopolo, G., Giovannini, O. and Pertot, I. (2014) *Lysobacter capsici* AZ78 can be combined with copper to effectively control *Plasmopara viticola* on grapevine. *Microbiological Research* 169, 633–642.
Qin, X., Xiao, H., Xue, C., Yu, Z., Yang, R., Cai, Z. and Si, L. (2015) Biocontrol of gray mold in grapes with the yeast *Hanseniaspora uvarum* alone and in combination with salicylic acid or sodium bicarbonate. *Postharvest Biology and Technology* 100, 160–167.
Rabosto, X., Carrau, M., Paz, A., Boido, E., Dellacassa, E. and Carrau, F.M. (2006) Grapes and vineyard soils as sources of microorganisms for biological control of *Botrytis cinerea*. *American Journal of Enology and Viticulture* 57, 332–338.
Raspor, P., Miklič-Milek, D., Avbelj, M. and Čadež, N. (2010) Biocontrol of grey mould disease on grape caused by *Botrytis cinerea* with autochthonous wine yeasts. *Food Technology and Biotechnology* 48, 336–343.
Reggiori, F., Aloi, C., Baleani, M., Benanchi, M., Bigot, G., Bortolotti, P., Bossio, D., Cotromino, M., Di Marco, S., Faccini, F. *et al.* (2014) Remedier® (*Trichoderma asperellum e Trichoderma gamsii*): nuova opportunità di contenimento del complesso del mal dell'esca della vite. Risultati di quattro anni di sperimentazione in Italia [Remedier® (*Trichoderma asperellum e Trichoderma gamsii*): a new opportunity to control the esca disease complex. Four-year results of field trials in Italy]. *Atti Giornate Fitopatologiche* 2, 363–372. [English abstract.]
Reglinski, T., Elmer, P.A.G., Taylor, J.T., Wood, P.N. and Hoyte, S.M. (2010) Inhibition of *Botrytis cinerea* growth and suppression of botrytis bunch rot in grapes using chitosan. *Plant Pathology* 59, 882–890.
Reino, J.L., Guerrero, R.F., Hernández-Galán, R. and Collado, I.G. (2008) Secondary metabolites from species of the biocontrol agent *Trichoderma*. *Phytochemistry Reviews* 7, 89–123.
Repka, V. (2001) Elicitor-stimulated induction of defense mechanisms and defense gene activation in grapevine cell suspension cultures. *Biologia Plantarum* 44, 555–565.
Repka, V., Fischerová, I. and Silhárová, K. (2004) Methyl jasmonate is a potent elicitor of multiple defense responses in grapevine leaves and cell-suspension cultures. *Biologia Plantarum* 48, 273–283.
Repka, V., Čarná, M. and Pavlovkin, J. (2013) Methyl jasmonate-induced cell death in grapevine requires both lipoxygenase activity and functional octadecanoid biosynthetic pathway. *Biologia* 68, 896–903.
Reuveni, M., Zahavi, T. and Cohen, Y. (2001) Controlling downy mildew (*Plasmopara viticola*) in field-grown grapevine with beta-aminobutyric acid (BABA). *Phytoparasitica* 29, 125–133.
Reymond, P., Grünberger, S., Paul, K., Müller, M. and Farmer, E.E. (1995) Oligogalacturonide defense signals in plants: large fragments interact with the plasma membrane *in vitro*. *Proceedings of the National Academy of Sciences of the United States of America*, 4145–4149.
Reynolds, A.G, Veto, L.J., Sholberg, P.L., Wardle, D.A. and Haag, P. (1996) Use of potassium silicate for the control of powdery mildew [*Uncinula necator* (Schwein) Burrill] in *Vitis vinifera* L. cultivar Bacchus. *American Journal of Enology and Viticulture* 47, 421–428.
Roatti, B., Perazzolli, M., Gessler, C. and Pertot, I. (2013) Abiotic stresses affect *Trichoderma harzianum* T39-induced resistance to downy mildew in grapevine. *Phytopathology* 103, 1227–1234.

Romanazzi, G., Lichter, A., Gabler, F.M. and Smilanick, J.L. (2012) Recent advances on the use of natural and safe alternatives to conventional methods to control postharvest gray mold of table grapes. *Postharvest Biology and Technology* 63, 141–147.
Romanazzi, G., Mancini, V., Feliziani, E., Bastianelli, M., Servili, A., Nardi, S. and Flamini, L. (2014) Efficacia di prodotti alternativi nella difesa antiperonosporica della vite [Effectiveness of alternative compounds in the control of grapevine downy mildew]. *Atti Giornate Fitopatologiche* 2, 363–372. [English abstract.]
Ronseaux, S., Clément, C. and Ait Barka, E. (2013) Interaction of *Ulocladium atrum*, a potential biological control agent, with *Botrytis cinerea* and grapevine plantlets. *Agronomy* 3, 632–647.
Rühmann, S., Pfeiffer, J., Brunner, P., Szankowski, I., Fischer, T.C., Forkmann, G. and Treutter, D. (2013) Induction of stilbene phytoalexins in grapevine (*Vitis vinifera*) and transgenic stilbene synthase-apple plants (*Malus domestica*) by a culture filtrate of *Aureobasidium pullulans. Plant Physiology and Biochemistry* 72, 62–71.
Sackenheim, R., Weltzien, H.C. and Kast, W.K. (1994) Effects of microflora composition in the phyllosphere on biological regulation of grapevine fungal diseases. *Vitis* 33, 235–240.
Sahariah, P., Gaware, V.S., Lieder, R., Jónsdóttir, S., Hjálmarsdóttir, M.Á., Sigurjonsson, O.E. and Másson, M. (2014) The effect of substituent, degree of acetylation and positioning of the cationic charge on the antibacterial activity of quaternary chitosan derivatives. *Marine Drugs* 12, 4635–4658.
Salmon, J.M. (2009) Valutazione della vaporizzazione del lievito come strumento per ridurre le malattie fungine sulla vite. In: Trioli, G. and Hofmann, U. (eds) *Codice di Buone Pratiche per la Viticoltura e l'Enologia Biologica*. ORWINE – Policy-oriented Research (SSP) – Progetto Nr. 022769 per la Commissione Europea and ECOVIN – German Organic Winegrowers Association, pp. 216–218. Available at: http://www.agrinnovazione.regione.sicilia.it/reti/Viticoltura/pubblicazioni/allegati/viticoltura_biologica.pdf (accessed 15 September 2015).
Sanchez, L., Courteaux, B., Hubert, J., Kauffmann, S., Renault, J., Clément, C., Baillieul, F. and Dorey, S. (2012) Rhamnolipids elicit defense responses and induce disease resistance against biotrophic, hemibiotrophic, and necrotrophic pathogens that require different signaling pathways in *Arabidopsis* and highlight a central role for salicylic acid. *Plant Physiology* 160, 1630–1641.
Santos, A. and Marquina, D. (2004) Killer toxin of *Pichia membranifaciens* and its possible use as a biocontrol agent against grey mould disease of grapevine. *Microbiology* 150, 2527–2534.
Sawant, S.D. (2014) *Trichoderma* – foliar pathogen interactions. *The Open Mycology Journal* 8, 58–70.
Sawant, S.D. and Sawant, I.S. (2008) Use of potassium bi-carbonates for the control of powdery mildew in table grapes. *Acta Horticulturae* 785, 285–291.
Sawant, S.D., Sawant, I.S., Shetty, D., Shinde, M., Jade, S. and Waghmare, M. (2011) Control of powdery mildew in vineyards by Milastin K, a commercial formulation of *Bacillus subtilis* (KTBS). *Journal of Biological Control* 25, 26–32.
Schena, L., Ippolito, A., Zahavi, T., Cohen, L., Nigro, F. and Droby, S. (1999) Genetic diversity and biocontrol activity of *Aureobasidium pullulans* isolates against postharvest rots. *Postharvest Biology and Technology* 17, 189–199.
Schena, L., Nigro, F., Pentimone, I., Ligorio, A. and Ippolito, A. (2003) Control of postharvest rots of sweet cherries and table grapes with endophytic isolates of *Aureobasidium pullulans*. *Postharvest Biology and Technology* 30, 209–220.
Schmidt, C.S., Lorenz, D., Wolf, G.A. and Jäger, J. (2001) Biological control of the grapevine dieback fungus *Eutypa lata*, II: influence of formulation additives and transposon mutagenesis on the antagonistic activity of *Bacillus subtilis* and *Erwinia herbicola*. *Journal of Phytopathology* 149, 437–445.

Schmitt, A. (2006) Induced resistance with extracts of *Reynoutria sachalinensis*: crucial steps behind the scene. *IOBC/WPRS Bulletin* 29(8), 85–90.
Schoonbeek, H.J., Jacquat-Bovet, A.C., Mascher, F. and Métraux, J.P. (2007) Oxalate-degrading bacteria can protect *Arabidopsis thaliana* and crop plants against *Botrytis cinerea*. *Molecular Plant-Microbe Interactions* 20, 1535–1544.
Sendhilvel, V., Marimuthu, T. and Samiappan, R. (2007) Talc-based formulation of *Pseudomonas fluorescens*-induced defense genes against powdery mildew of grapevine. *Archives of Phytopathology and Plant Protection* 40, 81–89.
Spagnolo, A., Magnin-Robert, M., Larignon, P., Clément, C. and Fontaine, F. (2012) Preliminary screening for the possible biological control activity of a *Chaetomium* sp. isolate on grapevine trunk disease agents. *Phytopathologia Mediterranea* 2, 449–450.
Sticher, L., Mauch-Mani, B. and Métraux, J.P. (1997) Systemic acquired resistance. *Annual Review of Phytopathology* 35, 235–270.
Sutton, J.C., Li, D.W., Peng, G., Yu, H., Zhang, P. and Valdebenito-Sanhueza, R.M. (1997) *Gliocladium roseum*: a versatile adversary of *Botrytis cinerea* in crops. *Plant Disease* 81, 316–328.
Tamm, L., Thürig, B., Fliessbach, A., Goltlieb, A.E., Karavani, S. and Cohen, Y. (2011) Elicitors and soil management to induce resistance against fungal plant diseases. *NJAS – Wageningen Journal of Life Sciences* 58, 131–137.
Tasin, M., Sicher, C., Zeqir, A., Simeone, V. and Pertot, I. (2009) Double role of *Metschnikowia fructicola*: biocontrol agent of *Botrytis cinerea* and repellant to *Lobesia botrana*. *IOBC-WPRS Bulletin* 43, 263–266.
Thuerig, B., Binder, A., Boller, T., Guyer, U., Jimenez, S., Rentsch, C. and Tamm, L. (2006) An aqueous extract of the dry mycelium of *Penicillium chrysogenum* induces resistance in several crops under controlled and field conditions. *European Journal of Plant Pathology* 114, 185–197.
Thuerig, B., Slaughter, A., Marouf, E., Held, M., Mauch-Mani, B. and Tamm, L. (2011) Site-specific field resistance of grapevine to *Plasmopara viticola* correlates to altered gene expression and was not modulated by the application of organic amendments. *European Journal of Plant Pathology* 129, 255–265.
Trotel-Aziz, P., Couderchet, M., Vernet, G. and Aziz, A. (2006) Chitosan stimulates defense reactions in grapevine leaves and inhibits development of *Botrytis cinerea*. *European Journal of Plant Pathology* 114, 405–413.
Trotel-Aziz, P., Couderchet, M., Biagianti, S. and Aziz, A. (2008) Characterization of new bacterial biocontrol agents *Acinetobacter*, *Bacillus*, *Pantoea* and *Pseudomonas* spp. mediating grapevine resistance against *Botrytis cinerea*. *Environmental and Experimental Botany* 64, 21–32.
Trouvelot, S., Varnier, A.L., Allègre, M., Mercier, L., Baillieul, F., Arnould, C., Gianinazzi-Pearson, V., Klarzynski, O., Joubert, J.M., Pugin, A. and Daire, X. (2008) A β-1,3 glucan sulfate induces resistance in grapevine against *Plasmopara viticola* through priming of defense responses, including HR-like cell death. *Molecular Plant-Microbe Interactions* 21, 232–243.
Van Aubel, G., Buonatesta, R. and Van Cutsem, P. (2014) COS-OGA: a novel oligosaccharidic elicitor that protects grapes and cucumbers against powdery mildew. *Crop Protection* 65, 129–137.
Van Loon, L.C., Bakker, P.A.H.M. and Pieterse, C.M.J. (1998) Systemic resistance induced by rhizosphere bacteria. *Annual Review of Phytopathology* 36, 453–483.
Varnier, A.L., Sanchez, L., Vatsa, P., Boudesocque, L., Garcia-Brugger, A., Rabenoelina, F., Sorokin, A., Renault, J.H., Kauffmann, S., Pugin, A. *et al.* (2009) Bacterial rhamnolipids are novel MAMPs conferring resistance to *Botrytis cinerea* in grapevine. *Plant, Cell and Environment* 32, 178–193.
Vatsa, P., Sanchez, L., Clement, C., Baillieul, F. and Dorey, S. (2010) Rhamnolipid biosurfactants as new players in animal and plant defense against microbes. *International Journal of Molecular Sciences* 11, 5095–5108.

Vera, J., Castro, J., Gonzalez, A. and Moenne, A. (2011) Seaweed polysaccharides and derived oligosaccharides stimulate defense responses and protection against pathogens in plants. *Marine Drugs* 9, 2514–2525.

Verhagen, B.W.M., Trotel-Aziz, P., Couderchet, M., Höfte, M. and Aziz, A. (2010) *Pseudomonas* spp.-induced systemic resistance to *Botrytis cinerea* is associated with induction and priming of defence responses in grapevine. *Journal of Experimental Botany* 61, 249–260.

Verhagen, B.W.M., Trotel-Aziz, P., Jeandet, P., Baillieul, F. and Aziz, A. (2011) Improved resistance against *Botrytis cinerea* by grapevine-associated bacteria that induce a prime oxidative burst and phytoalexin production. *Phytopathology* 101, 768–777.

Verma, M., Brar, S.K., Tyagi, R.D., Surampalli, R.Y. and Valéro, J.R. (2007) Antagonistic fungi, *Trichoderma* spp.: panoply of biological control. *Biochemical Engineering Journal* 37, 1–20.

Vezzulli, S., Civardi, S., Ferrari, F. and Bavaresco, L. (2007) Methyl jasmonate treatment as a trigger of resveratrol synthesis in cultivated grapevine. *American Journal of Enology and Viticulture* 58, 530–533.

Vinale, F., Sivasithamparam, K., Ghisalberti, E.L., Marra, R., Barbetti, M.J., Li, H., Woo, S.L. and Lorito, M. (2008) A novel role for *Trichoderma* secondary metabolites in the interactions with plants. *Physiological and Molecular Plant Pathology* 72, 80–86.

Walters, D.R., Ratsep, J. and Havis, N.D. (2013) Controlling crop diseases using induced resistance: challenges for the future. *Journal of Experimental Botany* 64, 1263–1280.

Whiteman, S.A. and Stewart, A. (1998) Suppression of *Botrytis cinerea* sporulation on irradiated grape leaf tissue by the antagonistic bacterium *Serratia liquefaciens*. *New Zealand Journal of Crop and Horticultural Science* 26, 325–330.

Woo, S.L., Scala, F., Ruocco, M. and Lorito, M. (2006) The molecular biology of the interactions between *Trichoderma* spp., phytopathogenic fungi, and plants. *Phytopathology* 96, 181–185.

Woo, S.L., Ruocco, M., Vinale, F., Nigro, M., Marra, R., Lombardi, N., Pascale, A., Lanzuise, S., Manganiello, G. and Lorito, M. (2014) *Trichoderma*-based products and their widespread use in agriculture. *The Open Mycology Journal* 8, 71–126.

Youssef, K. and Ruffo Roberto, S. (2014a) Applications of salt solutions before and after harvest affect the quality and incidence of postharvest gray mold of 'Italia' table grapes. *Postharvest Biology and Technology* 87, 95–102.

Youssef, K. and Ruffo Roberto, S. (2014b) Salt strategies to control *Botrytis* mold of 'Benitaka' table grapes and to maintain fruit quality during storage. *Postharvest Biology and Technology* 95, 95–102.

Zahavi, T., Cohen, L., Weiss, B., Schena, L., Daus, A., Kaplunov, T., Zutkhi, J., Ben-Arie, R. and Droby, S. (2000) Biological control of *Botrytis*, *Aspergillus* and *Rhizopus* rots on table and wine grapes in Israel. *Postharvest Biology and Technology* 20, 115–124.

Zanzotto, A., Bellotto, D. and Borgo, M. (2005) Efficacia antioidica di AQ_{10} ed effetti sulla formazione di cleistoteci in vigneto [Efficacy of AQ_{10} against powdery mildew and effects on cleistothecia formation in vineyard]. *Informatore Fitopatologico* 7/8, 68–76. [English abstract.]

Zerbetto, F., Saracchi, M., Maffi, D., Jucker, C. and Vercesi, A. (2002) Possibilità di impiego del bicarbonato di sodio nella difesa della vite [Possible use of sodium bicarbonate in grapevine disease management]. *Atti Giornate Fitopatologiche* 2, 395–402. [English abstract.]

Zerbetto, F., Maffi, D., Sburlino, P. and Fremiot, P. (2004) Attività antioidica su vite di alcuni olii vegetali e minerali, utilizzati da soli o con bicarbonato di sodio [Effectiveness of vegetable and mineral oils, used alone or with sodium bicarbonate, against grapevine powdery mildew]. *Atti Giornate Fitopatologiche* 2, 225–226. [English abstract.]

2 Grapevine Endophytes and Plant Health: A Culture-Independent Approach

S. Yousaf,[1*] M. Anees[2] and A. Campisano[3*]

[1]Department of Environmental Sciences, Quaid-i-Azam University, Islamabad, Pakistan; [2]Department of Biochemistry, Quaid-i-Azam University, Islamabad, Pakistan; [3]IASMA Research and Innovation Centre, Fondazione Edmund Mach, San Michele all'Adige, Italy

Introduction

An endophyte is a bacterial (including actinomycetal) or fungal microorganism that spends the whole or part of its life cycle colonizing healthy tissues of the host plant, either intercellularly and/or intracellularly, and typically causes no apparent disease symptoms (Wilson, 1995; Sturz *et al.*, 2000). The endophytic population of a given species varies from several to a few hundred bacterial and fungal strains (Strobel and Long, 1998).

The ecological role of these organisms is still not well determined but most of them have positive effects on host plants, which include promoting plant growth, improving resistance to multiple stresses, maintaining a reliable supply of nutrients, biocontrol of plant parasites and microbial synthesis of metabolites antagonistic to predators (Schulz *et al.*, 2002; Schulz and Boyle, 2005), as well as protection against diseases and insects (Rodriguez *et al.*, 2008). Endophytic microorganisms are also presumed to play an active role in enhancing the host defence response against infection by phytoplasmas (Musetti *et al.*, 2011). Factors responsible for improving plant growth and health are the microbial synthesis of phytohormones (Tudzynski, 1997; Kobayashi and Palumbo, 2000; Tudzynski and Sharon, 2002), access to minerals and other nutrients from the soil (Caldwell *et al.*, 2000; Barrow, 2003) and bacterial atmospheric nitrogen fixation (Reinhold-Hurek and Hurek, 1998).

In this chapter, we address the potential of culture-independent tools for the study of endophytes of the grapevine (*Vitis vinifera* L.) and the potential of as yet uncultured endophytes for plant health. The role of endophytic microbial communities in plant health in general is an exciting frontier for microbiologists. Fascinating interactions among pathogens and microbial communities, such as those observed in disease-suppressive soils, provide inspiration for developing a better understanding of the ecology of plant microbial communities. However, technical limitations have, in the past, made it difficult to assess with any precision the composition of these communities. The low frequency of occurrence of most community components complicates the identification of organisms that contribute towards disease suppression or other ecosystem services (Garrett *et al.*, 2013). New metagenomic tools, such as next-generation

*Corresponding authors: syousaf@qau.edu.pk; andrea.campisano@fmach.it

sequencing (NGS) and microarray technologies, have greatly contributed to overcoming this limitation. These novel approaches support much more complete analyses of microbial communities, but are still hindered by the lagged development of reference databases and experiments that allow the elucidation of gene function. The rapidly growing generation of broad data sets allows for data mining at previously unmatched global scales (Chakraborty and Newton, 2011; Garrett *et al.*, 2011; Juroszek and Von Tiedemann, 2011; Luck *et al.*, 2011).

Metagenomic information bears the promise to support plant health management under different conditions, including spillover for new experimental and analytical methods. New sequencing tools have created a revolution in biologists' ability to study the taxonomic composition and, to a degree, the functional composition of microbial communities. A similar revolution is needed in experimental techniques to evaluate the effects of microbial community structure on outcomes such as plant productivity and health (Alston *et al.*, 2009). The identification of endophytic microbes that contribute to grapevine health is analogous to the identification of genes that contribute to the productivity of particular plant genotypes.

Role of Endophytic Communities in Grapevine Plant Health

We consider beneficial microorganisms to be those known as biocontrol agents and/or growth promoters. Endophytes may represent especially efficient biocontrol agents as they colonize the same ecological niches as phytopathogens do (Ryan *et al.*, 2008). There are several modes of action by which endophytes can be beneficial to grapevine plant health. Their positive effects can be exerted directly or indirectly. Endophytic microorganisms have indirect positive effects on plants by affecting adversely the population density, dynamics and metabolic activities of pathogens, mainly through competition, antibiosis, lysis and hyperparasitism. Competition for space and nutrients against soil-borne pathogens takes place at the root surface; competitive colonization of the endosphere and successful establishment in the root zone and upper plant parts is a prerequisite for effective biocontrol. Antagonistic microorganisms can often produce a range of different antimicrobial secondary metabolites, and/or extracellular lytic enzymes. Direct positive effects on plants are exerted through phytostimulation and biofertilization of plants; these processes involve the production of phytohormones, non-symbiotic nitrogen fixation, and the increased availability of phosphate and other nutrients in the soil (Burdman *et al.*, 2001).

There are many potential roles for endophytes in epidemics. They may alter the uptake of plant nutrients, thereby potentially affecting nutrient stress as well as susceptibility to infectious disease. They may alter plant susceptibility through processes such as induced systemic resistance (ISR) and systemic acquired resistance (SAR) (Vallad and Goodman, 2004; Van Loon, 2007), or through potentially complex network structures related to microbial associations (Garrett *et al.*, 2013).

Endophytic Bacteria

Endophytes are known to reside in the roots, stems and leaves of plants, and their diversity encompasses a wide part of bacterial taxonomy. A survey dating back to 2002 enumerated 82 genera found in a broad range of plants (Lodewyckx *et al.*, 2002). The frequent occurrence of endophytic bacteria belonging to the genera *Pseudomonas*, *Enterobacter* and *Bacillus* in grapevines has been previously reported (Bell *et al.*, 1995; Thomas, 2004). Bulgari *et al.* (2011) described the bacterial diversity of healthy grapevines, flavescence dorée-affected grapevines and recovered grapevines in order to investigate the possible relation between recovery from this phytoplasmal disease and the presence of endophytic bacteria. The authors used length heterogeneity polymerase chain reaction (LH-PCR) to analyse the endophytic bacterial diversity in the grapevines under examination. In this study, a less diverse bacterial community was observed in diseased plants than in healthy and recovered plants. This could be the result of (i) a direct interaction between phytoplasmas and endophytic bacteria, or (ii) a phytoplasma-mediated plant response that restructured the endophytic bacterial community. In grapevines, phytoplasmas could compete with endophytic bacteria for carbon sources and the favourable niche that they both inhabit, thereby leading to a reduction of bacterial richness in infected plants (Musetti *et al.*, 2007).

Burkholderia sp., *Bacillus pumilus*, *Paenibacillus pasadenensis* and uncultured *Bacillus* were identified only in recovered plants. These taxa are all known as resistance inducers. *Burkholderia* sp. is capable of controlling several plant pathogens owing to its ability to produce antifungal compounds (El-Banna and Winkelmann, 1998). Moreover, *Bacillus pumilus* and other *Bacillus* strains such as *B. subtilis*, *B. cereus*, *B. pasteuri*, *B. sphaericus* and *Bacillus* sp. elicited a significant reduction of disease severity on different hosts by inducing ISR (Choudhary and Johri, 2009), a process that protects a plant from pathogens for a considerable part of its lifetime. Consequently, plant susceptibility, disease severity and the number of diseased plants can be reduced (Van Loon, 2007).

In the case of grapevine yellows (GY) diseases, it was observed that recovered plants maintained their healthy condition for at least 2 years and were not easily reinfected by phytoplasmas. The evidence that recovery was stable and that ISR-inducing bacteria were present in recovered grapevines suggested a possible involvement of endophytes in recovery from GY. In the GY-altered grapevine endophytic bacterial community after phytoplasmal infection, bacterial taxa with known tolerance towards reactive oxygen species (ROS) and with the ability to trigger plant defences were enriched. These bacteria could ultimately be responsible for disease recovery. This view is supported by the findings of some other studies which showed that recovered grapevine plants have higher amounts of ROS than diseased and healthy plants (Musetti *et al.*, 2007).

Bulgari *et al.* (2014) recently described the evolution of endophytic bacterial communities in grapevines affected by flavescence dorée during the crucial months when symptoms become evident and until later in the season. They supported their LH-PCR data with taxon-specific real-time PCR, and highlighted the possible role of the bacterial taxa *Burkholderia*, *Methylobacterium* and *Pantoea* in interacting with the phytoplasmas. Interestingly, the effect of phytoplasmas in diseased plants induced restructuring of the bacterial endophyte community in a season-dependent manner. Among those taxa in which biocontrol activity has been reported, those belonging to the genus *Pantoea* were limited to the months of July, August and October in disease-recovered plants, but not in healthy or diseased plants. Endophytes belonging to the genus *Burkholderia* were detected late in the season (September) in healthy plants, but had already appeared in July in diseased and recovered plants. These authors also suggested that the pathogen may alter the impact of seasonality on the microbial endophyte community dynamics, and proposed further investigation of the role of endophytes in the biocontrol of phytoplasmal disease.

Recently, a combination of methods that are cultivation dependent and independent were used to expose a highly uncommon event of host switching for the bacterium *Propionibacterium acnes* (Campisano *et al.*, 2014a), from humans to the grapevine. Interestingly, 16 rDNA (nuclear ribosomal DNA) sequences obtained through 454 pyrosequencing showed that the bacteria that had switched hosts also had a slightly different sequence. Conventional Sanger sequencing was used to study additional genes (*recA* and *tly*) in this symbiont in order to understand its potential for pathogenesis on the grapevine, as *P. acnes* is known to cause post-surgical infections on human hosts. The host switch was associated with loss of *recA* function and possibly the establishment of endocellular symbiosis in plant cells, while no detrimental health effect was associated with the presence of this bacterium in the plant's tissues.

Some bacteria that are potentially harmful to humans can also adopt an endophytic habit in grapevines, as shown by Yousaf *et al.* (2014). At least four taxa, including well-known human and animal pathogens, are present in grapevine, both as wood and leaf endophytes: *Staphylococcus*, *Propionibacterium*, *Clostridium* and *Burkholderia*. While their presence in plants does not seem to be associated with disease or other detectable effects, they may represent a threat to human health if they are detected in berries.

Bacterial distribution and biodiversity in grapevine plants

An understanding of the population dynamics of bacterial endophytes will enable more use to be made of their beneficial characteristics (Hallmann *et al.*, 1997). Endophytic bacteria belonging to various different genera have been reported to be present in grapevines. For example, the following genera have been detected: (i) the *Gammaproteobacteria* genera *Enterobacter*, *Pectobacterium*, *Sphingomonas* and *Stenotrophomonas*; (ii) the *Alphaproteobacteria* genera *Methylobacterium*, *Brevundimonas* and *Agrobacterium*; (iii) the *Betaproteobacteria* genus *Burkholderia*; (iv) the *Firmicutes* genera *Brevibacillus*, *Staphylococcus*, *Paenibacillus*, *Oceanobacillus* and *Bacillus*; and (v) the *Actinobacteria* genera *Acaricomes* and *Streptomyces* (Bulgari *et al.*, 2011). West *et al.* (2010) investigated the diversity and distribution of the non-culturable bacterial endophyte populations in grapevine and explored endophyte entry into vines in an effort to understand their impact on plant health.

The diversity of the bacteria associated with grapevines and endophytes was studied extensively using pyrosequencing by Pinto *et al.* (2014) and Campisano *et al.* (2014b). Both studies highlighted the variability of bacterial communities. While the former study broadly assessed the factors affecting plant-associated microorganisms as affected by a number of factors (including seasonality and treatment with fungicides), the latter study specifically addressed endophytic bacteria and how they differ between two well-known and common grapevine cultivars (Chardonnay and Merlot), and between vineyards where organic production or integrated pest management (IPM) are used. The most prevalent bacterial taxa at the time of sampling (*Ralstonia*, *Staphylococcus*, *Burkholderia* and *Pseudomonas*) were identified, and taxa *Mesorhizobium*,

Caulobacter and *Staphylococcus* were indicated as occurring more frequently in plants from organic production.

Endophytic Fungi

A large number of fungal endophytes have been reported in grapevines. Pancher *et al.* (2012) isolated many different endophytic fungal species, including *Alternaria*, *Epicoccum nigrum*, *Cladosporium*, *Xylaria*, *Botryosphaeria obtusa*, *B. dothidea*, *Truncatella angustata*, *Neofusicoccum parvum*, *Phoma herbarum* and *Davidiella tassiana*. Increasing amounts of evidence suggest that endophytic fungus–plant interactions are an important determinant of plant health, evolution and biodiversity (Davitt *et al.*, 2010; Gundel *et al.*, 2010). The colonization of host plants by endophytic fungi is believed to contribute to plant genotype adaptability to biotic and abiotic stress factors (Waller *et al.*, 2005; Bae *et al.*, 2009). The hyphomycete *Acremonium byssoides* colonizes the host grapevine without any visible effect and has been described as an antagonist of the downy mildew agent *Plasmopara viticola* (Burruano *et al.*, 2008). In fact, *A. byssoides* is able to parasitize *P. viticola* and to produce secondary metabolites, named acremines A–F, which inhibit sporangial germination of this oomycete (Assante *et al.*, 2005).

Bruez *et al.* (2014) studied the variability of healthy and esca-diseased grapevines by both single strand conformation polymorphism (SSCP) and the identification of fungal isolates from various surface and internal plant tissues. SSCP data were analysed by multivariate statistics and grouped together fungal communities into three clusters: those inhabiting the bark, those in the white rot tissue and those dwelling in the healthy inner tissues. Fungal communities from different seasons often formed different clusters (those in January and April, and a third cluster including samples from June and September). Overall, a high abundance of Hypocreales and Botryosphaeriales fungi were detected, but no difference in fungal richness and diversity was detected between esca-diseased and healthy plants over the four sampling seasons. Similarly to alpha diversity, differences in community composition between the healthy tissues of plants with or without grapevine trunk disease symptoms were not detected.

Endophytic fungi as biological control agents

Fungal endophytes have been successfully used as biological control agents against various grapevine diseases. It is recognized that these endophytes are of great importance for the hosts, and their role in the protection of the plants against insects, nematodes and pathogenic microorganisms has been well demonstrated (Saikkonen *et al.*, 1998; Arnold, 2007; Kavroulakis *et al.*, 2007). However, the role of endophytic microorganisms in relation to systemic, incurable diseases, such as those associated with phytoplasmas, is still to be investigated. Spontaneous recovery from phytoplasma-induced symptoms has been reported in grapevines and fruit crops, indicating that a type of resistance could occur in previously infected plants (Musetti *et al.*, 2007). It has been hypothesized that endophytic microorganisms may take part in this resistance phenomenon (Fleet *et al.*, 2002).

In order to determine the distribution and frequency of endophytes in host tissues, accurate identification of the organisms is required. Because endophytes do not cause symptoms in their hosts, they are difficult to detect. Conventional methods for identifying fungal endophytes rely on isolation, culturing and laboratory identification. However, some fungi develop slowly or do not readily grow in culture (Malvick and Grunden, 2005). Other methods used to detect endophytes include histochemical, immunological and molecular methods (Doss *et al.*, 1998).

Martini *et al.* (2009) examined the potential role of *Aureobasidium pullulans* and *Epicoccum* as biological control agents against grapevine pathogens using an efficient DNA extraction method and a sensitive PCR protocol. *A. pullulans* and *Epicoccum* have been isolated as endophytes from different crops, including grapevines (Schena *et al.*, 1999; Prasongsuk *et al.*, 2005; Campanile *et al.*, 2006; Elmer and Reglinski, 2006; Rodolfi *et al.*, 2006). In particular, they have been found in leaves (mainly near the midribs; Andrews *et al.*, 2002; Rodolfi *et al.*, 2006), berries (Schena *et al.*, 2003; Rodolfi *et al.*, 2006) and dormant buds (Dugan *et al.*, 2002). It was observed that *A. pullulans* and *E. nigrum* were present at high frequency in grapevines in north-eastern Italy. The occurrence of both *A. pullulans* and *E. nigrum* in the same tissue has also been reported (Lacava *et al.*, 2004; Rodolfi *et al.*, 2006), indicating that these endophytes may coexist in a host (Gimenez *et al.*, 2007). For these studies, specific primers were designed to amplify DNA from all tested *E. nigrum* and *A. pullulans* strains, including isolates from other sources. The primers did not amplify DNA from other fungal endophytes that are usually encountered during isolation attempts from grapevine tissues.

The majority (78%) of healthy and recovered plants hosted a strain of *E. nigrum*, which clearly shows the potential of this strain as a biocontrol agent. Endophytic fungi may influence the microbial community both within and on grapevines. For example, they may beneficially delay the establishment of wound pathogens such as *B. cinerea* (Dugan *et al.*, 2002), and some strains of *A. pullulans* are able to reduce ochratoxin accumulation in wine grapes by inhibiting *Aspergillus* sp. (De Felice *et al.*, 2008). Because microorganisms in grapes may influence wine quality, the association of endophytes such as *A. pullulans* and *E. nigrum* may play an important role in wine processing.

Endophytic fungi in relation to plant defence mechanisms

Schmid *et al.* (2011) showed that the differing plant protection measures adopted in conventional and organic vineyards influenced the grape-associated microorganisms found. The most interesting fact was that not only the structure but also the function of the fungal community was affected. In organically managed grapevine plants, the number of *in vitro* antagonists was enhanced as a result of an enrichment of *A. pullulans*. The findings of this study support a key role for this yeast-like fungus in explaining the structural and functional differences between the two agricultural systems. Interestingly, *Aureobasidium* can utilize inorganic sulfur (Killham *et al.*, 1981) and is able to absorb, and in this way detoxify, copper (Gadd and de Rome, 1988). These properties can explain the enrichment that occurs under organic farming conditions, Furthermore, *A. pullulans* is a well-studied potent antagonist of several fungal pathogens, and biocontrol products against *Botrytis* are already on the market (e.g. Botector from bio-ferm of Tulln, Austria). The mechanisms of *A. pullulans* antagonism against fungi include competition for nutrients and space and the production of cell wall-degrading enzymes (Chi *et al.*, 2009). Interestingly, the typical flavour components of wine were detected as being produced by *A. pullulans* (Verginer *et al.*, 2010), which could explain the better taste of organically produced wine (Yildirim *et al.*, 2004).

Molecular Techniques Used to Study Grapevine Endophytes

Denaturing gradient gel electrophoresis (DGGE)

Fungal endophytes have been reported to have an antagonistic effect against fungi, bacteria and other adversities affecting plants. Grisan *et al.* (2011) described the diversity of fungal endophytes in either phytoplasma-infected, recovered or healthy grapevines from organic vineyards. First, they identified 21 morphospecies via a culture-dependent method; these were subsequently analysed by internal transcribed spacer (ITS) region amplification and cloning, and grouped into 29 operational taxonomic units (OTUs) by restriction fragment length polymorphism (RFLP) analysis, of which 16 were subsequently shown to be unique. A similar but culture-independent analysis was then applied to the total genomic DNA extracted from pooled shoots of the grapevine cultivars. This resulted in the definition of 39 (26 unique) OTUs associated with the non-culturable fraction of the fungal endophytic community. Primer pairs amplifying the ITS1 region of all representative culturable endophytic isolates were used to generate amplicons suitable for DGGE analysis of the culture-independent OTUs. From the data collected, it appeared that more than 90% of the isolates obtained by the culture-dependent method belonged to seven main genera. The same seven genera represented 82% of the total OTUs defined by the culture-independent method. The application of PCR-based DGGE analysis was shown to be a valuable culture-independent approach for the rapid and reliable identification of fungal endophytic species. Using this approach, it may be possible in the future to discover fungal endophytes as potential biocontrol agents acting as inducers of recovery.

West *et al.* (2010) also used DGGE to examine the molecular profiles of bacterial endophyte communities in grapevine wood. In this investigation, DGGE profiles were produced for bacterial endophyte communities from the whole grapevine plants ('Chardonnay') from a single site. In most profiles, three or four major bands were visible and appeared to indicate that these were the community members present in the highest numbers. In addition, up to 20 other bands were clearly visible in some samples. The three or four major bands present were sequenced and given different identities: chloroplast DNA, uncultured bacterial clones or bacteria including *Enterobacter amnigenus*, *Vibrio salmonicida*, *Escherichia coli* and *Halomonas* sp. These results suggest that bacterial endophyte populations are highly homogenous but also unique to each vine and vine location. Bacterial endophytes are potentially important because they might provide benefits to plant health, help to control pests and diseases, and ultimately lead to minimizing the use of chemicals in agricultural management techniques (Hallmann *et al.*, 1997).

Automated ribosomal intergenic spacer analysis (ARISA)

Automated ribosomal intergenic spacer analysis (ARISA) is a community fingerprinting method based on the analysis of length polymorphisms of the rDNA region containing the two ITS regions and the gene for 5.8S for ribosomal RNA (rRNA). It is used in microbial community fingerprinting owing to its high accuracy in describing the diversity of the microbial community (Danovaro *et al.*, 2006), as well as its ease of use and the precision that the capillary electrophoresis system which is used by the method offers (Crosby and Criddle, 2003). Pancher *et al.* (2012) the first time, applied ARISA to the study of fungal endophytic communities in grapevines to compare organic and integrated pest management (IPM) and investigate the cultivar effect. The findings of this study showed that the mycota in grapevines from organic farms form communities that are significantly different from those in grapevines from IPM farms. This DNA-dependent approach was also found to be more powerful than that based on the analysis of culturable fungi, although they both indicate that IPM has an impact on the composition of endophytic fungal communities. A likely factor behind this could be the long-term use of synthetic fungicides in IPM or the use of organic fertilizers in organic farming.

Length heterogeneity (LH)-PCR

The development of cultivation-independent fingerprinting molecular methods based on 16S rRNA gene automated analysis has allowed a more specific, replicable and detailed description to be obtained of the diversity in complex bacterial communities. LH-PCR can be successfully used to describe bacterial communities (Brusetti *et al.*, 2006), and has previously been applied for characterization of the endophytic community in potatoes (Sessitsch *et al.*, 2002) and in sugarbeet (Dent *et al.*, 2004). Bulgari *et al.* (2009) studied the endophytic bacterial communities associated with healthy and phytoplasma-infected grapevine leaves characterized by LH-PCR and the sequencing of 16S rRNA gene libraries from the total leaf DNA. In this work, 98% of the 159 clones yielded best matches with bacteria of the family *Enterobacteriaceae* within the *Gammaproteobacteria*, which has been previously reported as a prevalent division associated with plant tissues (Chelius and Triplet, 2001; Kaiser *et al.*, 2001; Idris *et al.*, 2004). Among these, 137 sequences (86%) were more than 97% identical with *Pantoea agglomerans*, which was the dominant bacterium in the libraries. This microorganism has been associated with plants and insects. It was previously isolated as an endophyte from grapevine xylem (Bell *et al.*, 1995). Some other studies have emphasized the potential of *P. agglomerans* as a biocontrol agent against a range of plant pathogens through its secretion of antibacterial molecules (Ishimaru *et al.*, 1988) and/or by the activation of plant ISR (Ortmann *et al.*, 2006). Previously, *P. agglomerans* had also been successfully used for the control of the fire blight bacterial disease of apple, pear and other plants of the family Rosaceae (Wright *et al.*, 2001; Pusey *et al.*, 2008), and of *Rhizopus stolonifer* and *Monilinia laxa* in the postharvest storage of apricots and nectarines (Bonaterra *et al.*, 2003). Other bacterial species found in the libraries were *Ewingella americana*, *Erwinia persicina* and *P. ananatis*.

Fluorescence *in situ* hybridization (FISH)

Fluorescence *in situ* hybridization (FISH) with 16S rRNA-targeted oligonucleotide probes allows the detection, enumeration and identification of culturable and non-culturable microorganisms within their natural microhabitat; thus, it is widely applied in microbial ecology (Moter and Gobel, 2000; Amann *et al.*, 2001). Piccolo *et al.* (2010) used the FISH protocol in combination with confocal laser scanning microscopy (CLSM) for the visualization, localization and quantification of endophytic bacteria in grapevine leaf tissues with and without colonization by *A. byssoides*. The FISH/CLSM approach demonstrated endophytic colonization of grapevine leaves by bacteria. In fact, applying the microscopic *in situ* detection technique made the spatial distribution of bacteria in different layers of leaf tissue much clearer.

Following the observation of the high occurrence of microbial colonies in grapevine leaf tissues, a strain of *Burkholderia phytofirmans* was isolated. In particular, *B. phytofirmans* has been reported to be a plant growth-promoting bacterium known for establishing rhizosphere and endophytic associations with various hosts (Pillay and Nowak, 1997; Bensalim *et al.*, 1998; Compant *et al.*, 2005). Furthermore, potato and tomato plants inoculated with *B. phytofirmans* strain PsJN showed enhanced resistance to low levels of pathogens (Nowak *et al.*, 1995; Sharma and Nowak, 1998). This strain, formerly named *Pseudomonas* sp. PsJN (Frommel *et al.*, 1991), was also able to reduce *in vitro* infection by *Botrytis cinerea* as well as to enhance chilling resistance in grapevine (Ait Barka *et al.*, 2002, 2006).

Next generation sequencing (NGS)

Novel metagenomic data are emerging with the development of pyrosequencing-based approaches. It is hoped that these will bridge the gaps left unfilled by previous metagenomic (culture-independent) approaches and will help to address the issues associated with grapevine endophytes and plant health. New and much more complete views of the structure and function of microbial endophytic communities associated with grapevine are now becoming available. This information has the potential to be a groundbreaking tool to understand plant disease ecology and to develop innovative strategies for plant protection (Garrett *et al.*, 2013). Novel NGS-based metagenomic techniques provide a huge advance in the study of microbial communities by supporting the study of non-culturable microbes and the study of a previously unapproachable number of taxa simultaneously through high-throughput sequencing and microarray technologies (Handelsman, 2004; Acosta-Martínez *et al.*, 2008, 2010; Das *et al.*, 2008; Mendes *et al.*, 2011; Studholme *et al.*, 2011; Rosenzweig *et al.*, 2012). Previously, microbial community assessments relied on cursory evaluation of shallowly sampled cloned nucleotide fragments (Amann *et al.*, 1995), and the results obtained rarely came as close as newer approaches to defining the broader organismal diversity that is present (Jumpponen and Johnson, 2005).

Microbial ecology is now expanding and has the potential to produce extensive, if not perhaps comprehensive, lists of resident microbial entities (species or OTUs). The volumes of data from metagenomic next-generation sequencing studies enable microbial ecologists to characterize microbial communities in greater detail than before. At the same time, while these data are powerful, the new tools must be developed hand in hand with reference databases to permit molecular identification (Handelsman, 2004; Woyke *et al.*, 2006).

The study of endophytic microbial metagenomes has been evolving in multiple ways. First, more sophisticated experimental designs are being employed that support better statistical hypothesis testing and the construction of confidence intervals. Some of the first studies using high-throughput techniques tended to focus on comparisons of samples with little or no replication (Sogin *et al.*, 2006; Roesch *et al.*, 2007). The use of DNA tags for sample identification has permitted wider multiplexing and thus helped to make it more practical to use replicated designs. Unique DNA tags can be added to each sample prior to the pooling of samples, so that each sequence obtained can be assigned to its original sample based on that DNA tag sequence. Researchers can now determine a balance between sequencing depth per sample and the total number of samples to sequence: increase in sequencing depth inevitably leads to lower replication and vice versa. Higher levels of replication are often necessary for understanding ecological processes, where many factors may influence system variation, so DNA tagging makes it practical to incorporate metagenomic information in studies of disease ecology.

Pinto *et al.* (2014) provided the first comprehensive study of the composition of grapevine microbiota, both fungal and bacterial, using 454 pyrosequencing. Although they did not explicitly concentrate on endophytes, this was the first report on NGS-based community analysis in grapevine. The authors described copious Ascomycota and early diverging fungi, while Basidiomycota were less abundant. The diversity of plant-associated fungi decreased from spring to July, while bacterial diversity was unchanging. These authors also correlated the chemical treatment of plants for the control of fungal pathogens with the abundance of *Aureobasidium* and *Alternaria*.

Experimental data obtained through pyrosequencing using novel the Roche 454 genome sequencing FLX instrument (GS FLX) + titanium system indicate that complex and defined endophyte communities can be identified both in grapevine roots and stems. The taxa dominating these communities, as analysed in our laboratory, varied following changes along the environmental variables investigated. Despite this complexity, it was possible to identify clear trends in community profiles and identity of the largest taxa (see Fig. 2.1).

Conclusions and Future Prospects

The current knowledge of the structure of endophytic communities in grapevine plants is based on both cultivation-dependent and cultivation-independent studies. The combination of both approaches in the same study is recommended because cultivation-based techniques enable the recovery and testing of isolates, whereas cultivation-independent techniques enable screening for variations in the total endophytic communities (Van Overbeek and Van Elsas, 2008). Collectively, these studies have indicated that endophytic communities are dynamic over time, with endophytes showing a development that coincides with plant growth and development. The endophytic bacterial communities observed are less complex than soil bacterial communities, encompassing usually hundreds of different bacterial types. Hence, it seems that plants can function as true ‘filters’ of soil organisms, selecting only those that are successful and competent endophytes (Sessitsch *et al.*, 2002).

Many plant-associated microorganisms are well known for their capacity to confer plant growth promotion to the host and to increase resistance towards various

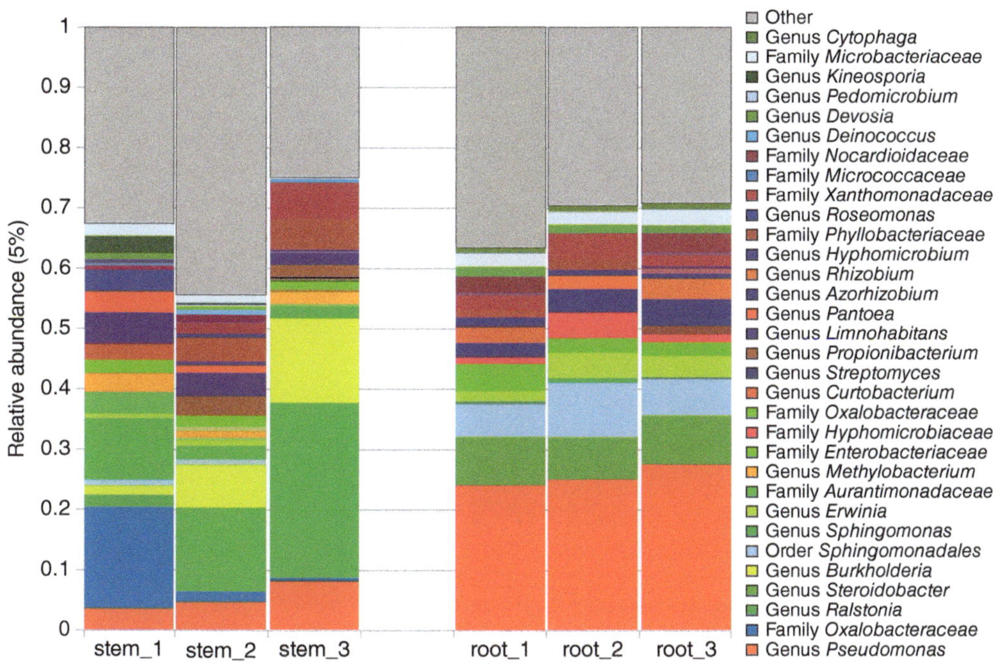

Fig. 2.1. Taxonomic assignment at genus level of 454 sequences of bacterial 16S amplicons from grapevine endophytes. The figure shows three replicates from the plant canopy (left) and three from the roots (right). Courtesy of Dr Livio Antonielli, AIT (Austrian Institute of Technology GmbH, Austria).

diseases as well as abiotic stresses. Research performed so far has been mostly related to grapevine plant growth promotion and/or rhizosphere or root endophytic colonization. The functioning of and contribution to grapevine plant health by endophytes localized in different plant parts is rather poorly understood. Exploring the microbial community structure by using DNA-dependent molecular and metagenomic techniques is helping towards a better understanding of the role of these endophytes in plant health. Further analysis of sequenced genomes, the characterization of yet unknown genes and the identification of genes expressed during plant growth promotion and plant defence against different diseases and/or pathogens will help to improve our understanding of grapevine endophytes and their role in plant health. It would not be surprising if some new factors and functions, as well as genes that are required for the endophytic lifestyle of microorganisms were to be identified in the near future.

Acknowledgements

The authors wish to acknowledge Dr Livio Antonielli (Fondazione Edmund Mach, San Michele all'Adige, Italy) for providing the image in Fig. 2.1. They also wish to thank MECAGraFIc, a project funded by Autonomous Province of Trento, Call 2 Team 2009 Incoming and COSTAction FA1103 for supporting this work.

References

Acosta-Martínez, V., Dowd, S., Sun, Y. and Allen, V. (2008) Tag-encoded pyrosequencing analysis of bacterial diversity in a single soil type as affected by management and land use. *Soil Biology and Biochemistry* 40, 2762–2770.

Acosta-Martínez, V., Dowd, S.E., Bell, C.W., Lascano, R., Booker, J.D., Zobeck, T.M. and Upchurch, D.R. (2010) Microbial community composition as affected by dryland cropping systems and tillage in a semiarid sandy soil. *Diversity* 2, 910–931.

Ait Barka, E., Gognies, S., Nowak, J., Audran, J.C. and Belarbi, A. (2002) Inhibitory effect of endophyte bacteria on *Botrytis cinerea* and its influence to promote the grapevine growth. *Biological Control* 24, 135–142.

Ait Barka, E., Nowak, J. and Clément, C. (2006) Enhancement of chilling resistance of inoculated grapevine plantlets with a plant growth promoting rhizobacterium, *Burkholderia phytofirmans* strain PsJN. *Applied and Environmental Microbiology* 72, 7246–7252.

Alston, J.M., Beddow, J.M. and Philip G. Pardey, P.G. (2009) Agricultural research, productivity, and food prices in the long run. *Science* 325, 1209–1210.

Amann, R.I., Ludwig, W. and Schleifer, K.H (1995) Phylogenetic identification and *in situ* detection of individual microbial cells without cultivation. *Microbiological Reviews* 59, 143–169.

Amann, R., Fuchs, B.M. and Behrens, S. (2001) The identification of microorganisms by fluorescence *in situ* hybridisation. *Current Opinion in Biotechnology* 12, 231–236.

Andrews, J.H., Harris, R.F. and Nordheim, E.V. (2002) Population biology of *Aureobasidium pullulans* on apple leaf surface. *Canadian Journal of Microbiology* 48, 500–513.

Arnold, A.E. (2007) Understanding the diversity of foliar endophytic fungi: progress, challenges, and frontiers. *Fungal Biology Reviews* 21, 51–66.

Assante, G., Dallavalle, S., Malpezzi, L., Nasini, G., Burruano, S. and Torta, L. (2005) Acremines A–F, novel secondary metabolites produced by a strain of an endophytic *Acremonium*, isolated from sporangiophores of *Plasmopara viticola* in grapevine leaves. *Tetrahedron* 61, 7686–7692.

Bae, H., Sicher, R.C., Kim, M.S., Kim, S.H., Strem, M.D., Melnick, R.L. and Bailey, B.A. (2009) The beneficial endophyte *Trichoderma hamatum* isolate DIS 219b promotes growth and delays the onset of the drought response in *Theobroma cacao*. *Journal of Experimental Botany* 60, 3279–3295.

Barrow, J.R. (2003) A typical morphology of dark septate fungal root endophytes of *Bouteloua* in arid south western USA rangelands. *Mycorrhiza* 13, 239–247.

Bell, C.R., Dickie, G.A., Harvey, W.L.G. and Chan, J.W.Y.F. (1995) Endophytic bacteria in grapevine. *Canadian Journal of Microbiology* 41, 46–53.

Bensalim, S., Nowak, J. and Asiedu, S.K. (1998) A plant growth promoting rhizobacterium and temperature effects on performance of 18 clones of potato. *American Journal of Potato Research* 75, 145–152.

Bonaterra, A., Mari, M., Casalini, L. and Montesinos, E. (2003) Biological control of *Monilinia laxa* and *Rhizopus stolonifer* in postharvest of stone fruit by *Pantoea agglomerans* EPS125 and putative mechanisms of antagonism. *International Journal of Food Microbiology* 84, 93–104.

Bruez, E., Vallance, J., Gerbore, J., Lecomte, P., Da Costa, J.-P., Guerin-Dubrana, L. and Rey, P. (2014) Analyses of the temporal dynamics of fungal communities colonizing the healthy wood tissues of esca leaf-symptomatic and asymptomatic vines. *PLoS ONE* 9(5): e95928.

Brusetti, L., Borin, S., Mora, D., Rizzi, A., Raddadi, N., Sorlini, C. and Daffonchio, D. (2006) Usefulness of length heterogeneity-PCR for monitoring lactic acid bacteria succession during maize ensiling. *FEMS Microbiology Ecology* 56, 154–164.

Bulgari, D., Casati, P., Brusetti, L., Quaglino, F., Brasca, M., Daffonchio, D. and Bianco, P.A. (2009) Endophytic bacterial diversity in grapevine (*Vitis vinifera* L.) leaves described by 16S rRNA gene sequence analysis and length heterogeneity-PCR. *Journal of Microbiology* 47, 393–401.

Bulgari, D., Casati, P., Crepaldi, P., Daffonchio, D., Quaglino, F., Brusetti, L. and Bianco, P.A. (2011) Endophytic bacterial community is restructured in grapevine yellows-diseased and recovered *Vitis vinifera* L. plants. *Applied and Environmental Microbiology* 77, 5018–5022.

Bulgari, D., Casati, P., Quaglino, F. and Bianco, P.A. (2014) Endophytic bacterial community of grapevine leaves influenced by sampling date and phytoplasma infection process. *BMC Microbiology* 14, 198.

Burdman, S., Jurkevitch, E. and Okon, Y. (2001) Recent advances in the use of plant growth promoting rhizobacteria (PGPR) in agriculture. In: Subba Rao, N.S. and Dommergues, Y.R. (eds) *Microbial Interactions in Agriculture and Forestry, Vol. II.* Science Publishers, Enfield, New Hampshire, pp. 229–250.

Burruano, S., Alfonzo, A., Lo Piccolo, S., Conigliaro, G., Mondello, V., Torta, L., Moretti, M. and Assante, G. (2008) Interaction between *Acremonium byssoides* and *Plasmopara viticola* in *Vitis vinifera*. *Phytopathologia Mediterranea* 47, 122–131.

Caldwell, B.A., Jumpponen, A. and Trappe, J.M. (2000) Utilization of major detrital substrates by dark-septate, root endophytes. *Mycologia* 92, 230–232.

Campanile, G., Campanile, D., Nigro, M., Mannerucci, F., Ruscelli, A. and Luisi, N. (2006) Prove preliminari di antagonismo di funghi endofiti contro *Diplodia corticola* e indicazioni gestionali per il contenimento del deperimento delle querce. *Foresta* 3, 78–85.

Campisano, A., Ometto, L., Compant, S., Pancher, M., Antonielli, L., Yousaf, S., Varotto, C., Anfora, G., Pertot, I., Sessitsch, A. and Rota-Stabelli, O. (2014a) Interkingdom transfer of the acne-causing agent, *Propionibacterium acnes*, from human to grapevine. *Molecular Biology and Evolution* 31, 1059–1065.

Campisano, A., Antonielli, L., Pancher, M., Yousaf, S., Pindo, M. and Pertot, I. (2014b) Bacterial endophytic communities in the grapevine depend on pest management. *PLoS ONE* 9(11): e112763.

Chakraborty, S. and Newton, A.C. (2011) Climate change, plant diseases and food security: an overview. *Plant Pathology* 60, 2–14.

Chelius, M.K. and Triplett, E.W. (2001) The diversity of archaea and bacteria in association with the roots of *Zea mays* L. *Microbial Ecology* 41, 252–263.

Chi, Z., Wang, F., Chi, Z., Yue, L., Liu, G. and Zhang, T. (2009) Bioproducts from *Aureobasidium pullulans*, a biotechnologically important yeast. *Applied Microbiology and Biotechnology* 82, 793–804.

Choudhary, D.K. and Johri, B.N. (2009) Interactions of *Bacillus* sp. and plants – with special reference to induced systemic resistance (ISR). *Microbiological Research* 164, 493–513.

Compant, S., Reiter, B., Sessitsch, A., Nowak, J., Clément, C. and Ait Barka, E. (2005) Endophytic colonization of *Vitis vinifera* L. by plant growth-promoting bacterium *Burkholderia* sp. strain PsJN. *Applied and Environmental Microbiology* 71, 1685–1693.

Crosby, L.D. and Criddle, C.S. (2003) Understanding bias in microbial community analysis techniques due to RRN operon copy number heterogeneity. *Biotechniques* 34, 790–794.

Danovaro, R., Luna, G.M., Dell'Anno, A. and Pietrangeli, B. (2006) Comparison of two fingerprinting techniques, terminal restriction fragment length polymorphism and automated ribosomal intergenic spacer analysis, for determination of bacterial diversity in aquatic environments. *Applied and Environmental Microbiology* 72, 5982–5989.

Das, S., Bhat, P.E., Sudhakar, C., Ehlers, J.D., Wanamaker, S., Roberts, P.A., Cui, X. and Close, T.J. (2008) Detection and validation of single feature polymorphisms in cowpea (*Vigna unguiculata* L. Walp) using a soybean genome array. *BMC Genomics* 9, 107.

Davitt, A., Stansberry, M. and Rudgers, J.A. (2010) Do the costs and benefits of fungal endophyte symbiosis vary with light availability? *New Phytologist* 188, 824–834.

De Felice, D.V., Solfrizzo, M., De Curtis, F., Lima, G., Visconti, A. and Castoria, R. (2008) Strains of *Aureobasidium pullulans* can lower ochratoxin A contamination in wine grapes. *Phytopathology* 98, 1261–1270.

Dent, K.C., Stephen, J.R. and Finch-Savage, W.E. (2004) Molecular profiling of microbial community associated with seeds of *Beta vulgaris* subsp. vulgaris (sugar beet). *Journal of Microbiological Methods* 56, 17–26.

Doss, R.P., Clement, S.L., Kuy, S.R. and Welty, R.E. (1998) A PCR-based technique for detection of *Neotyphodium* endophytes in diverse accessions of tall fescue. *Plant Disease* 82, 738–740.

Dugan, F.M., Lupien, S.L. and Grove, G.G. (2002) Incidence, aggressiveness and *in planta* interactions of *Botrytis cinerea* and other filamentous fungi quiescent in grape berries and dormant buds in central Washington State. *Journal of Phytopathology* 150, 375–381.

El-Banna, N. and Winkelmann, G. (1998) Pyrrolnitrin from *Burkholderia cepacia*: antibiotic activity against fungi and novel activities against *Streptomyces*. *Journal of Applied Microbiology* 85, 69–78.

Elmer, P.A.G. and Reglinski, T. (2006) Biosuppression of *Botrytis cinerea* in grapes. *Plant Pathology* 55, 155–177.

Fleet, G.H., Prakitchaiwattana, C., Beh, A.L. and Heard, G. (2002) The yeast ecology of wine grapes. In: Ciani, M. (ed.) *Biodiversity and Biotechnology of Wine Yeasts*. Research Signpost, Trivandrum, Kerala, India, pp. 1–17.

Frommel, M.I., Nowak, J. and Lazarovits, G. (1991) Growth enhancement and developmental modifications of *in vitro* grown potato (*Solanum tuberosum* ssp. *tuberosum*) as affected by a nonfluorescent *Pseudomonas* sp. *Plant Physiology* 96, 928–936.

Gadd, G.M. and de Rome, L. (1988) Biosorption of copper by fungal melanin. *Applied Microbiology and Biotechnology* 29, 610–617.

Garrett, K.A., Forbes, G.A., Savary, S., Skelsey, P., Sparks, A.H., Valdivia, C. and Van Bruggen, A.H.C. (2011) Complexity in climate-change impacts: an analytical framework for effects mediated by plant disease. *Plant Pathology* 60, 15–30.

Garrett, K.A., Dobson, A.D.M., Kroschel, J., Natarajan, B., Orlandini, S., Tonnang, H.E.Z. and Valdivia, C. (2013) The effects of climate variability and the color of weather time series on agricultural diseases and pests, and decision-making for their management. *Agricultural and Forest Meteorology* 170, 216–227.

Gimenez, C., Cabrera, R., Reina, M. and Gonzales-Coloma, A. (2007) Fungal endophytes and their role in plant protection. *Current Organic Chemistry* 11, 707–720.

Grisan, S., Martini, M., Musetti, M. and Osler, R. (2011) Development of a molecular approach to describe the diversity of fungal endophytes in either phytoplasma infected, recovered or healthy grapevines. *Bulletin of Insectology* 64(Supplement), S207–S208.

Gundel, P.E., Omacini, M., Sadras, V.O. and Ghersa, C.M. (2010) The interplay between the effectiveness of the grass–endophyte mutualism and the genetic variability of the host plant. *Evolutionary Applications* 3, 538–546.

Hallmann, J., Quadt Hallmann, A., Mahaffee, W.F. and Kloepper, J.W. (1997) Bacterial endophytes in agricultural crops. *Canadian Journal of Microbiology* 43, 895–914.

Handelsman, J. (2004) Metagenomics: application of genomics to uncultured microorganisms. *Microbiology and Molecular Biology Reviews* 68, 669–685.

Idris, R., Trifonova, R., Puschenreitr, M., Wenzel, W.W. and Sessitch, A.A. (2004) Bacterial communities associated with flowering plants of the Ni hyperaccumulator *Thlaspi goesingense*. *Applied and Environmental Microbiology* 70, 2667–2677.

Ishimaru, C.A.E., Klobs, E. and Brubaker, R.R. (1988) Multiple antibiotic production by *Erwinia herbicola*. *Phytopathology* 78, 746–750.

Jumpponen, A. and Johnson, L.C. (2005) Can rDNA analyses of diverse fungal communities in soil and roots detect effects of environmental manipulations – a case study from tallgrass prairie. *Mycologia* 97, 1177–1194.

Juroszek, P. and von Tiedemann, A. (2011) Potential strategies and future requirements for plant disease management under a changing climate. *Plant Pathology* 60, 100–112.
Kaiser, O., Pühler, A. and Selbitschka, W. (2001) Phylogenetic analysis of microbial diversity in the rhizoplane of oilseed rape (*Brassica napus* cv. Westar) employing cultivation-dependent and cultivation-independent approaches. *Microbial Ecology* 42, 136–149.
Kavroulakis, N., Ntougias, S., Zervakis, G.I., Ehaliotis, C., Haralampidis, K. and Papadopoulou, K.K. (2007) Role of ethylene in the protection of tomato plants against soil-borne fungal pathogens conferred by an endophytic *Fusarium solani* strain. *Journal of Experimental Botany* 29, 1–12.
Killham, K., Lindley, N.D. and Wainwright, M. (1981) Inorganic sulfur oxidation by *Aureobasidium pullulans*. *Applied and Environmental Microbiology* 42, 629–631.
Kobayashi, D.Y. and Palumbo, J.D. (2000) Bacterial endophytes and their effects on plants and uses in agriculture. In: Bacon, C.W. and White, J.F. (eds) *Microbial Endophytes*. Dekker, New York, pp. 199–236.
Lacava, P.T., Araújo, W.L., Marcon, J., Maccheroni, W. and Azevedo, J.L. (2004) Interaction between endophytic bacteria from citrus plants and the phytopathogenic bacteria *Xylella fastidiosa*, causal agent of citrus variegated chlorosis. *Letters in Applied Microbiology* 39, 55–59.
Lodewyckx, C., Vangronsveld, J., Porteous Moore, F., Moore, E.R.B., Taghavi, S. and Van der Lelie, D. (2002) Endophytic bacteria and their potential applications. *Critical Reviews in Plant Sciences* 21, 583–606.
Luck, J. Spackman, M., Freeman, A., Trębicki, P., Griffiths, W., Finlay, K. and Chakraborty, S. (2011) Climate change and diseases of food crops. *Plant Pathology* 60, 113–121.
Malvick, D.K. and Grunden, E. (2005) Isolation of fungal DNA from plant tissues and removal of DNA amplification inhibitors. *Molecular Ecology Notes* 5, 958–960.
Martini, M., Musetti, R., Grisan, S., Polizzotto, R., Borselli, S., Pavan, F. and Osler, R. (2009) DNA-dependent detection of the grapevine fungal endophytes *Aureobasidium pullulans* and *Epicoccum nigrum*. *Plant Disease* 93, 993–998.
Mendes, R., Kruijt, M., De Bruijn, I., Dekkers, E., Van der Voort, M., Schneider, J.H.M., Piceno, Y.M., DeSantis, T.Z., Andersen, G.L., Bakker, P.A.H.M. and Raaijmakers. J.M. (2011) Deciphering the rhizosphere microbiome for disease-suppressive bacteria. *Science* 332, 1097–1100.
Moter, A. and Gobel, U.B. (2000) Fluorescence *in situ* hybridization (FISH) for direct visualization of microorganisms. *Journal of Microbiological Methods* 41, 85–112.
Musetti, R., Marabottini, R., Badiani, M., Martini, M., Sanità di Toppi, L., Borselli, S., Borgo, M. and Osler, R. (2007) On the role of H_2O_2 in the recovery of grapevine (*Vitis vinifera*, cv. Prosecco) from flavescence dorée disease. *Functional Plant Biology* 34, 750–758.
Musetti, R., Grisan, S., Polizzotto, R., Martini, M., Paduano, C. and Osler, R. (2011) Interactions between '*Candidatus* Phytoplasma mali' and the apple endophyte *Epicoccum nigrum* in *Catharanthus roseus* plants. *Journal of Applied Microbiology* 110, 746–756.
Nowak, J., Asiedu, S.K. and Lazarovits, G. (1995) Enhancement of *in vitro* growth and transplant stress tolerance of potato and vegetable plants co-cultured with a plant growth promoting rhizobacterium. In: Carré, F. and Chagvardieff, P. (eds) *Ecophysiology and Photosynthetic In Vitro Cultures*. CEA, Aix-en-Provence, France, pp. 173–180.
Ortmann, I., Conrath, U. and Moerschbacher, B.M. (2006) Exopolysaccharides of *Pantoea agglomerans* have different priming and eliciting activities in suspension-cultured cells of monocots and dicots. *FEBS Letters* 580, 4491–4494.
Pancher, M., Ceol, M., Corneo, P.E., Longa, C.M.O., Yousaf, S., Pertot, I. and Campisano, A. (2012) Fungal endophytic communities in grapevines (*Vitis vinifera* L.) respond to crop management. *Applied and Environmental Microbiology* 78, 4308–4317.
Piccolo, S.L., Ferraro, V., Alfonzo, A., Settanni, L., Ercolini, D., Burruano, S. and Moschetti, G. (2010) Presence of endophytic bacteria in *Vitis vinifera* leaves as detected by fluorescence *in situ* hybridization. *Annals of Microbiology* 60, 161–167.

Pillay, V.K. and Nowak, J. (1997) Inoculum density, temperature, and genotype effects on *in vitro* growth promotion and epiphytic and endophytic colonization of tomato (*Lycopersicon esculentum* L.) seedlings inoculated with a pseudomonad bacterium. *Canadian Journal of Microbiology* 43, 354–361.

Pinto, C., Pinho, D., Sousa, S., Pinheiro, M., Egas, C. and Gomes, A.C. (2014) Unravelling the diversity of grapevine microbiome. *PLoS ONE* 9(1): e85622.

Prasongsuk, S., Sullivan, R.F., Kuhirun, M., Eveleigh, D.E. and Punnapayak, H. (2005) Thailand habitats as source of pullulan producing strains of *Aureobasidium pullulans*. *World Journal of Microbiology and Biotechnology* 21, 393–398.

Pusey, P.L., Stockwell, V.O. and Rudell, D.R. (2008) Antibiosis and acidification by *Pantoea agglomerans* strain E325 may contribute to suppression of *Erwinia amylovora*. *Phytopathology* 98, 1136–1143.

Reinhold-Hurek, B. and Hurek, T. (1998) Life in grasses: diazotrophic endophytes. *Trends in Microbiology* 6, 139–144.

Rodriguez, J.R., White, J.F. Jr, Arnold, A.E. and Redman, R.S. (2008) Fungal endophytes: diversity and functional roles. *New Phytologist* 182, 314–330.

Rodolfi, M., Legler, S.E. and Picco, A.M. (2006) Endofiti fungini in *Vitis vinifera* in Oltrepò Pavese. *Micologia Italiana* 35, 25–31.

Roesch, L.F.W., Fulthorpe, R.R., Riva, A., Casella, G., Hadwin, A.K.M., Kent, A.D., Daroub, S.H., Camargo, F.A.O., Farmerie, W.G. and Triplett, E.W. (2007) Pyrosequencing enumerates and contrasts soil microbial diversity. *The ISME Journal* 1, 283–290.

Rosenzweig, N., Tiedje, J.M., Quensen, J.F., Meng, Q. and Hao, J.J. (2012) Microbial communities associated with potato common scab-suppressive soil determined by pyrosequencing analyses. *Plant Disease* 96, 718–725.

Ryan, R.P., Germaine, K., Franks, A., Ryan, D.J. and Dowling, D.N. (2008) Bacterial endophytes: recent developments and applications. *FEMS Microbiology Letters* 278, 1–9.

Saikkonen, K., Faeth, S.H., Helander, M. and Sullivan, T.J. (1998) Fungal endophytes: a continuum of interactions with host plants. *Annual Review of Ecology and Systematics* 29, 319–343.

Schena, L., Ippolito, A., Zahavi, T., Cohen, L., Nigro, F. and Droby, S. (1999) Genetic diversity and biocontrol activity of *Aureobasidium pullulans* isolates against postharvest rots. *Postharvest Biology and Technology* 17, 189–199.

Schena, L., Nigro, F., Pentimone, I., Ligorio, A. and Ipolito, A. (2003) Control of postharvest rots of sweet cherries and table grapes with endophytic isolates of *Aureobasidium pullulans*. *Postharvest Biology and Technology* 30, 209–220.

Schulz, B. and Boyle, C. (2005) The endophytic continuum. *Mycological Research* 109, 661–687.

Schulz, B., Boyle, C., Draeger, S., Römmert, A.K. and Krohn, K. (2002) Endophytic fungi: a source of biologically active secondary metabolites. *Mycological Research* 106, 996–1004.

Schmid, F., Moser, G., Müller, H. and Berg, G. (2011) Functional and structural microbial diversity in organic and conventional viticulture: organic farming benefits natural biocontrol agents. *Applied and Environmental Microbiology* 77, 2188–2191.

Sessitsch, A., Reiter, B., Pfeifer, U. and Wilhelm, E. (2002) Cultivation-independent population of bacterial endophytes in three potato varieties based on eubacterial and *Actinomycetes*-specific PCR of 16S rRNA genes. *FEMS Microbiology Ecology* 39, 23–32.

Sharma, V. and Nowak, J. (1998) Enhancement of verticillium wilt resistance in tomato transplants by *in vitro* co-culture of seedlings with a plant growth promoting rhizobacterium (*Pseudomonas* sp. strain PsJN). *Canadian Journal of Microbiology* 44, 528–536.

Sogin, M.L., Morrison, H.G., Huber, J.A., Welch, D.M., Huse, S.M., Neal, P.R., Arrieta, J.M. and Herndl, G.J. (2006) Microbial diversity in the deep sea and the underexplored 'rare biosphere'. *Proceedings of the National Academy of Sciences of the United States of America* 103, 12115–12120.

Sturz, A.V., Christie, B.R. and Nowak, J. (2000) Bacterial endophytes: potential role in developing sustainable systems of crop production. *Critical Reviews in Plant Sciences* 19, 1–30.
Strobel, G.A. and Long, D.M. (1998) Endophytic microbes embody pharmaceutical potential. *ASM News* 64, 263–268.
Studholme, D.J., Wasukira, A., Paszkiewicz, K., Aritua, V., Thwaites, R., Smith, J. and Grant, M. (2011) Draft genome sequences of *xanthomonas sacchari* and two banana-associated xanthomonads reveal insights into the *Xanthomonas* Group 1 clade. *Genes* 2, 1050–1065.
Thomas, P. (2004) Isolation of *Bacillus pumilus* from *in vitro* grapes as a long-term alcohol-surviving and rhizogenesis inducing covert endophyte. *Journal of Applied Microbiology* 97, 114–123.
Tudzynski, B. (1997) Fungal phytohormones in pathogenic and mutualistic associations. In: Carroll, G.C. and Tudzynski, P. (eds) *The Mycota V. Plant Relationships: Part A*. Springer, Berlin/Heidelberg, Germany/New York, pp. 167–184.
Tudzynski, B. and Sharon, A. (2002) Biosynthesis, biological role and application of fungal hormones. In: Osiewacz, H.D. (ed.) *The Mycota X. Industrial Applications*. Springer, Berlin/Heidelberg, Germany/New York, pp. 183–211.
Vallad, G.E. and Goodman, R.M. (2004) Systemic acquired resistance and induced systemic resistance in conventional agriculture. *Crop Science* 44, 1920–1934.
Van Loon, C.L. (2007) Plant responses to plant growth-promoting rhizobacteria. *European Journal of Plant Pathology* 119, 243–254.
Van Overbeek, L. and Van Elsas, J.D. (2008) Effects of plant genotype and growth stage on the structure of bacterial communities associated with potato (*Solanum tuberosum* L.). *FEMS Microbiology Ecology* 64, 283–296.
Verginer, M., Leitner, E. and Berg, G. (2010) Production of volatile metabolites by grape-associated microorganisms. *Journal of Agricultural and Food Chemistry* 58, 8344–8350.
Waller, F., Achatz, B., Baltruschat, H., Fodor, J., Becker, K., Fischer, M., Heier, T., Hückelhoven, R., Neumann, C., von Wettstein, D. *et al.* (2005) The endophytic fungus *Piriformospora indica* reprograms barley to salt-stress tolerance, disease resistance, and higher yield. *Proceedings of the National Academy of the United States of America* 102, 386–391.
West, E.R., Cother, E.J., Steel, C.C. and Ash, G.J. (2010) The characterization and diversity of bacterial endophytes of grapevine. *Canadian Journal of Microbiology* 56, 209–216.
Wilson, D. (1995) Endophyte: the evolution of a term, and clarification of its use and definition. *Oikos* 73, 274–276.
Woyke, T., Teeling, H., Ivanova, N.N., Huntemann, M., Richter, M., Gloeckner, F.O., Boffelli, D., Anderson, I.J., Barry, K.W., Shapiro, H.J. *et al.* (2006) Symbiosis insights through metagenomic analysis of a microbial consortium. *Nature* 443, 950–955.
Wright, S.A.I., Zumoff, C.H., Schneider, L. and Beer, S.V. (2001) *Pantoea agglomerans* strain EH318 produces two antibiotics that inhibit *Erwinia amylovora in vitro*. *Applied and Environmental Microbiology* 67, 284–292.
Yildirim, H.K., Akçay, Y.D., Güvenç, Y. and Sözmen, E.Y. (2004) Protection capacity against low-density lipoprotein oxidation and antioxidant potential of some organic and non-organic wines. *International Journal of Food Sciences and Nutrition* 5, 351–362.
Yousaf, S., Bulgari, D., Bergna, A., Pancher, M., Quaglino, F., Casati, P. and Campisano, A. (2014) Pyrosequencing detects human and animal pathogenic taxa in the grapevine endosphere. *Frontiers in Microbiology* 5, 327.

3 Ecology and Function of Grape-Associated Microorganisms with a Special Focus on Biocontrol of *Botrytis cinerea*

M. CARDINALE[1]* AND G. BERG[2]

[1]*Institute of Applied Microbiology, Justus-Liebig-University, Giessen, Germany;* [2]*Institute of Environmental Biotechnology, Graz University of Technology, Graz, Austria*

Introduction

Grapevine plants are affected by numerous pests, most of which are fungal pathogens. Significant yield losses are caused by fungal diseases and these thus represent an economic threat for grapevine production and for the wine industry. One of the most aggressive and destructive pathogens is *Botrytis cinerea*, an ascomycete fungus. It infects the plant through wounds inflicted, for example, by insects, and causes the bunch rot disease also known as grey mould (Elad *et al.*, 2004). This pathogen can bypass disinfestation procedures as a result of its ability both to develop resistance (Elad *et al.*, 1992) and to survive in a latent stage on non-host plants (Williamson *et al.*, 2007). As the use of fungicides has been limited by international regulations owing to the environmental and health problems that they can cause (Spadoro and Gullino, 2005), new environmentally friendly alternatives are now required to control *Botrytis* and other pathogens. One promising technology is represented by the induction of systemic resistance in plants based on the addition of certain elicitors, such as oligogalacturonides (Aziz *et al.*, 2003) and chitosan (Ait Barka *et al.*, 2004; Trotel-Aziz *et al.*, 2006). The most common alternative to chemical pesticides is biocontrol, a method that uses naturally occurring microorganisms as so-called biocontrol agents (BCAs) that specifically inhibit the growth of pathogens by various means. The major advantage of this approach is sustainability, i.e. the absence of environmental pollutants and chemical residues.

Most known modes of action of BCAs act through a combination of diverse mechanisms including competition for space and nutrients between the nutrient and the BCA, direct inhibition of the pathogen by the secretion of antifungal substances, and induced systemic resistance (ISR) in the plant (Compant *et al.*, 2005; Berg, 2009; Verhagen *et al.*, 2009; Jacometti *et al.*, 2010; Muzammil *et al.*, 2012). Most BCAs act by a combination of these mechanisms (Berg *et al.*, 2006). Even though they have such diverse components, the modes of action of BCAs are usually investigated under extremely

*massimiliano.cardinale@umwelt.uni-giessen.de

simplified simulations of natural conditions, such as in a laboratory environment, in microcosms or in axenic cultures. Field conditions are much more variable and complex, so the results of BCA treatments in the field are highly unpredictable. This is primarily due to the fact that the colonization capabilities of the BCA strains – the basic prerequisite for any successful treatment – are outcompeted by the native microbial species that are already present (Lugtenberg and Kamilova, 2009). Moreover, even after a possible colonization of the host, the antimicrobial activity of a BCA can be strongly reduced or totally inhibited by the intricate net of physical and metabolic interactions with the species of the plant-associated microbiome (Cardinale *et al.*, 2015). The development of successful BCA treatments should, therefore, carefully consider the microbial ecology of the host (Köberl *et al.*, 2011, 2012). Despite such evidence, only a few studies have investigated this aspect in grapevines. In this chapter, we present a comparative review focusing on the biocontrol of *B. cinerea*. We also attempt to draw general conclusions and highlight future perspectives for biocontrol in the grapevine.

Ecology of Grape-Associated Bacteria and Natural Biocontrol

Grapevine-associated microorganisms act in tripartite interactions between pathogens, plants and non-pathogenic microbes. Natural microbial communities are characterized by seasonal shifts as well as changes related to the growth stage of the plant (Smalla *et al.*, 2001; Houlden *et al.*, 2008). Only a certain number of microbial strains are able to colonize the plant tissues and hosts endophytically, which is of particular importance because endophytism is a key feature of successful BCAs (Compant *et al.*, 2008, 2011). Endophytic strains can exert a plant growth-promoting effect which, in turn, can enhance the global fitness of the plant either directly or indirectly (Compant *et al.*, 2005). The plant itself is a puzzle of different habitats and niches; within (and around) the same plant, the rhizosphere, endorhiza (i.e. the root endosphere), phyllosphere, carposphere and anthosphere can harbour different microbial communities at the strain level (Berg *et al.*, 2005). For the grapevine, Compant *et al.* (2011) found that some bacterial endophytes of flowers and berry pulp (*Bacillus*, *Paenibacillus* and *Pseudomonas* spp.) were different from those associated with the rhizosphere and the endorhiza (Fig. 3.1).

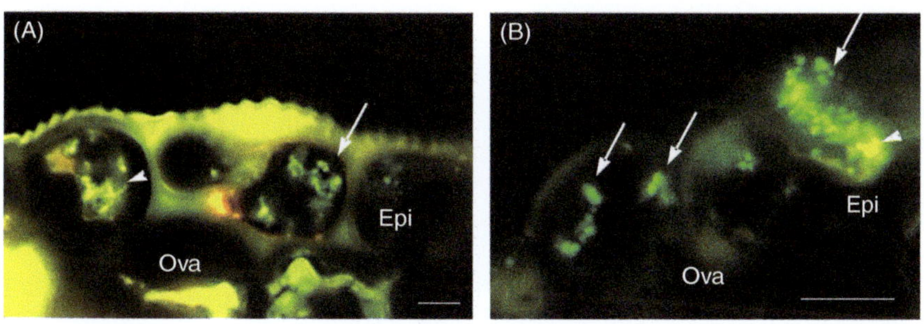

Fig. 3.1. Bacterial colonization of grapevine flowers analysed by fluorescence *in situ* hybridization (FISH) with specific probes. (A) *Gammaproteobacteria* appear yellow (arrowhead) and other bacteria green (arrows). (B) *Firmicutes* appear yellow (arrowhead) and other bacteria green (arrows). Key: Ova, ovary; Epi, epidermis. Scale bars 10 μm. Reproduced with permission from Compant *et al.* (2011).

This finding suggests that the soil or rhizosphere soil is not the only source of endophytic bacteria in grapevine: some of them could derive, for example, from epiphytic colonizers of the anthosphere. This hypothesis is also supported by the evidence that bacteria associated with xylem vessels are different from those of the rhizosphere and are attached to the vessel walls (Bell *et al.*, 1995).

While many ubiquitous bacteria as well as BCAs are able to colonize the whole plant (Compant *et al.*, 2008), the microenvironmental specificity at strain level implies that inoculated BCAs can be active only within a specific habitat of the plant as a result of partial or complete inhibition in other plant parts.

The farming method is another driver for the grapevine microbiome. Organic farming has become more widespread in viticulture in recent years (Willer, 2008), and it generally leads to a higher microbial abundance and diversity in the farming system (Gasser and Berg, 2011). It has been shown that microorganisms in organic farming systems colonize leaves more than do those in conventional farming systems, although bacteria and yeasts were more abundant in conventionally treated grapes (Schmid and Berg, 2009). Although endophytes always occur two to three orders of magnitude lower than epiphytes, they were, likewise, more abundant in plants from organic farms; similarly, more bacteria with antagonistic potential against fungal pathogens were found in cultivated grapes than in wild ones (Schmid and Berg, 2009). Thus, organically managed grapevines demonstrate the richest source of potentially antagonistic bacteria (Table 3.1).

Table 3.1. Abundance of microorganisms on and around grapevines of different species according to habitat and farming method, given as colony forming units (CFU) per gram of fresh weight (FW).

	Plant habitat				
Microorganism type	Grape	Stem ectosphere	Leaf ectosphere	Stem endosphere	Leaf endosphere
***Vitis vinifera sylvestris* (wild variety)**					
Bacteria	2×10^4	2×10^5	4×10^5	$<10^2$	2×10^2
Yeasts	5×10^3	2×10^4	9×10^4	$<10^2$	$<10^2$
Fungi	9×10^3	6×10^3	1×10^4	$<10^2$	2×10^2
***Vitis riparia* (wild variety)**					
Bacteria	[a]	5×10^5	5×10^5	2×10^2	2×10^2
Yeasts	[a]	4×10^4	5×10^4	$<10^2$	$<10^2$
Fungi	[a]	1×10^4	3×10^4	$<10^2$	$<10^2$
***Vitis vinifera* cv. Sauvignon Blanc, organically treated**					
Bacteria	1×10^4	2×10^5	5×10^5	3×10^3	2×10^3
Yeasts	9×10^3	6×10^4	2×10^5	8×10^2	5×10^2
Fungi	7×10^3	1×10^4	2×10^5	5×10^2	6×10^2
***Vitis vinifera* cv. Sauvignon Blanc, conventionally treated**					
Bacteria	1×10^5	1×10^5	7×10^4	5×10^2	$<10^2$
Yeasts	5×10^4	3×10^4	4×10^4	3×10^2	$<10^2$
Fungi	7×10^3	8×10^3	1×10^4	$<10^2$	$<10^2$

[a]No grapes available as the wild variety of *Vitis riparia*; they are dioecious plants and in this case pollination was not possible.

Bacteria with antagonistic potential isolated from the endosphere of organic grapevine included both Gram-positive and Gram-negative species, such as *Bacillus*, diverse *Actinobacteria* and *Pantoea* spp. (Bulgari *et al.*, 2009; West *et al.*, 2010; Gasser *et al.*, 2012).

Grapevine is grown in an extremely long-term monoculture, so the age of the vineyard should also be considered in studies of microbial ecology. Svercel *et al.* (2009) detected higher numbers and percentages of rhizosphere pseudomonads in soil from long-term vineyards producing 2,4-diacetylphloroglucinol and hydrogen cyanide. This suggests that the soil in old vineyards has highly suppressive potential, as has also been shown for other suppressive soils after long-term monoculture (Ramette *et al.*, 2006; Mendes *et al.*, 2011).

Grapevine-associated microbial communities have demonstrated notable stability and resilience. Enrichment by black fungi, which usually affects host–microbe interactions, did not significantly shape the native community (Grube *et al.*, 2011). Nevertheless, after a strong perturbation, such as disease, the grapevine microbial community can experience a plastic change. Two years after recovering from a phytoplasma infection, grapevine plants harboured a different bacterial community that was enriched in species resistant to reactive oxygen species (ROS) and those able to elicit plant defence responses, including ROS production (Bulgari *et al.*, 2011). Such microbes included *Burkholderia*, *Bacillus* and *Paenibacillus* strains. However, in other less drastic cases, changes in microbial communities can be very specific and limited only to certain strains. *Aureobasidium*, a fungal genus harbouring antagonistic species, was highly abundant in organic farming (Grube *et al.*, 2011; Fig. 3.2) and its enrichment was correlated with a higher antagonistic potential of organically managed plants (Schmid *et al.*, 2011). In contrast to the fungal community, the bacterial community profile of *Firmicutes* and *Pseudomonas* was

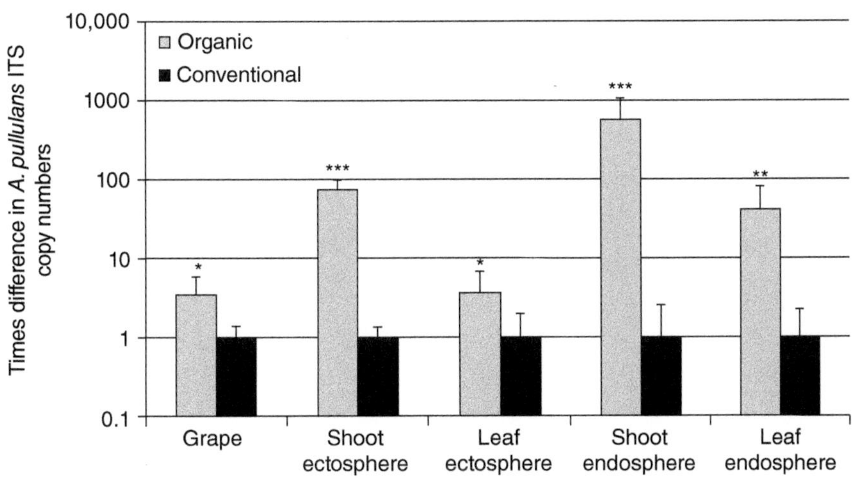

Fig. 3.2. Internal transcribed spacer (ITS) copy numbers of *Aureobasidium pullulans* per gram of fresh weight determined by quantitative PCR. The relative copy numbers of samples from organically managed plants is shown in relation to the copy numbers of samples from conventionally managed plants. Error bars indicate standard deviations and asterisks indicate statistical significance of differences (*, $P = 0.05$; **, $P = 0.01$; ***, $P = 0.001$). Reproduced with permission from Schmid *et al.* (2011).

not significantly affected by farming methods, although the cell numbers of cultivable bacteria were different.

Different fungal communities were also found between grapes growing under integrated pest management (IPM) and organic management systems (Pancher *et al.*, 2012). In the latter, a higher relative abundance of the genera *Alternaria* and *Epicoccum* was detected as endophytes in grapevines. Furthermore, some strains of *Alternaria* are known to play a role in biocontrol; for example, control of *Plasmopara viticola*. Conversely, the abundances of the genera *Aureobasidium* and *Cladosporium* were reduced (Pancher *et al.*, 2012; Fig. 3.3A). Interestingly, in this survey, the effect of the farming method on the fungal community structure overcame that of the plant cultivar (Pancher *et al.*, 2012; Fig. 3.3B).

New functions of grapevine-associated microorganisms can be utilized for biotechnological applications. Verginer *et al.* (2010) analysed volatile compounds (VOCs) made by native bacteria, fungi, and yeasts isolated from grapevine and produced the typical aromas of wines. These VOCs were found to be related to the type of organism (bacteria, fungi or yeasts; see Fig. 3.4) and to be vineyard specific. Thus, the native microflora associated with grapevines can also contribute to the unique features of the wine that is produced from them. Also, not only do volatiles have an impact on flavour, they are also effective biocontrol compounds. Moreover, they are already used in the control of postharvest *Botrytis* fruit rot (Huang *et al.*, 2011).

Biocontrol Using Antagonistic Microorganisms

In recent years, numerous microbes antagonistic to *Botrytis* have been isolated from grapevine microflora. While some have been tested *ad planta* under controlled conditions, others were developed into a product (Elmer and Reglinski, 2006). A few biocontrol agents against *Botrytis* are available for the European market; for example, Botector®, which is produced by bio-ferm at Tulln in Austria (http://www.bio-ferm.com) and Trichopulvin 25WP from the Research Institute of Plant Protection at Bucharest in Romania (Sesan *et al.*, 1999). A *Bacillus* sp. was able to suppress the growth of *Botrytis* in vitroplants of *Vitis vinifera* (Paul *et al.*, 1998), but was not tested in the field. This bacterium was able to produce resveratrol, a compound that is beneficial to humans, but which can also induce systemic resistance in plants.

Pantoea ananatis BLBT1-08 was isolated in Austria from an organically cultivated vine and was shown to exert effective biocontrol activity against *Botrytis*. In a 3-year field trial, disease symptoms of the bunches were significantly reduced (Gasser *et al.*, 2012). A detached leaves *in vitro* assay was performed to assess the effect on the pathogen *B. cinerea* at the cellular and molecular levels. In the leaves treated with the BCA, no symptoms of the disease appeared. Interestingly, the conidia of the pathogen germinated and no direct contact with the BCA was observed using confocal laser scanning microscopy of the interaction (Fig. 3.5). Moreover, the same copy number of the *Cut*A gene (a gene specific to *B. cinerea*) was found in BCA-treated leaves and in untreated leaves. To investigate the possible modes of action, dual culture tests were performed and the effect of amino acids was tested on minimal medium; methionine, threonine, tryptophan, leucine and ornithine inhibited the biocontrol activity against *Botrytis*. This suggests that antibiotics, whose action can be reversed by certain amino acids, could be responsible for the biocontrol by BLBT1-08. As a result, growth inhibition of

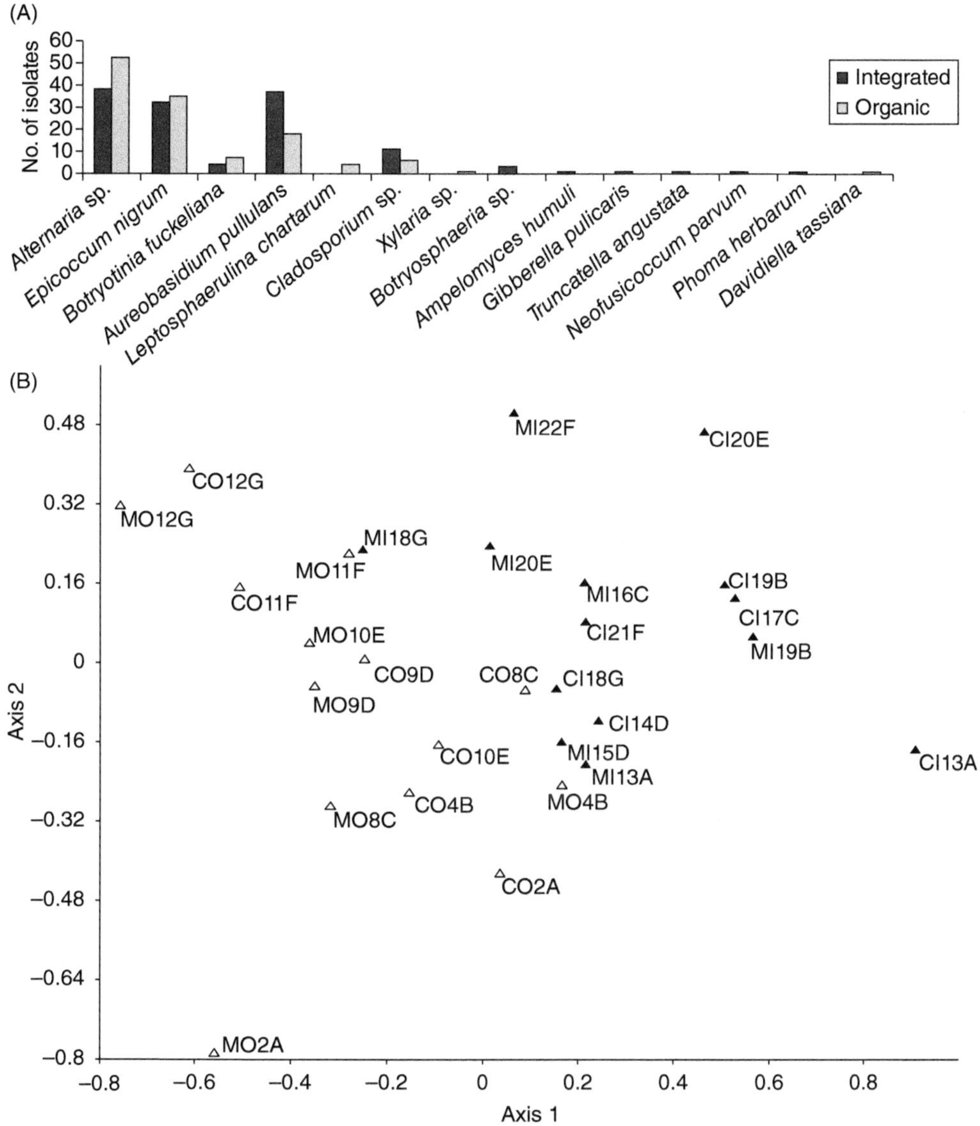

Fig. 3.3. (A) Frequency of fungal species isolated from vineyards according to phytosanitary regime (integrated pest management (IPM) vs organic). (B) Canonical correspondence analysis (CCA) plot of the fungal community structures according to farming method (▲, IPM; or Δ, organic) and plant variety (C, Chardonnay; or M, Merlot). Data from cultivation and automated ribosomal intergenic spacer analysis (ARISA) of fungal internal transcribed spacer (ITS) regions were combined. Reproduced with permission from Pancher *et al.* (2012).

Botrytis on rich medium indicates that biocontrol mechanisms other than those related to amino acid biosynthesis can occur. In addition, it also demonstrates the interaction of an abiotic condition with the modes of action.

Fungi as well as bacteria can also show biocontrol effects against grapevine pests. Yeast species belonging to the genus *Pichia* successfully controlled the development of

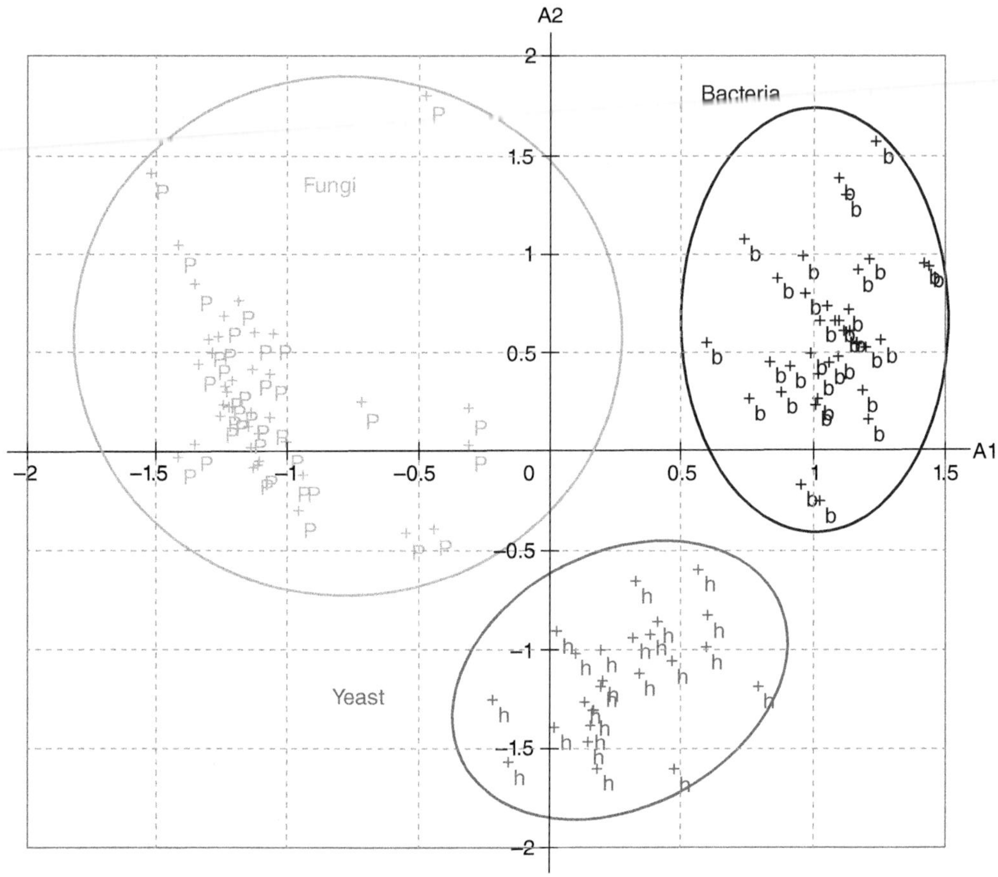

Fig. 3.4. Principle component analysis (PCA) showing the organism specificity of volatile compounds (VOCs) emitted from bacteria, fungi and yeast isolated from different vineyards. Reprinted with permission from Verginer *et al*. (2010). Copyright 2010 American Chemical Society.

grey mould symptoms *in vitro* (Masih *et al*., 2000, 2001). *P. membranifaciens* inhibited *Botrytis* growth most likely by secreting exo- and endo-β-1,3-glucanases (Masih *et al*., 2001). Santos and Marquina (2004) were able to purify an 18-kDa killer toxin from *P. membranifaciens* that was able to suppress *Botrytis* on grapevine leaves. *Trichoderma harzianum*, another fungus, interacts with plants by various mechanisms (Harman, 2006; Lorito *et al*., 2010). Inoculation of *T. harzianum* onto grapevine reduced the growth of *Phaeomoniella chlamydospora* and *Eutypa lata* on pruning wounds, although complete elimination of the pathogens was not achieved (Mutawila *et al*., 2011). These pathogenic fungi infect the plant by growing inside the xylem vessels. Interestingly, *T. harzianum* recognized the presence of antagonistic fungi that resulted in a deeper growth inside the canes. Against *B. cinerea*, treatment with *T. harzianum* was less efficient than the application of antifungal chemicals (Elad, 1994; O'Neill *et al*., 1996). The genus *Ulocladium* contains species with antagonistic effects on *Botrytis* (Kessel *et al*., 2005). *U. oudemansii* considerably reduced bunch rot, especially at the vintage and postharvest stages, when used in combination with the chemical

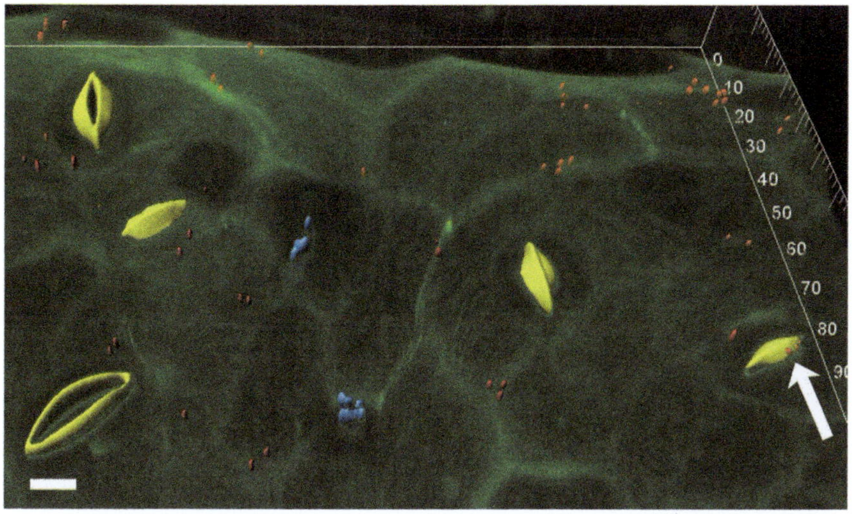

Fig. 3.5. Three-dimensional model of a confocal microscopy image series showing *Pantoea ananatis* strain BLBT1-08 and *Botrytis cinerea* on grapevine leaves. No direct contact between the antagonist and the pathogen appeared. Bacteria adhering to stomatal cells were also detected (arrow). Key: red, DsRed-fluorescently labelled bacteria; blue, conidia of *B. cinerea*; yellow, stomata; green, grapevine leaf surface. Scale bar: 10 μm. The confocal stack was acquired with a Leica TCS SPE confocal system (Leica Microsystems GmbH, Mannheim, Germany) and processed with the software Imaris 7.3 (Bitplane, Switzerland).

elicitor 5-chlorosalicylic acid (Reglinski *et al.*, 2005). *U. atrum* was also suggested for use as a possible co-application with other botryticides to increase the success of control in grapevines (Schoene *et al.*, 2000).

Conclusions and Future Perspectives

The biocontrol of *Botrytis* and other fungal pathogens by antagonistic microorganisms is promising and represents an environmentally friendly alternative to the use of chemicals. Results obtained thus far allow us to draw the following general conclusions: (i) the native microbiome associated with the grapevine is highly abundant and diverse and includes bacterial as well as fungal species, and (ii) this microbiome is highly stable but influenced by several factors, such as cultivar, plant stage, vineyard age, soil quality and climate, with a substantial impact from the management system used. Due to the observed cultivar/habitat-specificity of BCAs, colonization studies are important to evaluate potential BCA strains. Quantitative molecular analysis and microscopy observations *in situ* can shed light on to the complex network of interactions between beneficial microbes, the pathogen and the host plant (Cardinale, 2014). The microbiome can affect the behaviour of BCAs, so their interactions should be studied under field conditions using next-generation sequencing (NGS) techniques (Köberl *et al.*, 2012; Erlacher *et al.*, 2014) – we need to know the battlefield if we want to organize an efficient army and win the war against pests.

However, scientific research alone is not sufficient to achieve this goal; a better policy of cooperation is required between scientists, farmers, the wine industry and the biological products industry for agriculture. A pesticide-free environment will finally result in safer outdoor living and food which, in turn, will drastically reduce the costs of environmental sanitation and human health. As a first step, biocontrol measures can be integrated into other plant protection systems. For example, chemical elicitors of natural plant defence mechanisms could contribute to the success rate of biocontrol. Simultaneous or sequential application of more BCAs in combined approaches could also lead to higher efficiency. Moreover, the inoculation method used also affects the efficacy of biocontrol (Magnin-Robert *et al.*, 2007). A low biodiversity of microorganisms is often correlated with the occurrence of pathogens – saving diversity can, therefore, avoid diseases (Keesing *et al.*, 2010).

Acknowledgements

The help of Margaret Starcher (Virginia Polytechnic and State University, USA) is acknowledged for improving the language of the manuscript.

References

Ait Barka, E., Eullaffroy, P., Clement, C. and Vernet, G. (2004) Chitosan improves development, and protects *Vitis vinifera* L. against *Botrytis cinerea*. *Plant Cell Reports* 22, 608–614.

Aziz, A., Heyraud, A. and Lambert, B. (2003) Oligogalacturonide signal transduction, induction of defense-related responses and protection of grapevine against *Botrytis cinerea*. *Planta* 218, 767–774.

Bell, C.R., Dickie, G.A., Harvey, W.L.G. and Chan, J.W.Y.F. (1995) Endophytic bacteria in grapevine. *Canadian Journal of Microbiology* 41, 46–53.

Berg, G. (2009) Plant-microbe interactions promoting plant growth and health: perspectives for controlled use of microorganisms in agriculture. *Journal of Applied Microbiology and Biotechnology* 84, 11–18.

Berg, G., Krechel, A., Ditz, M., Sikora, R.A., Ulrich, A. and Hallmann, J. (2005) Endophytic and ectophytic potato-associated bacterial communities differ in structure and antagonistic function against plant pathogenic fungi. *FEMS Microbiology Ecology* 51, 215–229.

Berg, G., Opelt, K., Zachow, C., Lottmann, J., Götz, M., Costa, R. and Smalla, K. (2006) The rhizosphere effect on bacteria antagonistic towards the pathogenic fungus *Verticillium* differs depending on plant species and site. *FEMS Microbiology Ecology* 56, 250–261.

Bulgari, D., Casati, P., Brusetti, L., Quaglino, F., Brasca, M., Daffonchio, D. and Bianco, P.A. (2009) Endophytic bacterial diversity in grapevine (*Vitis vinifera* L.) leaves described by 16S rRNA gene sequence analysis and length heterogeneity-PCR. *Journal of Microbiology* 47, 393–401.

Bulgari, D., Casati, P., Crepaldi, P., Daffonchio, D., Quaglino, F., Brusetti, L. and Bianco, P.A. (2011) Restructuring of endophytic bacterial communities in grapevine yellows-diseased and recovered *Vitis vinifera* L. plants. *Applied Environmental and Microbiology* 77, 5018–5022.

Cardinale, M. (2014) Scanning a microhabitat: plant–microbe interactions revealed by confocal laser microscopy. *Frontiers in Microbiology* 5, 94.

Cardinale, M., Grube, M., Erlacher, A., Quehenberger, J. and Berg, G. (2015) Bacterial networks and co occurrence relationships in the lettuce root microbiota. *Environmental Microbiology* 17, 239–252.

Compant, S., Duffy, B., Nowak, J., Clément, C. and Ait Barka, E. (2005) Use of plant growth-promoting bacteria for biocontrol of plant diseases: principles, mechanisms of action, and future prospects. *Applied Environmental and Microbiology* 71, 4951–4959.

Compant, S., Kaplan, H., Sessitsch, A., Nowak, J., Ait Barka, E. and Clément, C. (2008) Endophytic colonization of *Vitis vinifera* L. by *Burkholderia phytofirmans* strain PsJN: from the rhizosphere to inflorescence tissues. *FEMS Microbiology Ecology* 63, 84–93.
Compant, S., Mitter, B., Colli-Mull, J.G., Gangl, H. and Sessitsch, A. (2011) Endophytes of grapevine flowers, berries, and seeds: identification of cultivable bacteria, comparison with other plant parts, and visualization of niches of colonization. *Microbial Ecology* 62, 188–197.
Elad, Y. (1994) Biological control of grape grey mould by *Trichoderma harzianum*. *Crop Protection* 13, 35–38.
Elad, Y., Yunis, H. and Katan, T. (1992) Multiple fungicide resistance to benzimidazoles, dicarboxymides, and diethofencarb in field isolates of *Botrytis cinerea* in Israel. *Plant Pathology* 41, 41–46.
Elad, Y., Williamson, B., Tudzynski, P. and Delen, N. (eds) (2004) *Botrytis: Biology, Pathology and Control*. Kluwer, Dordrecht, The Netherlands.
Elmer, P.A.G. and Reglinski, T. (2006) Biosuppression of *Botrytis cinerea* in grapes. *Plant Pathology* 55, 155–177.
Erlacher, A., Cardinale, M., Grosch, R., Grube, M. and Berg, G. (2014) The impact of the pathogen *Rhizoctonia solani* and its beneficial counterpart *Bacillus amyloliquefaciens* on the indigenous lettuce microbiome. *Frontiers in Microbiology* 5: 175.
Gasser, F. and Berg, G. (2011) Organic versus conventional agriculture: a review from a microorganism's point of view. *Current Trends in Microbiology* 7, 41–51.
Gasser, F., Cardinale, M., Schildberger, B. and Berg, G. (2012) Biocontrol of *Botrytis cinerea* by successful introduction of *Pantoea ananatis* in the grapevine phyllosphere. *International Journal of Wine Research* 4, 53–63.
Grube, M., Schmid, F. and Berg, G. (2011) Black fungi and associated bacterial communities in the phyllosphere of grapevine. *Fungal Biology* 115, 978–986.
Harman, G.E. (2006) Overview of mechanisms and uses of *Trichoderma* spp. *Phytopathology* 96, 190–194.
Houlden, A., Timms-Wilson, T.M., Day, M.J. and Bailey, M.J. (2008) Influence of plant developmental stage on microbial community structure and activity in the rhizosphere of three field crops. *FEMS Microbiology Ecology* 65, 193–201.
Huang, R., Li, G.Q., Zhang, J., Yang, L., Che, H.J., Jiang, D.H. and Huang, H.C. (2011) Control of postharvest *Botrytis* fruit rot of strawberry by volatile organic compounds of *Candida intermedia*. *Phytopathology* 101, 859–869.
Jacometti, M.A., Wratten, S.D. and Walter, M. (2010) Review: alternatives to synthetic fungicides for *Botrytis cinerea* management in vineyards. *Australian Journal of Grape and Wine Research* 16, 154–172.
Keesing, F., Belden, L.K., Daszak, P., Dobson, A., Harvell, C.D., Holt, R.D., Hudson, P., Jolles, A., Jones, K.E., Mitchell, C.E. *et al.* (2010) Impacts of biodiversity on the emergence and transmission of infectious diseases. *Nature* 468, 647–652.
Kessel, G.J., Köhl, J., Powell, J.A., Rabbinge, R. and Van der Werf, W. (2005) Modeling spatial characteristics in the biological control of fungi at leaf scale: competitive substrate colonization by *Botrytis cinerea* and the saprophytic antagonist *Ulocladium atrum*. *Phytopathology* 95, 439–448.
Köberl, M., Ramadan, E.M., Roßmann, B., Staver, C., Fürnkranz, M., Lukesch, B., Grube, M. and Berg, G. (2011) Using ecological knowledge and molecular tools to develop effective and safe biocontrol strategies. In: Stoytcheva, M. (ed.) *Pesticides in the Modern World: Pests Control and Pesticides Exposure and Toxicity Assessment*. InTech, Rijeka, Croatia, pp. 1–26.
Köberl, M., Schmidt, R., Ramadan, E.M., Bauer, R. and Berg, G. (2012) Chapter 16: Biocontrol strategies and next generation sequencing: organic desert agriculture in Egypt. In: iConcept Press (ed.) *Genomics 1 – Humans, Animals and Plants*. iConcept Press, Kowloon, Hong Kong [downloadable online only]. Available at: https://www.iconceptpress.com/books/genomics-i--humans-animals-and-plants/14410060/1205000413.pdf (accessed 17 September 2015).

Lorito, M., Woo, S.L., Harman, G.E. and Monte, E. (2010) Translational research on *Trichoderma*: from 'omics to the field. *Annual Review of Phytopathology* 48, 395–417.

Lugtenberg, B. and Kamilova, F. (2009) Plant-growth-promoting rhizobacteria. *Annual Review of Microbiology* 63, 541–556.

Magnin-Robert, M., Trotel-Aziz, P., Quantinet, D., Biagianti, S. and Aziz, A. (2007) Biological control of *Botrytis cinerea* by selected grapevine-associated bacteria and stimulation of chitinase and β-1,3 glucanase activities under field conditions. *European Journal of Plant Pathology* 118, 43–57.

Masih, E.I., Alie, I. and Paul, B. (2000) Can the grey mould disease of the grape-vine be controlled by yeast? *FEMS Microbiology Letters* 189, 233–237.

Masih, E.I., Slezack-Deschaumes, S., Marmaras, I., Ait Barka, E., Vernet, G., Charpentier, C., Adholeya, A. and Paul, B. (2001) Characterisation of the yeast *Pichia membranifaciens* and its possible use in the biological control of *Botrytis cinerea*, causing the grey mould disease of grapevine. *FEMS Microbiology Letters* 202, 227–232.

Mendes, R., Kruijt, M., De Bruijn, I., Dekkers, E., Van der Voort, M., Schneider, J.H., Piceno, Y.M., DeSantis, T.Z., Andersen, G.L., Bakker, P.A. and Raaijmakers, J.M. (2011) Deciphering the rhizosphere microbiome for disease-suppressive bacteria. *Science* 332, 1097–1100.

Mutawila, C., Fourie, P.H., Halleen, F. and Mostert, L. (2011) Histo-pathology study of the growth of *Trichoderma harzianum*. *Phytopathologia Mediterranea* 50(Suppl.), S46–S60.

Muzammil, S., Graillon, C., Mathieu, F., Lebrihi, A. and Compant, S. (2012) *Saccarothrix algeriensis*, an actinobacterium with high potential as biocontrol agent of *B. cinerea*, induces systemic resistance in grapevine. In: *COST Action FA1103: Current Aspects of European Endophyte Research. Endophytes in Biotechnology and Agriculture (www.endophytes.eu), Workshop 28–30 March 2012, University of Reims, France. Book of Abstracts*, p. 37. [No further details available.]

O'Neill, T.M., Elad, Y., Shtienberg, D. and Cohen, A. (1996) Control of grapevine grey mould with *Trichoderma harzianum* T39. *Biocontrol Science and Technology* 6, 139–146.

Pancher, M., Ceol, M., Corneo, P.E., Oliveira Longa, C.M., Yousaf, S., Pertot, I. and Campisano, A. (2012) Fungal endophytic communities in grapevines (*Vitis vinifera* L.) respond to crop management. *Applied Environmental and Microbiology* 78, 4308–4317.

Paul, B., Chereyathmanjiyil, A., Masih, I., Chapuis, L. and Benoît, A. (1998) Biological control of *Botrytis cinerea* causing grey mould disease of grapevine and elicitation of stilbene phytoalexin (resveratrol) by a soil bacterium. *FEMS Microbiology Letters* 165, 65–70.

Ramette, A., Moënne-Loccoz, Y. and Défago, G. (2006) Genetic diversity and biocontrol potential of fluorescent pseudomonads producing phloroglucinols and hydrogen cyanide from Swiss soils naturally suppressive or conducive to *Thielaviopsis basicola*-mediated black root rot of tobacco. *FEMS Microbiology Ecology* 55, 369–381.

Reglinski, T., Elmer, P.A.G., Taylor, J.T., Parry, F.J., Marsden, R. and Wood, P.N. (2005) Suppression of *Botrytis* bunch rot in Chardonnay grapevines by induction of host resistance and fungal antagonism. *Australasian Plant Pathology* 34, 481–488.

Santos, A. and Marquina, D. (2004) Killer toxin of *Pichia membranifaciens* and its possible use as a biocontrol agent against grey mould disease of grapevine. *Microbiology* 150, 2527–2534.

Schmid, F. and Berg, G. (2009) Grapevine-associated microorganisms: antagonistic potential towards *Botrytis cinerea* varies between habitats, cultivation methods and grapevine species. *IOBC-WPRS Bulletin* 43, 351–354.

Schmid, F., Moser, G., Müller, H. and Berg, G. (2011) Functional and structural microbial diversity in organic and conventional viticulture: organic farming benefits natural biocontrol agents. *Applied and Environmental Microbiology* 77, 2188–2191.

Schoene, P., Oerke, E.C. and Dehne, H.W. (2000) A new concept for integrated control of grey mould (*Botrytis cinerea*) in grapevine. In: *The BCPC Conference: Pests and Diseases, Volume 3. Proceedings of an International Conference Held at the Brighton Hilton Metropole Hotel, Brighton, UK, 13–16 November 2000*. British Crop Protection Council, Farnham UK, pp. 1031–1036.

Sesan, T., Oprea, M., Podosu, C.A., Tica, C. and Oancea, F. (1999) Biocontrol of *Botrytis cinerea* on grapevine with *Trichoderma* spp. and *Saccharomyces chevalieri*. *Bulletin of the Polish Academy of Sciences: Biological Sciences* 47, 197–205.
Smalla, K., Wieland, G., Buchner, A., Zock, A., Parzy, J., Kaiser, S., Roskot, N., Heuer, H. and Berg, G. (2001) Bulk and rhizosphere soil bacterial communities studied by denaturating gradient gel electrophoresis: plant-dependent enrichment and seasonal shifts revealed. *Applied and Environmental Microbiology* 67, 4742–4751.
Spadoro, D. and Gullino, M.L. (2005) Improving the efficacy of biocontrol agents against soil-borne pathogens. *Crop Protection* 24, 601–613.
Svercel, M., Christen, D., Moënne-Loccoz, Y., Duffy, B. and Défago, G. (2009) Effect of long-term vineyard monoculture on rhizosphere populations of pseudomonads carrying the antimicrobial biosynthetic genes *phl*D and/or *hcn*AB. *FEMS Microbiology Ecology* 68, 25–36.
Trotel-Aziz, P., Couderchet, M., Vernet, G. and Aziz, A. (2006) Chitosan stimulates defense reactions in grapevine leaves and inhibits development of *Botrytis cinerea*. *European Journal of Plant Pathology* 114, 405–413.
Verginer, M., Leitner, E. and Berg, G. (2010) Production of volatile metabolites by grape-associated microorganisms. *Journal of Agricultural and Food Chemistry* 58, 8344–8350.
Verhagen, B.W.M., Trotel-Aziz, P., Couderchet, M., Höfte, M. and Aziz, A. (2009) *Pseudomonas* spp.-induced systemic resistance to *Botrytis cinerea* is associated with induction and priming of defence responses in grapevine. *Journal of Experimental Botany* 61, 249–260.
West, E.R., Cother, E.J., Steel, C.C. and Ash, G.J. (2010) The characterization and diversity of bacterial endophytes of grapevine. *Canadian Journal of Microbiology* 56, 209–216.
Willer, H. (2008) Europe: statistics, support schemes and research. In: Willer, H. and Yussefi, M. (eds) *The World of Organic Agriculture – Statistics and Emerging Trends 2007*. International Federation of Organic Agriculture Movements (IFOAM), Bonn, Germany, and Research Institute of Organic Agriculture FiBL, Frick, Switzerland, pp. 134–142.
Williamson, B., Tudzynski, B., Tudzynski, P. and Van Kan, J.A.L. (2007) *Botrytis cinerea*: the cause of grey mould disease. *Molecular Plant Pathology* 8, 561–580.

4 Beneficial Bacteria Prime Local and Systemic Immunity Against *Botrytis cinerea* in Grapevine

A. Aziz,[1]* B. Verhagen,[1] S. Villaume,[1] M. Höfte,[2] F. Baillieul,[1] C. Clément[1] and P. Trotel-Aziz[1]

[1]*URVVC EA 4707, SDRP, University of Reims, Reims, France;*
[2]*Laboratory of Phytopathology, University of Ghent, Ghent, Belgium*

Introduction

Plants have evolved the ability to form beneficial relationships with mutualistic microbes that are important for plant growth and health (Mendes *et al.*, 2011; Berendsen *et al.*, 2012). The colonization of plant roots by mutualistic microbes is associated with an enhanced defensive capacity in the aerial parts of the plant, resulting in induced systemic resistance (ISR) (Pieterse *et al.*, 2012). This resistance may also result from a priming of the tissues to express basal defence mechanisms more rapidly and more strongly after subsequent pathogen challenge (Verhagen *et al.*, 2011; Zamioudis and Pieterse, 2012).

In grapevine, investigations on ISR using beneficial bacteria are limited to a few reports (Bordiec *et al.*, 2011; Verhagen *et al.*, 2011; Compant *et al.*, 2013; Magnin-Robert *et al.*, 2013). In recent years, we have identified and screened various beneficial bacteria isolated from vineyards for their ability to protect grapevines against *Botrytis cinerea* (Magnin-Robert *et al.*, 2007; Trotel-Aziz *et al.*, 2008; Verhagen *et al.*, 2011). Crude extracts from these bacteria also triggered immune responses in grapevine cell suspensions and leaves (Verhagen *et al.*, 2010, 2011), resembling those that occurred with various microbe-associated molecular patterns (MAMPs) that enhance resistance against *B. cinerea* and *Plasmopara viticola* (Aziz *et al.*, 2007; Trouvelot *et al.*, 2008; Varnier *et al.*, 2009). However, the relationships between ISR and local or systemic immune responses before or after pathogen infection are largely unknown. In this study, we examined the effectiveness of various beneficial bacteria, originating from different host plants, to trigger ISR in relation to their capacity to induce or prime immune responses in grapevine. We focused especially on oxidative burst in grapevine cell suspensions and the systemic production of stilbenic phytoalexins in plants.

Materials and Methods

The bacteria used in this study were from different plants or rhizospheres: *Pantoea agglomerans* PTA-AF2, *Bacillus subtilis* PTA-271, *Acinetobacter lwoffii* PTA-113 and

*Corresponding author: aziz.aziz@univ-reims.fr

Pseudomonas fluorescens PTA-CT2 from grapevine (Trotel-Aziz *et al.*, 2008), *P. fluorescens* CHA0 from tobacco, *P. fluorescens* WCS417 and Q2-87 from wheat, *P. putida* WCS358 from potato and *P. aeruginosa* 7NSK2 and its mutants KMPCH and KMPCH-567 from bean (see Verhagen *et al.*, 2010). Bacterial isolates were grown in Luria-Bertani (LB) medium and used at 1×10^7 colony forming units (cfu) ml^{-1} in 10 mM $MgSO_4$.

Plantlets of *Vitis vinifera* cv. Chardonnay were obtained *in vitro* by growing explants on modified Murashige and Skoog (MS) medium. Cells from *V. vinifera* cv. Gamay were cultivated in Nitsch medium under continuous light as described previously (Aziz *et al.*, 2007).

ISR was assayed on 4-week-old *in vitro* plantlets (Verhagen *et al.*, 2010). The roots were dipped into bacterial suspensions, and leaves were removed 4 weeks later and challenged with 7.5 μl of a conidial suspension of *B. cinerea* (10^6 conidia ml^{-1}). Necrotic lesions were measured at 7 days post-inoculation.

The production of H_2O_2 by the cells was assayed by chemiluminescence using luminol (Aziz *et al.*, 2003). Stilbenic phytoalexins were extracted in methanol and analysed by HPLC (high performance liquid chromatography) coupled to a fluorometer detector (Aziz *et al.*, 2006).

Results and Discussion

Induction of oxidative burst in grapevine cells by bacteria

The production of active oxygen species (AOS; also known as reactive oxygen species, ROS) has been assumed to be one of the earliest defence events in cell suspension cultures in response to various MAMPs (Aziz *et al.*, 2007). Here, we showed that selected bacteria induced a single and transient release of H_2O_2 in grapevine cell suspensions (Fig. 4.1A), with a maximum after 90–120 min. The burst was high with *P. fluorescens* PTA-CT2 and CHA0 and low with *P. aeruginosa* 7NSK2. This indicates that these bacteria can be perceived rapidly by grapevine cells. The potency of grapevine-associated bacteria to induce oxidative burst was compared with that of other bacterial isolates that have different modes of inducing ISR in other plants. The results showed that bacterial isolates were all capable of inducing an oxidative burst with similar kinetics, but to different extents (Fig. 4.1B–D).

High levels of H_2O_2 were released after treatment with PTA-CT2, PTA-113, CHA0 and Q2-87 (as producers of DAPG – 2,4-diacetylphloroglucinol). However, a lower response was observed with WCS417. WCS358 induced an intermediate level of H_2O_2 release similar to that induced by 7NSK2 and its mutants. This suggests that oxidative burst might play an important role in the recognition of bacterial signals that can initiate defence pathways to boost plant immunity. Different levels of accumulated H_2O_2 could be ascribed to different active compounds or the quantities released by each bacterium in the medium (Verhagen *et al.*, 2010).

Phytoalexin accumulation in grapevine cells by bacteria

The production of stilbenic phytoalexins is an important response that is usually associated with defence responses and direct resistance induction in plants. We have shown that all of the selected bacteria induced a slight accumulation of *trans*-resveratrol

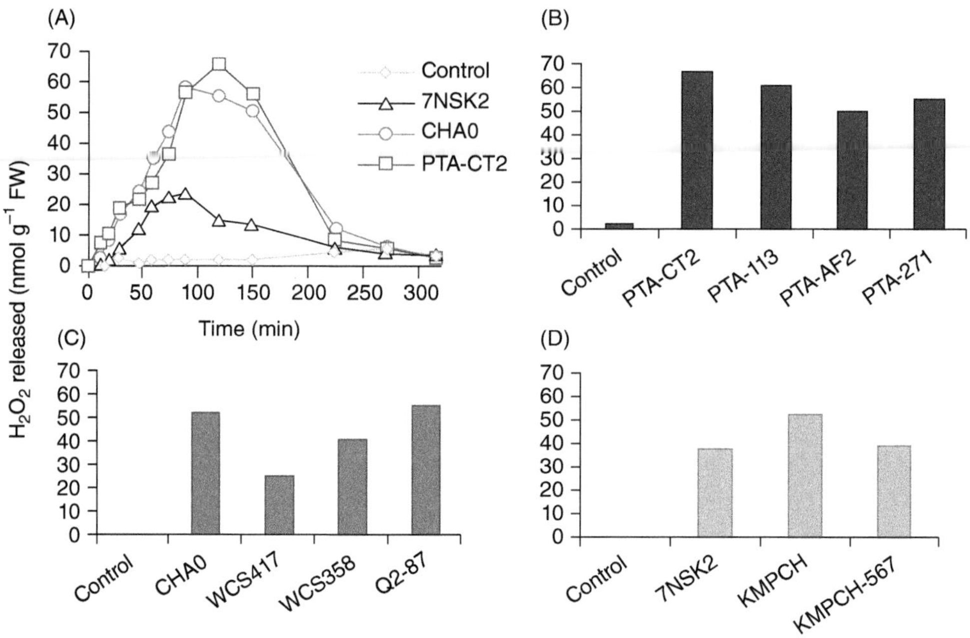

Fig. 4.1. H_2O_2 production by grapevine cells induced by bacteria. (A) Time course of H_2O_2 production in response *to Pseudomonas fluorescens* PTA-CT2 and CHA0 and *P. aeruginosa* 7NSK2. (B–D) Maximum H_2O_2 induced by different isolates: *Acinetobacter lwoffi* PTA-113, *Pantoea agglomerans* PTA-AF2, *Bacillus subtilis* PTA-271, *P. fluorescens* WCS417 and Q2-87, *P. putida* WCS358 and *P. aeruginosa* KMPCH and KMPCH-567. See text for details of host origin.

and *trans*-ε-viniferin in grapevine cells or in directly treated leaves. Elicitation with PTA-CT2 and PTA-113 (originating from vineyards) had the strongest effect (data not shown). Resveratrol production was lower with PTA-271 treatment. CHA0, Q2-87 and WCS358 also enhanced resveratrol and ε-viniferin production in grape cells, as did 7NSK2. However, lower responses were observed with WCS417 and the 7NSK2-derivative mutants KMPCH and KMPCH-567.

Priming for enhanced phytoalexin production after challenge inoculation

In order to investigate whether the induced resistance against *B. cinerea* is associated with the ability of bacteria to prime phytoalexin production, roots of grapevine plantlets were treated with bacteria, and leaves were removed and challenged with *B. cinerea*. Phytoalexins were determined during inoculation.

Priming by the fungus induced the accumulation of both resveratrol and ε-viniferin in PTA-CT2 and PTA-113-treated plants (Fig. 4.2A–B). However, PTA-AF2 and PTA-271 (to a lesser degree) did not affect the pathogen-induced production of phytoalexins. CHA0, Q2-85 and WCS358 also primed plants for enhanced production of both phytoalexins, whereas WCS417 did not.

Pretreatment with 7NSK2 resulted in a slight increase of phytoalexin levels before challenge, but strongly primed resveratrol and viniferin accumulation in the leaves

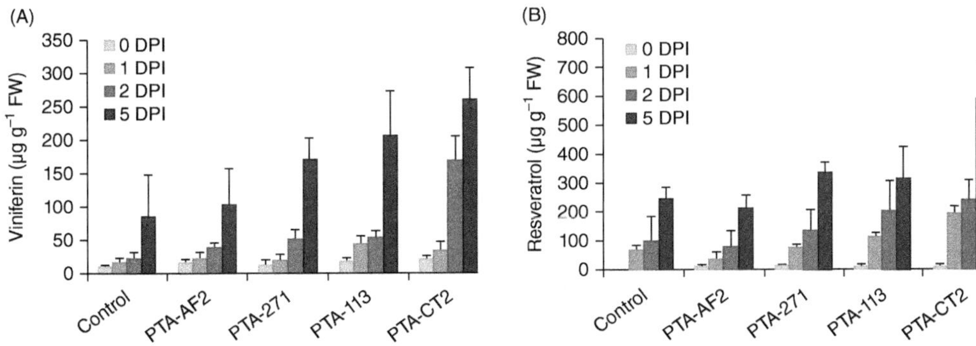

Fig. 4.2. Priming by various bacteria for the enhancement of *trans*-resveratrol (A) and *trans*-ε-viniferin (B) production in grapevine leaves upon challenge (inoculation) with *B. cinerea* (DPI, days post-inoculation). The bacteria used were *Pantoea agglomerans* PTA-AF2, *Bacillus subtilis* PTA-271, *Acinetobacter lwoffi* PTA-113 and *Pseudomonas fluorescens* PTA-CT2. See text for details of bacterial host origin.

after challenge. Resveratrol and viniferin in the *P. aeruginosa* 7NSK2-treated plantlets were 2–3-fold high than in the control. The mutant *P. aeruginosa* KMPCH also potentiated resveratrol and viniferin accumulation, but to a lower level than its wild type, while *P. aeruginosa* KMPCH-567 had no consistent effect on resveratrol production after 2 days of challenge. Similar results have been observed with growing media and bacterial extracts (Verhagen *et al.*, 2011). This pattern suggests that the plants had been primed by a sensitization of a defence arsenal rather than fully induced. Primed plants usually show no enhanced expression of phenotypic defence traits, such as callose deposition and H_2O_2 accumulation, but they respond faster or more strongly following a challenge inoculation.

Induction of systemic resistance in grapevine

Root treatment of grapevine plantlets with selected bacteria inhibited the development of disease caused by *B. cinerea* in the leaves by ≈30 to 50% (Fig. 4.3). PTA-CT2, -AF2 and -113 were the most active in enhancing resistance. PTA-271-induced resistance was lower (Fig. 4.3A). The difference between selected bacteria could be attributed to the different cellular MAMPs produced. *P. fluorescens* CHA0, WCS417 and Q2-87, and *P. putida* WCS358 significantly reduced grey mould disease in leaves (Fig. 4.3B). *P. aeruginosa* 7NSK2 and its mutant KMPCH-567 also reduced disease, but the reduction was smaller with KMPCH and KMPCH-567 (Fig. 4.3C).

Conclusions

Different bacterial species associated with grapevines are capable of inducing and, in some cases, priming a transient oxidative burst and the production of *trans*-resveratrol and *trans*-ε-viniferin in grapevine cells and leaves. These effects were associated with an induced resistance of grapevine leaves against *B. cinerea*. The early production of

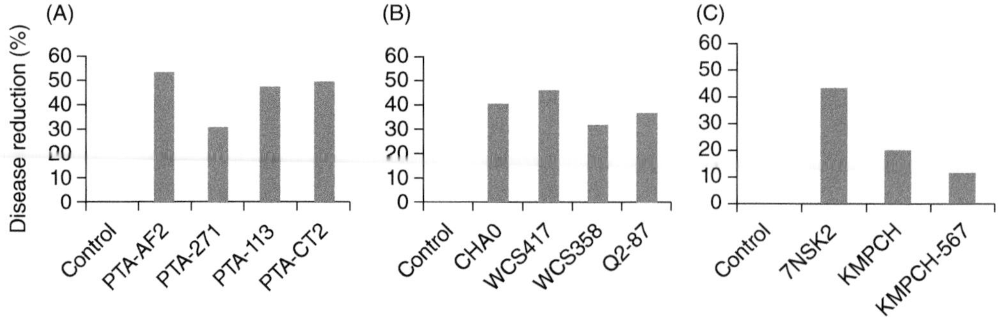

Fig. 4.3. Induced resistance of grapevine leaves to *Botrytis cinerea* after root treatment of plantlets with beneficial bacteria. *Pantoea agglomerans* PTA-AF2, *Bacillus subtilis* PTA-271, *Acinetobacter lwoffi* PTA-113, *Pseudomonas fluorescens* PTA-CT2, CHA0, WCS417 and Q2-87, *P. putida* WCS358, P. aeruginosa 7NSK2 and its mutants KMPCH and KMPCH-567. See text for details of bacterial host origin.

H_2O_2 by grapevine cells was not necessarily linked to enhanced resistance in the plants. However, induced resistance by *Pseudomonas fluorescens* PTA-CT2 and *Acinetobacter lwoffii* PTA-113, as well as by *P. fluorescens* CHA0, Q2-87, *P. putida* WCS358 and *P. aeruginosa* 7NSK2, appears to be linked to both the induction and the priming of phytoalexins. *Pantoea agglomerans* PTA-AF2 and *P. fluorescens* WCS417, which are effective in triggering ISR, can induce, but not prime, phytoalexin accumulation after challenge. The results also highlight that there are different mechanisms involved in the establishment of ISR in grapevine against *B. cinerea*.

Acknowledgements

We are grateful to Prof. C.M.J. Pieterse for providing bacterial strains WCS417 and WCS358, to Prof. D. Haas for providing CHA0 and to Prof. D.M. Weller for providing Q2-87. This research was supported by Champagne-Ardenne Region and Europol'Agro through the VINEAL programme.

References

Aziz, A., Poinssot, B., Daire, X., Adrian, M., Bezier, A., Lambert, B., Joubert, J.M. and Pugin, A. (2003) Laminarin elicits defense responses in grapevine and induces protection against *Botrytis cinerea* and *Plasmopara viticola*. *Molecular Plant-Microbe Interactions* 16, 1118–1128.

Aziz, A., Trotel-Aziz, P., Dhuicq, L., Jeandet, P., Couderchet, M. and Vernet, G. (2006) Chitosan oligomers and copper sulfate induce grapevine defense reactions and resistance to gray mold and downy mildew. *Phytopathology* 96, 1188–1194.

Aziz, A., Gauthier, A., Bézier, A., Poinssot, B., Pugin, A., Heyraud, J.M. and Baillieul, F. (2007) Elicitor and resistance-inducing activities of β-1,4 cellodextrins in grapevine, comparison with β-1,3 glucans and β-1,4 oligogalacturonides. *Journal of Experimental Botany* 58, 1463–1472.

Berendsen, R.L., Pieterse, C.M.J. and Bakker, P.A.H.M. (2012) The rhizosphere microbiome and plant health. *Trends in Plant Science* 17, 478–486.

Bordiec, S., Paquis, S., Lacroix, H., Dhondt, S., Ait Barka, E., Kauffmann, S., Jeandet, P., Mazeyrat-Gourbeyre, F., Clément, C., Baillieul, F. and Dorey, S. (2011) Comparative analysis of defence responses induced by the endophytic plant growth-promoting rhizobacterium *Burkholderia phytofirmans* strain PsJN and the non-host bacterium *Pseudomonas syringae* pv. *pisi* in grapevine cell suspensions. *Journal of Experimental Botany* 62, 595–603.

Compant, S., Brader, G., Muzammil, S., Sessitsch, A., Lebrihi, A. and Mathieu, F. (2013) Use of beneficial bacteria and their secondary metabolites to control grapevine pathogen diseases. *BioControl* 58, 435–455.

Magnin-Robert, M., Trotel-Aziz, P., Quantinet, D., Biagianti, S. and Aziz, A. (2007) Biological control of *Botrytis cinerea* by selected grapevine-associated bacteria and stimulation of chitinase and β-1,3 glucanase activities under field conditions. *European Journal of Plant Pathology* 118, 43–57.

Magnin-Robert, M., Quantinet, D., Couderchet, M., Aziz, A. and Trotel-Aziz, P. (2013) Differential induction of grapevine resistance and defense reactions against *Botrytis cinerea* by bacterial mixtures in vineyards. *BioControl* 58, 117–131.

Mendes, R., Kruijt, M., de Bruijn, I., Dekkers, E., Van der Voort, M., Schneider, J.H.M., Piceno, I.M., DeSantis, T.Z., Andersen, G.L., Bakker, P.A.H.M. and Raaijmakers, J.M. (2011) Deciphering the rhizosphere microbiome for disease-suppressive bacteria. *Science* 332, 1097–1100.

Pieterse, C.M.J., Van der Does, D., Zamioudis, C., Leon-Reyes, A. and Van Wees, S.C.M. (2012) Hormonal modulation of plant immunity. *Annual Review of Cell and Developmental Biology* 28, 489–521.

Trotel-Aziz, P., Couderchet, M., Biagianti, S. and Aziz, A. (2008) Characterization of new bacterial biocontrol agents *Acinetobacter*, *Bacillus*, *Pantoea* and *Pseudomonas* spp. mediating grapevine resistance against *Botrytis cinerea*. *Environmental and Experimental Botany* 64, 21–32.

Trouvelot, S., Varnier, A.-L., Allègre, M., Mercier, L., Baillieul, F., Arnould, C., Gianinazzi-Pearson, V., Klarzynski, O., Joubert, J.-M., Pugin, A. and Daire, X. (2008) A β-1,3 glucan sulfate induces resistance in grapevine against *Plasmopara viticola* through priming of defense responses, including HR-like cell death. *Molecular Plant-Microbe Interactions* 21, 232–243.

Varnier, A.-L., Sanchez, L., Vatsa, P., Boudesocque, L., Garcia-Brugger, A., Rabenoelina, F., Sorokin, A., Renault, J.-H., Kauffmann, S., Pugin, A. *et al.* (2009) Bacterial rhamnolipids are novel MAMPs conferring resistance to *Botrytis cinerea* in grapevine. *Plant, Cell and Environment* 32, 178–193.

Verhagen, B., Trotel-Aziz, P., Couderchet, M., Höfte, M. and Aziz, A. (2010) *Pseudomonas* spp.-induced systemic resistance to *Botrytis cinerea* is associated with induction and priming of defense responses in grapevine. *Journal of Experimental Botany* 61, 249–260.

Verhagen, B., Trotel-Aziz, P., Jeandet, P., Baillieul, F. and Aziz, A. (2011) Improved resistance against *Botrytis cinerea* by grapevine-associated bacteria that induce and prime oxidative burst and phytoalexin production. *Phytopathology* 101, 768–777.

Zamioudis, C. and Pieterse, C.M.J. (2012) Modulation of host immunity by beneficial microbes. *Molecular Plant-Microbe Interactions* 25, 139–150.

5 Potential of Two *Bacillus* Antagonists for Biocontrol of Grey Mould

S. Ben-Maachia,[1,2] R. Errakhi,[1,3,4]* F. Mathieu[1] and A. Lebrihi[1]

[1]LGC UMR 5503 (CNRS/INPT/UPS), Département Bioprocédés et Systèmes Microbiens, Université de Toulouse, Castanet-Tolosan, France; [2]Centre Régional des Recherches en Agriculture Oasienne à Dégache, Tozeur, Tunisia; [3]Plateforme de Biotechnologie, AGRONUTRITION, Carbonne, France; [4]Laboratoire d'Électrophysiologies des Membranes, Institut de Biologie des Plantes, Université Paris Diderot, Orsay, France

Introduction

Grapes are one of the most important fruit crops in the world. Most of the grapevine cultivars used for wine production are highly sensitive to pathogen attack and require many fungicidal treatments seasonally (Girault *et al.*, 2008). Currently, grey mould, which is caused by *Botrytis cinerea*, is controlled by preventive fungicides. The use of these chemicals has its limitations, and this has promoted consideration of the strategies of biological disease control and induction of plant resistance using non-pathogenic plant-associated microorganisms (Van Loon, 1997; Bargabus *et al.*, 2003; Tjamos *et al.*, 2005; Errakhi *et al.*, 2007). In this chapter, we describe a preliminary characterization of two bacterial strains isolated from the south of Tunisia and evaluate induced systemic resistance (ISR) as a mechanism of biological control of *B. cinerea* by these two bacterial strains. Additionally, molecular identification of the two *Bacillus* strains investigated is demonstrated.

Materials and Methods

Fungal and bacterial cultures

B. cinerea Pers. was isolated from infected leaves of grapevines in Tunisia and identified by its morphological characteristics. The fungus was grown routinely on potato dextrose agar (PDA) medium at 20°C. Conidia were scratched gently with a brush from 8-day-old cultures, suspended in distilled water, and diluted to 10^5 spores ml^{-1}.

*Corresponding author: r.errakhi@umi.ac.ma

The bacterial isolates B29 and B27 had been previously isolated from saline soil in the south of Tunisia and identified as *Bacillus* spp. These bacteria were routinely grown on nutrient agar at 30°C.

Grapevine treatment with *Bacillus* isolates and challenge with *Botrytis cinerea*

An 'inducing treatment' with the *Bacillus* strains on grapevine plants cv. COT-NOIR grown in a glasshouse was applied to a single leaf located at a central position of the shoot. Challenge inoculations with *B. cinerea* were applied 2 days later to the treated leaves. Twelve untreated leaves (three leaves on each of four vines), were also inoculated as controls. Leaf samples taken included the treated leaves and also the immediately adjacent leaves in the direction of the trunk and in the direction of the shoot tip.

Peroxidase and lipoxygenase activities

Peroxidase activity was assayed with a spectrophotometer at 470 nm according to the method described by El Hadrami *et al.* (1997). Lipoxygenase activity was measured according to the method of Axelrod *et al.* (1981).

16S rDNA gene sequencing of *Bacillus* isolates

A 16S rDNA gene sequence analysis of the *Bacillus* strains was done. The 16S rDNA was amplified using the PCR method with Taq DNA polymerase and primers 27f (AGAGTTTGATCCTGGCTCAG) and 1492r (GGTTACCTTGTTACGACTT). The conditions for thermal cycling were as follows: denaturation of the target DNA at 98°C for 3 min followed by 30 cycles at 94°C for 1 min, primer annealing at 53°C for 1 min and primer extension at 72°C for 5 min. At the end of the cycles, the reaction mixture was left at 72°C for 5 min and then cooled to 4°C. PCR amplification was detected by agarose gel electrophoresis and visualized by ultraviolet (UV) fluorescence after ethidium bromide staining. The PCR product obtained was also submitted to GenomExpress (Meylan, France) for sequence determination. The analysis of sequences was done using the BLAST (Basic Local Alignment Search Tool) software at the NCBI (US National Center for Biotechnology Information) server (http://www.ncbi.nlm.nih.gov/BLAST). All of the analysed sequences were deposited in the GenBank database.

Results and Discussion

Grapevines are highly sensitive to fungal attacks, such as the grey mould caused by *B. cinerea*. Current disease control methods are primarily based on chemical compounds, but the negative effects on the environment and humans that are associated with these fungicides have led to increasing interest in other methods of control, such as biological control by using beneficial microorganisms.

Microbiological activities

The study reported here investigated two new strains of *Bacillus* spp., B27 and B29, that had previously been selected and were highly effective *in vitro* against *B. cinerea*. Its aim was to determine the mode of action of these two bacteria.

The experimental data revealed the ability of the two selected *Bacillus* strains to degrade the (fungal) cell-wall of *B. cinerea*, as well as their ability to inhibit the growth of this fungus by the production of antibiotics and antifungal molecules in the media. Interestingly, previous results have demonstrated that the two strains not only showed antifungal activity against *B. cinerea* but also antimicrobial activity against many micro-organisms, including different fungi, yeasts, and Gram-positive and Gram-negative bacteria, thus demonstrating the wide range of the putative applications of these beneficial bacteria (Table 5.1).

Biocontrol effect of two strains

Pretreatment of grapevine plants with *Bacillus* strains B27 and/or B29 significantly reduced infection by *B. cinerea* inoculated 2 days later. Observations of the treated (and adjacent) leaves were conducted under an optical microscope 10 days after inoculation, and used to determine the impact of infection by *B. cinerea* on the host plant (see Fig. 5.1). Grapevine leaves inoculated with the fungus showed disorganized cellular tissues and cells that had been destroyed by the fungus. The observations also revealed the presence of dense mycelium of the fungus in the main veins of the leaf, as well as in other part of the leaves. An examination of the surface of the leaves attacked by *Botrytis* showed the development of the fungus and the appearance of fruiting bodies of the pathogen on the surfaces of inoculated control leaves that had not been pretreated with bacteria. In contrast, observations of vine leaves treated with the two bacterial strains but not inoculated showed no fungus in their tissues, and were well organized, showing clear ribs. It is interesting to note that under UV light (results not shown), both control leaves inoculated with *B. cinerea* and leaves treated with bacteria showed a blue fluorescence of the veins and in the cells of the tissues.

Table 5.1. Antimicrobial activity of *Bacillus* spp. strains B27 and B29.

	Diameter of inhibition (mm)	
Microorganism	B27	B29
Gram-positive bacteria		
Micrococcus luteus CIP A270	24.25 ± 0.85	25 ± 0.00
Gram-negative bacteria		
Escherichia coli CIP 102 400	20.5 ± 0.50	21.75 ± 0.75
Pseudomonas aeruginosa CIP 100720	20.5 ± 0.50	21.25 ± 0.47
Yeasts		
Saccharomyces cerevisiae ATCC 9226	22.5 ± 0.28	24 ± 0.40
Filamentous fungi		
Mucor ramannianus NRLL 1829	17.25 ± 0.28	19.25 ± 0.47
Botrytis cinerea	21.75 ± 0.62	24 ± 0.40

Induction of systemic resistance by two antagonistic strains

Our studies also demonstrated some of the means by which the two bacterial antagonists were able to induce grapevine defence mechanisms and thus allow a greater resistance to infection by *B. cinerea*. This was done by studying some of the metabolic changes associated with this resistance.

The results showed the ability of the two bacterial strains to induce ISR, which was clearly associated with the accumulation of phenolic compounds (particularly flavonoids) and the stimulation of peroxidase activity (Fig. 5.2A). The induction of ISR was further proven biochemically by an increase of lipoxygenase activity (Fig. 5.2B) in the leaves of plants treated with B27 and B29 compared with activity in the controls.

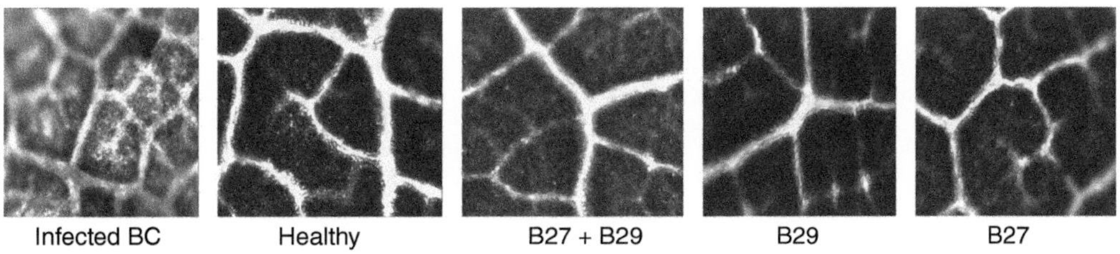

Fig. 5.1. Observations of grapevine leaves under the optical microscope according to different treatments. Infected BC: vine leaf infected with *Bacillus cinerea* showing the poorly organized cells, and revealing the development of the fungus. Healthy: healthy leaf showing organized cells with a clear absence of infection. B27 + B29: leaf pretreated with *Bacillus* strains B27 and B29, and inoculated with *B. cinerea* 2 days later. B29: leaf treated with *Bacillus* B29 strain and inoculated with *B. cinerea* 2 days later. B27: leaf treated with *Bacillus* B27 strain and inoculated with *B. cinerea* 2 days later.

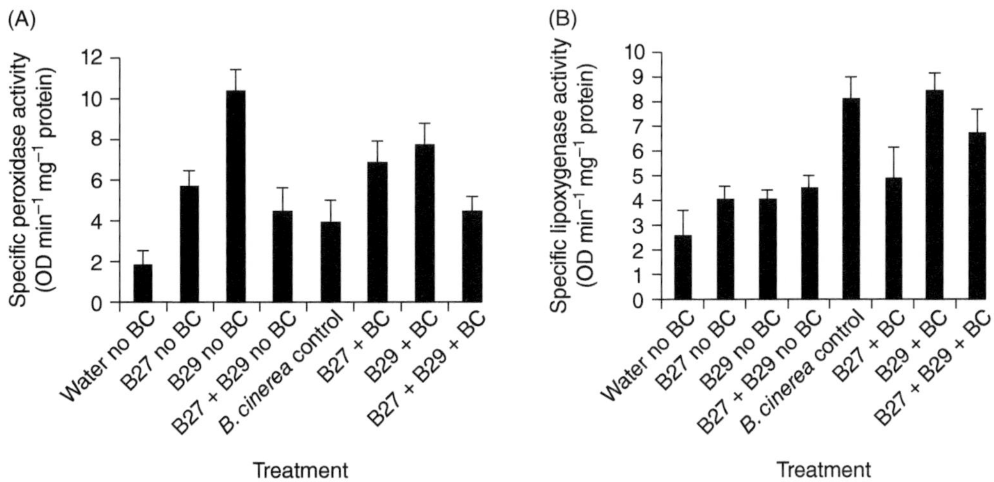

Fig. 5.2. The production of peroxidases (A) and lipoxygenases (B) in grapevines and the different amounts of activity of these enzymes in response to foliar pretreatments with *Bacillus* strains B27 and/or B29 and inoculation with *Botrytis cinerea* (BC). The error bars represent the standard error of the means of eight repetitions; OD, optical density.

In the presence of the pathogen, the host plant reacts strongly and quickly to implement its defence mechanisms when the pathogen is perceived, suggesting a phenomenon of priming of plant defences.

Molecular identification and phylogenetic analysis of bacteria

Data from the 16S rDNA gene provide a limited number of characters that are informative on the phylogeny of the *Bacillus* strains tested, and a phylogenetic tree is presented in Fig. 5.3. Based on the sequence homology and phylogenetic analysis, the B27 strain was found to be a *Bacillus* strain (accession number 1104513.1), which is closely related to *Bacillus* sp. Z19 (EU236746.1). The B29 strain was found to be a *Bacillus* strain (accession number 1104513.2), which was closely related to *B. amyloliquefaciens* strain EXWB3-03 (EU334107.1).

The partial sequence analysis of the 16S rDNA gene and BLAST sequence comparison in the GenBank database showed an alignment of 1414/1415 bases (99%

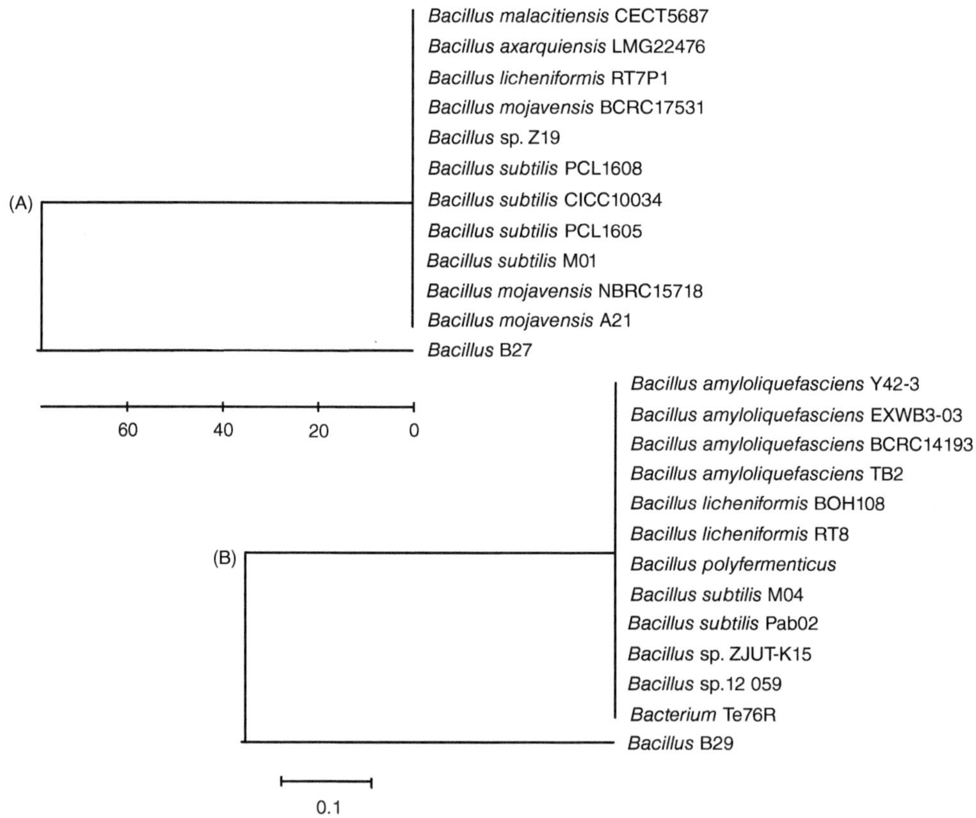

Fig. 5.3. Phylogenetic tree derived from parsimony analysis of 16S rDNA gene from (A) *Bacillus* sp. strain B27 (accession number 1104513.1) and (B) *Bacillus* strain B29 (accession number 1104513.2). The numbers at the nodes indicate the levels of bootstrap support based on neighbour-joining analyses of 1000 resampled data sets; only values over 50% are given. Bar, 0.0005 nucleotide (nt) substitutions per nt position.

similarity) of strain B27 (accession number 1104513.1) with the sequence of *B. subtilis* strain PCL 1605 (accession number DQ779882.1); comparable results were obtained for the similarity of strain B27 with *B. mojavensis* strain BCRC 17531 (accession number DQ993678.1). For strain B29 (accession number 1104513.2), a similarity level of 99% with an alignment of 1414/1415 bases was determined with *B. subtilis* strain Pab02 (accession number EU346662.1).

Conclusion

The work that has been reported here shows the potential of two new strains of *Bacillus* spp. isolated from the south of Tunisia to alleviate the infection of grapevines by *B. cinerea* through direct biocontrol activities and by inducing an ISR phenomenon. It has further demonstrated the taxonomy of these two members of the *Bacillaceae*, a prerequisite to better understanding of the potential of these new strains to reduce pathogen diseases.

References

Axelrod, B., Cheesbrough, T.M. and Laakso, S. (1981) Lipoxygenase from soybeans. *Methods in Enzymology* 71, 441–451.

Bargabus, R., Zidack, N., Sherwood, J. and Jacobsen, B. (2003) Oxidative burst elicited by *Bacillus mycoides* isolate Bac J, a biological control agent, occurs independently of hypersensitive cell death in sugar beet. *Molecular Plant-Microbe Interactions* 16, 1145–1153.

El Hadrami, I., Ramos, T., El Bellaj, M., El Idrissi-Tourane, A. and Macheix, J.J. (1997) A sinapic derivative as an induced defense compound of date palm against *Fusarium oxysporum* f. sp. *albedinis*, the agent causing bayoud disease. *Journal of Phytopathology* 145, 329–333.

Errakhi, R., Bouteau, F., Lebrihi, A. and Barakate, M. (2007) Evidences of biological control capacities of *Streptomyces* spp. against *Sclerotium rolfsii* responsible for damping-off disease in sugar beet (*Beta vulgaris* L.). *World Journal of Microbiology and Biotechnology* 23, 1503–1509.

Girault, T., François, J., Rogniaux, H., Pascal, S., Delrot, S., Coutos-Thévenot, P. and Gomès, E. (2008) Exogenous application of a lipid transfer protein–jasmonic acid complex induces protection of grapevine towards infection by *Botrytis cinerea*. *Plant Physiology and Biochemistry* 46, 140–149.

Tjamos, S.E., Flemetakis, E., Paplomatas, E.J. and Katinakis, P. (2005) Induction of resistance to *Verticillium dahliae* in *Arabidopsis thaliana* by the biocontrol agent K-165 and pathogenesis-related proteins gene expression. *Molecular Plant-Microbe Interactions* 18, 555–561.

Van Loon, L.C. (1997) Induced resistance in plants and the role of pathogenesis-related proteins. *European Journal of Plant Pathology* 103, 753–765.

6 Isolation and Characterization of Antibiotics Produced by *Streptomyces* J-2 and their Role in Biocontrol of Plant Diseases, Especially Grey Mould

R. Errakhi,[1,2,3*] F. Bouteau,[3] M. Barakate[4] and A. Lebrihi[2]

[1]*Plateforme de Biotechnologie, AGRONUTRITION, Carbonne, France;* [2]*LGC UMR 5503 (CNRS/INPT/UPS), Département Bioprocédés et Systèmes Microbiens, Université de Toulouse, Castanet-Tolosan, France;* [3]*Laboratoire d'Électrophysiologies des Membranes, Institut de Biologie des Plantes, Université Paris Diderot, Orsay, France;* [4]*Laboratoire de Microbiologie, Université Cadi Ayyad, Marrakech, Morocco*

Introduction

The biological control of plant pathogens is an integral component in maintaining ecological balance in the world of microorganisms in the rhizosphere of plants. It also helps to reduce the use of chemical pesticides to the minimum (Nautiyal, 2000; Nautiyal *et al.*, 2002). Whipps (2001) has reviewed the use of various biocontrol antagonists, including bacteria, in plant disease control. Mechanisms that have been proposed to explain how some microorganisms control plant diseases caused by fungal pathogens include mycoparasitism and antibiosis (Whipps, 2001). Mycoparasitism involves the production of extracellular enzymes that hydrolyse fungal cell-walls, while antibiosis involves the production of secondary metabolites in the rhizosphere. These metabolites inhibit the growth and/or differentiation of fungal pathogens.

Members of the genus *Streptomyces* are important soil microorganisms and well-known producers of antibiotics and extracellular enzymes (Crawford *et al.*, 1993). Some species exhibit biocontrol activity that correlates with their production of antibiotic(s) (Rothrock and Gottlieb, 1984; Raatikainen *et al.*, 1994). *S. violaceusniger* YCED9 is an isolate from a British soil that exhibits biocontrol activity against a variety of plant pathogenic fungi. The strain produces at least three antifungal antibiotics: nigericin, geldanamycin and a complex of polyenes that includes guanidylfungin A (Trejo-Estrada *et al.*, 1998).

*Corresponding author: r.errakhi@umi.ac.ma

 Biocontrol of Major Grapevine Diseases (eds S. Compant and F. Mathieu)

In this study, we present briefly our work on another *Streptomyces* species, *Streptomyces* J-2. Recently identified by our research team as a new species, *Streptomyces* J-2 is a bacterium belonging to the *Actinomycetales* that was isolated from Morocco and is of important agronomic interest. The bacterium produces metabolites with antifungal and antibacterial activities. The application of these molecules to seedlings of *Arabidopsis thaliana* enables the expression of the plant's defence genes (*PR1*, *PDF1.2* and *PAL*). Moreover, *Streptomyces* J-2 plays an important role in protecting plants against various phytopathogenic diseases. Of particular interest, and presented here, is its potential to control infection by and growth of *Botrytis cinerea* on the grapevine.

Materials and Methods

Isolation of *Streptomyces* sp.

The strain J-2 was isolated from agricultural soil in Morocco as described by Errakhi *et al.* (2009) and classified as belonging to the genus *Streptomyces*. The identification was based on morphological, cultural, physiological and biochemical characteristics, as described by Shirling and Gottlieb (1966) and Lechevalier and Lechevalier (1970). The isolate was maintained in glycerol 20% at –20°C until use.

DNA preparation

The selected antagonistic actinomycete was grown for 3–4 days at 28°C with agitation in 500 ml flasks containing 100 ml of ISP2 (International Streptomyces Project Yeast Malt Extract Agar) medium. The resulting biomass was harvested by centrifugation (7000 *g*, 10 min) and washed twice with double-distilled water. The mycelium (200 mg) was used for DNA extraction as follows: the sample was dispersed in 800 µl of the aqueous lysis solution (100 mmol Tris HCl, pH 7; 20 mmol EDTA; 250 mmol NaCl; 2% m/v SDS (sodium dodecyl sulfate); 1 mg ml^{-1} lysozyme); 5µl of 50 mg ml^{-1} RNase solution was added, and the suspension incubated at 37°C for 60 min. Proteinase K solution (10 µl at 20 mg ml^{-1}) was added and the lysis solution was re-incubated at 65°C for 30 min. The lysate was extracted with an equal volume of phenol and centrifuged (7000 *g*, 10 min). The aqueous layer was then re-extracted with phenol (50:50% v/v), then by chloroform (50:50% v/v). The DNA was recovered from the aqueous phase by the addition of NaCl (150 mmol as a final concentration) and two volumes of cold 95% (v/v) ethanol prior to centrifugation. The precipitated DNA was cleaned with 50 µl of 70% (v/v) ethanol, centrifuged (7000 *g*, 10 min), resuspended in 50 µl of TE buffer (10 mmol Tris HCl, pH 7; 1 mmol EDTA, pH 8), and stored at 20°C. The purity of the DNA solutions was checked with a spectrophotometer at 260 and 280 nm, and the quantities of DNA were measured at 260 nm.

16S rDNA sequence analysis of *Streptomyces* J-2

The 16S rDNA was amplified using the PCR method with Taq DNA polymerase and primers 27f (AGAGTTTGATCCTGGCTCAG) and 1492r (GGTTACCTTGTTACGACTT).

The conditions for thermal cycling were as follows: denaturation of the target DNA at 98°C for 3 min followed by 30 cycles at 94°C for 1 min; primer annealing at 53°C for 1 min; and primer extension at 72°C for 5 min. At the end of the cycling, the reaction mixture was left at 72°C for 5 min and then cooled to 4°C. PCR amplification was detected by agarose gel electrophoresis and visualized by ultraviolet (UV) fluorescence after ethidium bromide staining. The PCR product obtained was submitted to GenomExpress (Meylan, France) for sequence determination. The same primers as above and an automated sequencer were used for this purpose. The sequence was compared for similarity with the reference species of bacteria in genomic database banks, using the NCBI BLAST tool (US National Center for Biotechnology Information Basic Local Alignment Search Tool) tool (available at http://www.ncbi.nlm.nih/BLAST). The sequence analysis results of the 16S rDNA gene sequence were deposited in the GenBank database.

Test of antibiosis activity

The antimicrobial activity of the isolate was determined by the plate diffusion method (Errakhi *et al.*, 2009) against Gram-positive bacteria such as *Bacillus subtilis*, *Micrococcus luteus*, *S. scabies* and phytopathogenic fungi and oomycetes such as *Botrytis cinerea*, *Eutypa lata*, *Pythium ultimum* and *Verticillium dahliae*. For this part of the study, the isolate was grown on Bennett medium (beef extract 1 g l^{-1}; glucose 10 g l^{-1}; peptone 2 g l^{-1}; yeast extract 1 g l^{-1} and agar 15 g l^{-1}) for 14 days and three discs from *Streptomyces* agar plates (6 mm in diameter) were cut and placed on potato dextrose agar medium (PDA, Difco, USA) that was seeded with the test organism. Plates were first kept in a refrigerator (4°C) for at least 4 h to allow the diffusion of any antibiotics produced; they were then incubated at 30°C. Inhibition zones were determined after 48 h of incubation.

Purification of antibiotics from *Streptomyces* J-2

The culture broth (10 l) of *Streptomyces* J-2 was centrifuged at 12,000 rpm for 15 min, and the supernatant was extracted twice with an equal volume of ethyl acetate. The organic phase was dehydrated by Na_2SO_4 and dried *in vacuo* using a rotary evaporator. The red crude extract obtained was suspended in a minimum volume of methanol (MeOH), filtered through a 45-μm pore size filter and then subjected to high performance liquid chromatography (HPLC) using the following set-up to obtain the pure products: an Uptisphere UP5ODB C18 phase inverse column (250 × 7.8 mm inner diameter, Interchrom) with a linear gradient solvent system from 0% MeOH to 100% MeOH for 45 min and 20 min of isocratic gradient at 100% MeOH in H_2O. The flow rate was 1 ml min^{-1} and UV detection was at 220 nm. The active peaks were detected by the agar diffusion method against *B. subtilis* and *Mucor ramannianus*. Nuclear magnetic resonance (NMR) samples were prepared by dissolving 3 mg of C_1, C_2 and C_3 compounds in 600 μl CD_3OD (deuterated methanol) or in DMSO (dimethylsulfoxide). All spectra were recorded on a Bruker AV500 spectrometer equipped with a 5-mm triple resonance inverse probe operating at 500.33 MHz for ^1H and 125.81 MHz for ^{13}C. All chemical shifts for ^1H and ^{13}C were relative to tetramethylsilane (TMS) using

^{1}H (residual) or ^{13}C chemical shifts of the solvent as a secondary standard. The temperature was set at 298°K. The 2D NMR spectra were collected with 2048 t_2 data points and 256 t_1 increments (^{1}H-^{1}H gs-COSY45 – gradient-selected correlation spectroscopy 45°; ^{1}H-^{13}C gs-HSQC – gradient-selected heteronuclear single quantum correlation spectroscopy; and ^{1}H-^{13}C – gs-HMBC gradient-selected heteronuclear multiple bond correlation spectroscopy).

Biological control against *Botrytis cinerea*

Leaves were detached from grapevine plants aged 2 months and were of the variety COT-NOIR grown under glasshouse conditions. The leaves were removed at the same level from plants at the same stage of development and with the same orientation to the sun. They were disinfected with bleach, rinsed with sterile distilled water and placed in sterile Petri dishes containing filter paper and cotton soaked in sterile distilled water. Inoculation by the phytopathogenic fungus *B. cinerea* was performed 48 h after treatment of the leaves with suspensions of *Streptomyces* J-2. The leaves were placed under controlled temperature (20°C) and photoperiod (16 h) conditions.

Arabidopsis thaliana plant bioassay and RT-PCR analysis of SDN gene expression

A. thaliana COL seeds were germinated on Gamborg agar medium (1%) in square Petri dishes. Conditions in the growth room were 22°C with continuous white light (40 $\mu E\ m^{-2}\ s^{-1}$). After 2 weeks, plants were inoculated with pure metabolite from *Streptomyces* J-2 diluted to 1 in 10 with twice distilled water. RNA was extracted from the leaves 24 h later.

Total RNA samples were prepared with an RNA extraction Kit (Sigma). They were treated with the Deoxyribonuclease I kit (Sigma) and then quantified spectrophotometrically and their integrity checked on a denaturing agarose gel. The total RNAs were converted into first-strand cDNA using the Superscript II RNase H Reverse Transcriptase Kit (Invitrogen, Carlsbad, California) with oligo(dT)22. cDNA samples (1 µl) were amplified in 20 µl PCR mixture. Specific primers were used for the genes *PR1*, *PDF1.2a* and *PAL1*. Primers for the genes *VDAC* and *AOX1a* were designed (VDAC forward, CCTGCCCCTGGACTGAAAGTT; VDAC reverse, CAGTCGACGGGCTCCAATCT; AOX1a forward, TTTTCGATTTGAAACAATGATGA; AOX1a reverse, CCCAATAGCTCGCGATTCCTTTAT). Control PCR was performed using the constitutive 'housekeeping' gene *EF1α4*. Thermal cycling conditions comprised an initial denaturation step at 94°C for 2 min, followed by 34 cycles or by 26 cycles for *EF1α4*, 94°C for 30 s, 55 °C for 30 s, 72 °C for 1 min 30 s and ending at 72 °C for 10 min. The PCR products were then loaded on to a gel and visualized by ethidium bromide fluorescence. Representative results from three independent experiments are shown in this study.

Statistics

Significant differences between treatments were determined by the Mann–Whitney test and P values <0.05 were considered significant.

Results and Discussion

Isolation and identification of the strain

Streptomyces J-2 (Fig. 6.1) was isolated from rhizospheric Moroccan soil using the technique described by Errakhi *et al.* (2009). Its phenotypic and biochemical characteristics showed that the strain belongs to the genus *Streptomyces*. Physiological and molecular (16s rDNA sequencing and DNA–DNA hybridization) studies of the J-2 strain showed that it is a new species. It has now been deposited as a patented strain at two international collection centres of microorganisms.

Antibiosis activity of *Streptomyces* J-2 strain

Strain J-2 showed significant antimicrobial activities against Gram-positive bacteria such as *Micrococcus luteus*, *B. subtilis*, *B. cereus* and *S. scabies*, and against fungi such as *Mucor ramannianus* (see Fig. 6.2). However, it showed low activity against Gram-negative bacteria such as *Escherichia coli* and no activity against *Pseudomonas* sp.

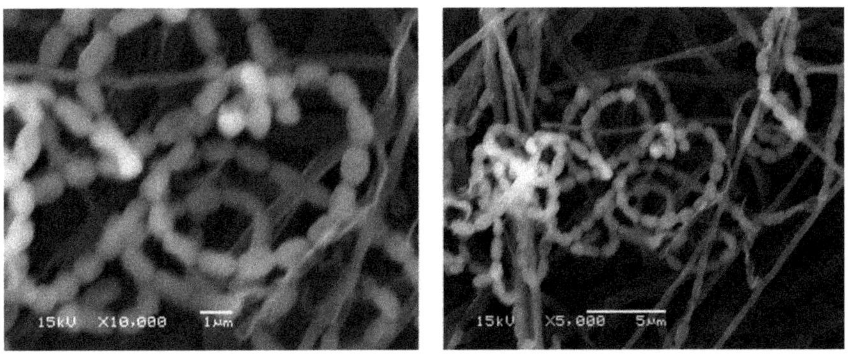

Fig. 6.1. Spore chains of *Streptomyces* J-2 obtained by observation under a scanning electron microscope of a culture of the strain on Bennett medium for 14 days. The spores are smooth and have a spiral channel (type S).

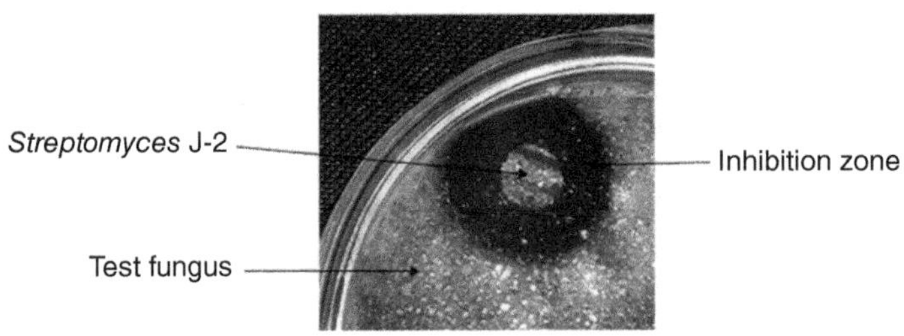

Fig. 6.2. The antimicrobial activity of *Streptomyces* J-2 against a test fungal plant pathogen as indicated by the fungal inhibition zone around a colony of J-2.

Antibiotic production by *Streptomyces* J-2

The bioactive molecules from strain J-2 are mainly extracted by ethyl acetate. The purification of these products was performed by HPLC (Fig. 6.3) using an increasing gradient of methanol in methanol/water. The chromatogram and biotests products show that the strain produced eight antibiotic molecules with different activities. One of the products was identified as a new molecule with a new structure. Three other products were also identified that had cyclopeptide structures and antifungal and antibacterial activities.

Biological control of plant pathogens

Streptomyces J-2 shows significant biological activity against various plant pathogens, namely: *Fusarium oxysporum* f. sp. *albedinis*, the causal agent of Bayaud disease; *B. cinerea* (Fig. 6.4), the causal agent of grey mould in grapes; *Pythium ultimum*, the causal agent of damping-off in many plants; *Verticillium dahliae*, the causal agent of Verticillium wilt of olive trees; *Sclerotium rolfsii*, the causal agent of damping off and root rot of sugarbeet; and *Streptomyces scabies*, which is responsible for common scab of potato.

Strain J-2 also exhibited a strong activity against the fungi causing diseases of the wood of the grapevine, such as *Phaeomoniella chlamydospora*, *Phaeoacremonium aleophilum*, *Eutypa lata*, *Fomitiporia mediterranea* and *Botryosphaeria obtusa*. Although in this study we have shown only the activity of *Streptomyces* J-2 against *B. cinerea*, these other data demonstrate its potential to be a biocontrol agent with a wide range of antibiosis properties.

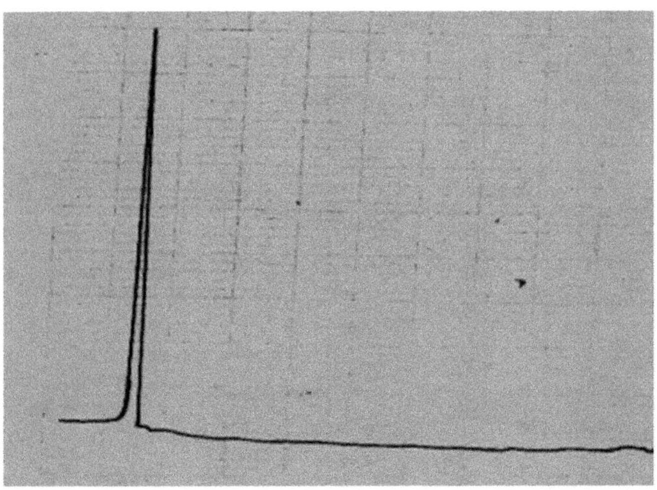

Fig. 6.3. Chromatogram of a pure product from *Streptomyces* strain J-2. This product has antifungal and antibacterial activities. The pure product was obtained after several injections into a C18 reverse phase high performance liquid chromatography column.

Induction of systemic defences in plants

The treatment of seedlings of *Arabidopsis thaliana* with the crude extract and pure compounds of strain J-2 activated the expression of genes that are related to plant defence. Indeed, RT-PCR analyses showed that some molecules from J-2 induce the expression of *PR1* (the gene responsible for the synthesis of antifungal compounds in plants) and *PDF1.2* (which activates the ethylene and jasmonate pathways) (see Fig. 6.5).

Conclusion

The need to reduce the use of chemical products in agriculture has increased research on new farming strategies that could be sustainable and protective of the environment. *Streptomyces* J-2 strain is a promising microbe that is able to play an important role not only as an antagonist agent against various diseases and in the induction of plant defence systems, but also as an effective biofertilizer agent that can stimulate plant growth and improve the quality of agricultural soils (unpublished results).

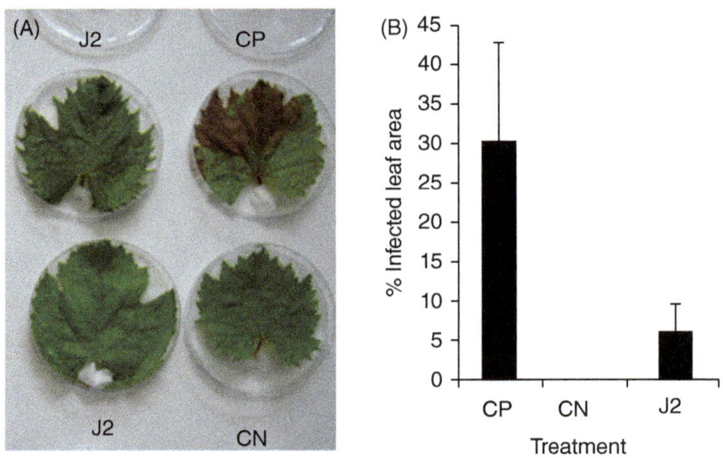

Fig. 6.4. Effect of treatment with spores of *Streptomyces* J-2 on the infection of grapevine leaves by *Botrytis cinerea*. (A) Photographs of leaves subjected to various treatments: CP, positive control; treatment with pathogenic fungi only; CN, negative control; no treatment; J-2, treatment with spores of *Streptomyces* J-2 strain. (B) Percentages of infected leaf area under the same three treatments.

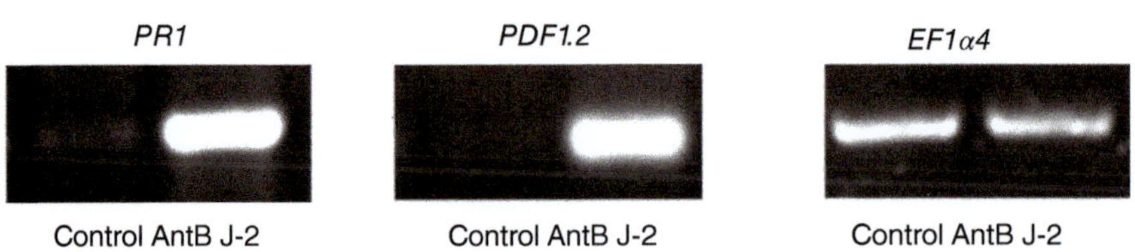

Fig. 6.5. Effect of antibiotics from *Streptomyces* J-2 strain on the activation of the defence genes *PR1* and *PDF1.2*, and the constitutive 'housekeeping' gene *EF1α4* in *Arabidopsis thaliana* as demonstrated by reverse-transcription PCR.

Acknowledgements

We thank Dr Stéphane Compant (AIT, Tulln, Austria) for the reading and correction of the manuscript.

References

Crawford, D.L., Lunch, J.M., Whipps, J.M. and Ousley, M.A. (1993) Isolation and characterization of actinomycete antagonists of a fungal root pathogen. *Applied and Environmental Microbiology* 59, 3899–3905.

Errakhi, R., Lebrihi, A. and Barakate, M. (2009) *In vitro* and *in vivo* antagonism of actinomycetes isolated from Moroccan rhizospherical soils against *Sclerotium rolfsii*: a causal agent of root rot on sugar beet (*Beta vulgaris* L.). *Journal of Applied Microbiology* 107, 672–681.

Lechevalier, H.A. and Lechevalier, M.P. (1970) Composition of whole-cell hydrolysates as a criterion in the classification of aerobic actinomycetes. In: Prauer, H., Fisher, G. and Verlag, G. (eds) *The Actinomycetales*. Gustav Fisher, Jena, Germany, pp. 393–404.

Nautiyal, C.S. (2000) Biocontrol of plant diseases for agricultural sustainability. In: Upadhyay, R.K., Mukherji, K.G. and Chamola, B.P. (eds) *Biocontrol Potential and its Exploration in Sustainable Agriculture, Volume I. Crop Diseases, Weeds, and Nematodes*. Kluwer/Plenum, New York, pp. 9–23.

Nautiyal, C.S., Johri, J.K. and Singh, H.B. (2002) Survival of the rhizosphere-competent biocontrol strain *Pseudomonas fluorescens* NBRI2650 in the soil and phytosphere. *Canadian Journal of Microbiology* 48, 588–601.

Raatikainen, O.J., Paivinen, T.H. and Tahvonen, R.T. (1994) HPLC separation and subsequent detection of aromatic heptaene polyenes in peat after treatment with *Streptomyces griseoviridis*. *Pesticide Science* 41, 149–154.

Rothrock, C.S. and Gottlieb, D. (1984) Role of antibiosis in antagonism of *Streptomyces hygroscopicus* var. *geldanus* to *Rhizoctonia solani* in soil. *Canadian Journal of Microbiology* 30, 1440–1447.

Shirling, E.B. and Gottlieb, D. (1966) Methods for characterization of *Streptomyces* species. *International Journal of Systematic Bacteriology* 16, 313–340.

Trejo-Estrada, S.R., Paszczynski, A. and Crawford, D.L. (1998) Antibiotics and enzymes produced by the biocontrol agent *Streptomyces violaceusniger* YCED-9. *Journal of Industrial Microbiology and Biotechnology* 21, 81–90.

Whipps, J.W. (2001) Microbial interactions and biocontrol in rhizosphere. *Journal of Experimental Botany* 52, 487–511.

7 Use of *Saccharothrix algeriensis* NRRL B-24137 to Control *Botrytis cinerea*?

S. Muzammil,[1*] R. Saria,[1*] Z. Yu,[1*] C. Graillon,[1*] F. Mathieu,[1] A. Lebrihi[1] and S. Compant[1,2†]

[1]*LGC UMR 5503 (CNRS/INPT/UPS), Département Bioprocédés et Systèmes Microbiens, Université de Toulouse, Castanet-Tolosan France;* [2]*AIT Austrian Institute of Technology GmbH, Health and Environment Department, Bioresources Unit, Tulln an der Donau, Austria*

Introduction

Beneficial bacteria are known to help their hosts by increasing plant growth and/or protecting them from several pathogenic diseases (Bakker *et al.*, 2007; Lugtenberg and Kamilova, 2009; Zamioudis and Pieterse, 2011). Some of these bacteria can be isolated from the phyllosphere. Others can be isolated from the anthosphere and the carposphere, as well as from the caulosphere. The majority of these bacterial microsymbionts are however known to colonize the rhizosphere, which is a rich zone of microbial interactions with their hosts (Lugtenberg and Kamilova, 2009). Some of the rhizosphere microflora can also enter into plants and establish subpopulations in various plant parts (Rosenblueth and Martínez-Romero, 2006; Hallmann and Berg, 2007; Compant *et al.*, 2010a; Reinhold-Hurek and Hurek, 2011). It is still not known though whether some specific bacteria can protect plants from diseases. This is especially the case with microbes that have been isolated from harsh environments (Compant *et al.*, 2010b). Such environments could provide a new source of beneficial bacteria such as growth-promoting bacteria or biocontrol agents (Compant *et al.*, 2010b). Indeed, it is becoming increasingly evident that microbes from soil and/or plants growing in harsh environments may represent an enormous potential for plant improvement. The transfer of these microbes from native plants to non-host plants could promise a revolutionary biotechnology to rapidly improve plant germplasm (Barrow *et al.*, 2008).

In relation to the search for beneficial microbes from harsh environments, a member of the *Actinobacteria* was isolated from a desert soil in a palm grove (at Adrar) in Algeria. This strain, *Saccharothrix algeriensis* NRRL B-24137, is known to secrete various secondary metabolites with broad bioactive activities (Zitouni, 1995; Lamari *et al.*, 2002a,b; Zitouni *et al.*, 2004, 2005). However, it is not yet known if this strain could help grapevine plants to counteract the fungal infections that are responsible for the different kinds of diseases that they acquire. Grapevines can suffer from

*These authors contributed equally.
†Corresponding author: stephane.compant@ait.ac.at

several phytopathogenic infections, and there is an urgent need to replace the chemical pesticides currently used to control these, i.e. for sustainable management of viticulture (Compant *et al.*, 2013a). In this study, we evaluated whether the strain *S. algeriensis* NRRL B-24137 can control *Botrytis cinerea* infection on grapevine. This was evaluated directly by assays of antagonists and indirectly on grapevine plants. We further used the model plant *Arabidopsis thaliana*, as well as two mutants of this plant in which ethylene or jasmonate signalling was reduced, to evaluate part of the mechanisms of resistance that are induced by strain NRRL B-24137 and thereby allow a reduction of infection by *B. cinerea*. This type of study is a prerequisite for better understanding of the interaction between a bacterium isolated from a harsh environment and plants, as well as aiding in finding a new solution for the control of *B. cinerea*.

Materials and Methods

Bacterial and fungal cultures

S. algeriensis NRRL B-24137 was used in this work. It was grown on International Streptomyces Project 2 (ISP2) agar plates (pH 7.0). This medium contained, per litre of distilled water: 4 g D(+)-glucose (Acros Organics), 10 g malt extract (Fluka), 4 g yeast extract (Fluka) and 18 g agar (Sigma). The bacterium was grown at 30°C.

B. cinerea strain BC1 (isolated by S. Compant from grapevine plants in Illmitz, Burgenland, Austria in 2008) was grown on potato dextrose agar (PDA) medium at 25°C under light conditions.

Antifungal assay on Petri dishes

For the direct biocontrol test, a medium was prepared using 14 g of nutrient agar (Fluka), 12 g of potato dextrose broth (PDB; Bacto) and 7 g of agar (Bacto), pH 5.1. Plugs of *S. algeriensis* NRRL B-24137 (grown on ISP2 medium for 7 days) were then placed on one side of the Petri dish and *B. cinerea* BC1 (grown on PDA for 9 days) was placed on the opposite side. Petri dishes without *S. algeriensis* NRRL B-24137 or *B. cinerea* BC1 served as controls. All the plates were allowed to grow at 25°C for 10 days. The results were visualized 10 days post-inoculation.

Grapevine plant material

Grapevine plants with the cv. Cabernet Sauvignon clone 15 as the scion and 44-53 M (Malègue) as the rootstock were used to analyse whether the strain NRRL B-24137 can protect plants from *B. cinerea*. Plant material was provided by 'Pépinières Colombie Vendries' (Camparnaud, France), and treated with Cryptonol® at 0.05% before being placed in a dark cold chamber at 4°C for at least 2 weeks. Plants were then surface sterilized with 1.6% bleach (10 min) and 70% ethanol (30 min) before rinsing twice with sterile tap water. They were then planted in a twice autoclaved soil containing

$\frac{1}{3}$ perlite, $\frac{1}{3}$ potting soil and $\frac{1}{3}$ sand. The plants were then grown in a phytotronic growth chamber (16 h photoperiod, 20:25°C night:day, 70% relative humidity) and watered with sterile tap water.

Grapevine plant Inoculation with *S. algeriensis* NRRL B-24137 and challenge with *B. cinerea*

One month after planting, the grapevine plants were carefully separated from their soils. The root systems were immersed for 3 min in a bacterial suspension of *S. algeriensis* NRRL B-24137. This was prepared by growing strain NRRL B-24137 on a plate for 8 days, harvesting the aerial mycelium + spores in phosphate-buffered saline (PBS) and adjusting the concentration to 5×10^7 colony forming units (cfu) ml^{-1}. Control plants were inoculated with PBS only. The plants were then replaced in pots filled with soil (as previously described) and allowed to grow for 10 days before challenging them with the phytopathogen. The leaves were infected with *B. cinerea* by inoculating five leaves per plant with 3 µl of a spore suspension of *B. cinerea* (spores were harvested in half diluted PDB from a 9-day-old *B. cinerea* plate culture to give a spore suspension with a concentration of 6.5×10^5 conidia ml^{-1}). The leaf inoculation was carried out five times per leaf or, in the case of control, the leaves were mock inoculated. To ensure infection, the inoculated plants were then kept at 100% relative humidity during the infection process under the growth chamber conditions described before (16 h photoperiod, 20:25°C night:day and 70% relative humidity). Plant susceptibility to *B. cinerea* was evaluated by the percentage of leaves with necrosis 3 days post-infection. Leaves that had been inoculated or not (mock) with *B. cinerea* were additionally photographed 3 days post-inoculation.

Arabidopsis plant material

Plants of *A. thaliana* were also used in this work. Seeds of wild type (WT) *A. thaliana* accession Columbia (Col. parent) and of the mutants *ein2-1* and *jar1-1* were obtained from the Nottingham *Arabidopsis* stock centre at Loughborough, UK. The seeds of *Arabidopsis* were surface sterilized with 70% ethanol (5 min) and 2.6% NaClO (2 min) before rinsing three times with sterilized distilled water. They were then placed on plates containing Murashige and Skoog (MS) media (Sigma, France) amended with 2% saccharose and 0.8% agar (pH 5.7). Seedlings were then allowed to grow *in vitro* for 16 days in a phytotronic growth chamber (16 h photoperiod, 20:25°C night:day and 70% relative humidity) on vertically oriented plates.

Induction treatment

Following 16 days of seedling growth, seedlings corresponding to the WT or mutants were taken from the *in vitro* plates, rinsed with sterilized distilled water and the roots were dipped in a suspension of the strain NRRL B-24137 (as for grapevine) or with PBS (control) for 3 min. The seedlings were then planted in Petri dishes perforated at the bottom and amended with twice sterilized soil ($\frac{1}{3}$ perlite, $\frac{1}{3}$ sand, $\frac{1}{3}$ potting soil).

The seedlings were allowed to grow horizontally for 10 days in a box covered with a transparent perforated film; sterilized tap water was added daily with a syringe.

Challenge inoculation

Five leaves of each *A. thaliana* seedling (WT or mutants), root inoculated or not with NRRL B-24137 (10 days post-inoculation), were infected with a suspension of 9-day-old *B. cinerea* culture by inoculating with 3 µl of 6.5×10^5 conidia ml^{-1}. To ensure infection, the inoculated plants were then kept at 100% relative humidity during the infection process in celled boxes (with a transparent film) under the phytotronic chamber conditions described before. Plant susceptibility to *B. cinerea* was then evaluated by the percentage of necrotic leaves, as for grapevine, 3 days post-infection. Photos of leaves inoculated or not (mock) with *B. cinerea* were additionally taken 3 days post-inoculation.

Statistics

Experiments were done three times independently with eight dual plates for each plate assay experiment. Grapevine and *Arabidopsis* resistances were evaluated independently three times on ten plants for each treatment. Statistical analyses were performed using Min–Max normalization before using Student's t test (http://studentsttest.com/).

Results

In vitro biocontrol activity by *S. algeriensis* NRRL B-24137

S. algeriensis NRRL B-24137 was evaluated for its activity against the phytopathogenic fungus *B. cinerea* BC1 by a dual-culture *in vitro* assay. Strain NRRL B-24137 exhibited a strong ability to inhibit the growth of *B. cinerea* growth (Fig. 7.1). During the interaction between *S. algeriensis* NRRL B-24137 and *B. cinerea* BC1, the development of the fungal strain was prevented over an area at least 1 cm wide around *S. algeriensis* NRRL B-24137 (Fig. 7.1). A red/orange pigmented compound was further observed in the area of antibiosis (Fig. 7.1B). This secretion was, however, also observed when strain NRRL B-24137 was cultivated alone on the same medium (data not shown).

S. algeriensis NRRL B-24137 induces a systemic resistance in grapevine against *B. cinerea*

Once challenged with *B. cinerea*, control grapevine plants and plants treated by inoculating with *S. algeriensis* strain NRRL B-24137 at root level showed different degrees of infection by *B. cinerea* (Fig. 7.2A). In the plants inoculated with bacteria, some of the leaves showed only a few necrotic patches in comparison with control plants

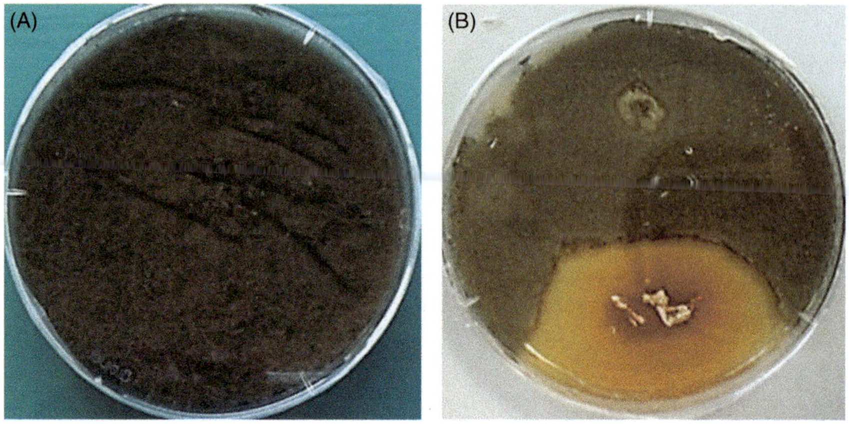

Fig. 7.1. Biocontrol test in Petri dishes showing the activity of *Saccharothrix algeriensis* strain NRRL B-24137 against *Botrytis cinerea* BC1. (A) *B. cinerea* BC1 alone; (B) *S. algeriensis* NRRL B-24137 with *B. cinerea* BC1, showing an inhibition zone.

Fig. 7.2. The protection of grapevine (*Vitis vinifera* L. cv. Cabernet Sauvignon) against infection by *Botrytis cinerea*. Comparison of symptoms due to *B. cinerea* infection on control plants and plants previously inoculated with *Saccharothrix algeriensis* strain NRRL B-24137 and then challenged with *B. cinerea*. (A) Leaves from control and pre-inoculated plants showing necrotic spots; arrows indicate *B. cinerea* infection; (B) % leaves with symptoms (* indicates significantly different at $P < 0.05$). Neither control plants nor plants inoculated with bacterial strain NRRL B-24137 but not infected with *B. cinerea* showed any necrosis (data not shown).

(Fig. 7.2A). The leaves of control plants were 62 ± 17.51% infected, whereas those of plants previously inoculated with strain NRRL B-24137 showed 42 ± 23.94% infection (significantly different at $P < 0.05$; Fig. 7.2B). Neither control plants nor those treated with strain NRRL B-24137 showed any disease symptoms on leaves if not challenged with *B. cinerea* (data not shown).

S. algeriensis NRRL B-24137 induces ethylene and jasmonate dependent systemic resistance in *A. thaliana*

When *Arabidopsis* was used as the test plant instead of grapevine, the results showed that *B. cinerea* infection on the WT Col. parent was lower in leaves from plants that had been root inoculated with *S. algeriensis* strain NRRL B-24137 compared with control non-inoculated plants (Fig. 7.3A). In control plants, 68.18 ± 13.97% of the leaves were infected, whereas in plants that had been inoculated with NRRL B-24137, 30.71 ± 14.47% of the leaves were infected (significantly different at $P < 0.05$; Fig. 7.3B).

Two *Arabidopsis* mutants with either reduced ethylene signalling (*ein2*) or reduced jasmonate signalling (*jar1-1*), were used to determine whether these types of signalling are required for the resistance to *B. cinerea* induced by strain NRRL B-24137 and the consequent reduction of *B. cinerea* infection. The results showed that the mutants *ein2* and *jar1-1* have no more resistance against *B. cinerea*, 3 days post-infection, when the plants had been pre-inoculated with NRRL B-24137, than did the wild type plants and mutant plants that had not been pre-inoculated with the beneficial strain (Fig. 7.4).

Discussion and Conclusions

In this study, we demonstrated that the desert soil isolate *S. algeriensis* NRRL B-24137 can reduce the fungal growth of *B. cinerea*. This was previously demonstrated in a large study by Lamari *et al.* (2002a,b). The strain NRRL B-24137 is known to secrete various secondary metabolites with broad bioactive activities (Lamari *et al.*, 2002a,b; Zitouni *et al.*, 2004) and can reduce the growth of various fungal agents (Lamari *et al.*, 2002a). Although the nature of the antifungal agent is still unknown, several bioactive metabolites with antifungal properties have been characterized as secreted by the beneficial strain, such as dithiolopyrrolones (Lamari *et al.*, 2002b). Although we still do not know what kinds of metabolites are responsible for the growth reduction

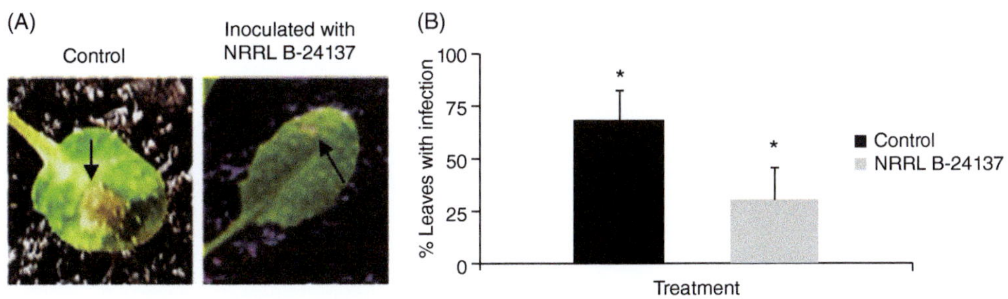

Fig. 7.3. The protection of wild type *Arabidopsis thaliana* against infection by *Botrytis cinerea*. Comparison of symptoms due to *B. cinerea* infection on control plants and on plants previously inoculated with *Saccharothrix algeriensis* strain NRRL B-24137 and then challenged with *B. cinerea*. (A) Leaves from control and pre-inoculated plants, arrows indicate symptoms due to *B. cinerea*; (B) % leaves with symptoms (* significantly different at $P < 0.05$). Neither control plants nor plants inoculated with bacterial strain NRRL B-24137 but not infected with *B. cinerea* showed any necrosis (data not shown). Results from Muzammil *et al.* (2014) with permission.

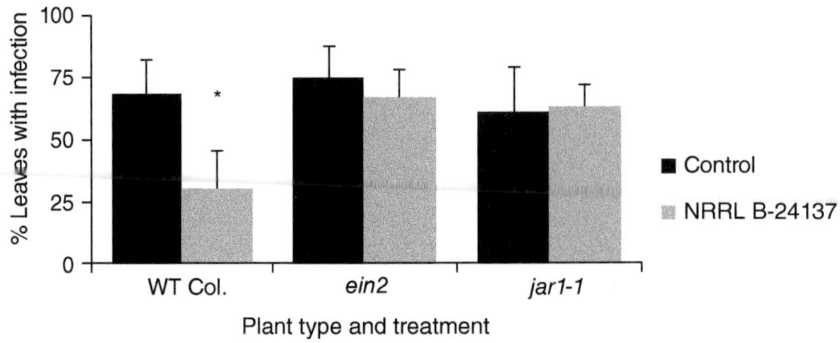

Fig. 7.4. The protection of *Arabidopsis thaliana* against infection by *Botrytis cinerea*. Comparison of symptoms due to *B. cinerea* infection in wild type Columbia (WT Col.) plants or mutant plants (*ein2* and *jar1-1*) previously inoculated or not with *Saccharothrix algeriensis* strain NRRL B-24137 and then challenged with *B. cinerea*. Results from Muzammil *et al.* (2014) with permission.

of *B. cinerea*, we have shown in this study that the beneficial strain of *S. algeriensis* can directly reduce its growth. Interestingly, a red/orange compound seems to be linked to the area where the growth of *B. cinerea* is prevented. Further experiments will give more information on which kind of metabolites are secreted by the beneficial strain that enable the growth reduction of *B. cinerea*.

We showed additionally that *S. algeriensis* NRRL B-24137 can protect grapevine plants from the fungus *B. cinerea*, which is a necrotrophic agent. The beneficial actinobacterium may only be present at the root level. Indeed, the strain NRRL B-24137 has not been recovered from leaves either 10 or 13 days post-inoculation, but was found colonizing the rhizosphere, the root surface and the inner tissues of grapevine roots (Compant *et al.*, 2013b). Therefore, there could have been no direct interaction between the fungal pathogen and the beneficial bacterium, which suggests that a systemic resistance is induced by strain NRRL B-24137 when it is applied to grapevine roots that allows protection against *B. cinerea*. Other bacteria, such as members of the *Bacillaceae*, *Pseudomonodaceae* and *Enterobacteriaceae*, as well as *Actinobacteria*, have been demonstrated as inducing protection against the grey mould agent in grapevine following inoculation (reviewed in Compant *et al.*, 2011, 2013a). However, this has never previously been demonstrated with a member of *Actinomycetales* such as *Saccharothrix*, and in this study we demonstrated that the strain NRRL B-24137 can protect grapevine cv. Cabernet Sauvignon from *B. cinerea* irrespective of its natural environment (i.e. desert soil).

Using seedlings of the *A. thaliana* mutants *ein2* and *jar1-1*, we also showed that the resistance of plants induced by strain NRRL B-24137 that allows reduction of infection by *B. cinerea* is ethylene and jasmonate dependent. In neither of the mutants used in this study was systemic resistance enabled against phytopathogens after they had been inoculated with a beneficial bacterium (Pieterse *et al.*, 1996, 1998; Kloepper *et al.*, 2004; Bakker *et al.*, 2007). However, as induced systemic resistance (ISR) stimulated by beneficial bacteria (both epiphytic and endophytic strains) has been shown to be linked to ethylene and jasmonate signalling (Kloepper *et al.*, 2004; Bakker *et al.*, 2007), it is not surprising that ethylene and jasmonate seem to be involved in the resistance induced by the strain NRRL B-24137. More work on this aspect is now needed as several other mutants involved in ethylene or jasmonate signalling could also be tested. Nevertheless, we have already demonstrated in a recent study that the resistance induced by strain NRRL B-24137 is ethylene/jasmonate dependent, that it also requires AtrbohF oxidase (an *Arabidopsis* respiratory burst oxidase homologue) and UPS1 (underinducer after pathogen and stress 1) functionalities, and that some crosstalks with salicylic acid (SA) signalling can take place as well (Muzammil *et al.*, 2014).

Although more work is required to understand better how the strain NRRL B-24137 could have an impact on grapevine plants leading to a systemic resistance against *B. cinerea*, we have now shown that this strain can reduce *B. cinerea* infection on grapevine plants. Interestingly, we know that this beneficial strain can reduce *B. cinerea* infection on *Arabidopsis* too and that this requires ethylene and jasmonate signalling. The strain NRRL B-24137 can also induce reduction of infection by *B. cinerea* on grapevine plants under high temperature stress and that this protection seems to be greater under high temperature conditions (Muzammil *et al.*, 2012). This suggests that the study of the beneficial interaction between the strain NRRL B-24137 and plants has great potential for the future sustainable management of agriculture.

Acknowledgements

This work was supported by a grant from the Higher Education Commission (HEC) Pakistan. Additional funding came from a 'Bonus Qualité Recherche' (BQR) from the Institut Nationale Polytechnique de Toulouse (INPT). We are grateful to Dr Brian Murphy (Trinity College Dublin, Irish Republic) for correction of the English of this manuscript.

References

Bakker, P.A.H.M., Pieterse, C.M.J. and Van Loon, L.C. (2007) Induced systemic resistance by fluorescent *Pseudomonas* spp. *Phytopathology* 97, 239–243.

Barrow, J.R., Lucero, M.E., Reyes-Vera, I. and Havstad, K.M. (2008) Do symbiotic microbes have a role in plant evolution, performance and response to stress? *Communicative and Integrative Biology* 1, 69–73.

Compant, S., Clément, C. and Sessitsch, A. (2010a) Plant growth-promoting bacteria in the rhizo- and endosphere of plants. Their role, colonization, mechanisms involved and prospects for utilization. *Soil Biology and Biochemistry* 42, 669–678.

Compant, S., Van der Heijden, M. and Sessitsch, A. (2010b) Climate change effects on beneficial plant–microorganism interactions. *FEMS Microbiology Ecology* 73, 197–214.

Compant, S., Mathieu, F. and Lebrihi, A. (2011) Biocontrol de *Botrytis cinerea* par des bactéries bénéfiques de la rhizosphère ou de l'endosphère. *Revue des Œnologues* 141, 15–16.

Compant, S., Brader, G., Muzammil, S., Sessitsch, A., Lebrihi, A. and Mathieu, F. (2013a) Use of beneficial bacteria and their secondary metabolites to control grapevine pathogen diseases. *BioControl* 58, 435–455.

Compant, S., Muzammil, S., Lebrihi, A. and Mathieu, F. (2013b) Visualization of grapevine root colonization by the Saharan soil isolate *Saccharothrix algeriensis* NRRL B-24137 using DOPE-FISH microscopy. *Plant and Soil* 370, 583–591.

Hallmann, J. and Berg, B. (2007) Spectrum and population dynamics of bacterial root endophytes. In: Schulz, B.J.E., Boyle, C.J.C. and Sieber, T.N. (eds) *Microbial Root Endophytes*. Springer, Berlin/Heidelberg, Germany, pp. 15–31.

Kloepper, J.W., Ryu, C.-M. and Zhang, S. (2004) Induced systemic resistance and promotion of plant growth by *Bacillus* spp. *Phytopathology* 94, 1259–1266.

Lamari, L., Zitouni, A., Boudjella, H., Badji, B., Sabaou, N., Lebrihi, A., Lefebvre, G., Seguin, E. and Tillequin, F. (2002a) New dithiolopyrrolones antibiotics from *Saccharothrix* sp. SA 233. I. Taxonomy, fermentation, isolation and biological activities. *The Journal of Antibiotics* 55, 696–701.

Lamari, L., Zitouni, A., Dob, T., Sabaou, N., Lebrihi, A., Germain, P., Seguin, E. and Tillequin, F. (2002b) New dithiolopyrrolone antibiotics from *Saccharothrix* sp. SA 233. II. Physiochemical properties and structure elucidation. *The Journal of Antibiotics* 55, 702–707.

Lugtenberg, B. and Kamilova, F. (2009) Plant growth-promoting rhizobacteria. *Annual Review of Microbiology* 63, 541–556.

Muzammil, S., Compant, S., Yu, Z., Mathieu, F. and Lebrihi, A. (2012) *Saccharothrix algeriensis* NRRL B-24137, a new endophyte with high potential to protect grapevine towards *Botrytis cinerea* in case of high temperature conditions. In: Faculté d'Oenologie de Bordeaux (ed.) *Oeno 2011: Actes de Colloques du 9e Symposium International d'Oenologie de Bordeaux*. Dunod, Paris, France.

Muzammil, S., Graillon, C., Saria, R., Mathieu, F., Lebrihi, A. and Compant, S. (2014) The Saharan isolate *Saccharothrix algeriensis* NRRL B-24137 induces systemic resistance in *Arabidopsis thaliana* seedlings against *Botrytis cinerea*. *Plant and Soil* 374, 423–434.

Pieterse, C.M.J., Van Wees, S.C.M., Hoffland, E., Van Pelt, J.A. and Van Loon, L.C. (1996) Systemic resistance in *Arabidopsis* induced by biocontrol bacteria is independent of salicylic acid accumulation and pathogenesis-related gene expression. *The Plant Cell* 8, 1225–1237.

Pieterse, C.M.J., Van Wees, S.C.M., Van Pelt, J.A., Knoester, M., Laan, R., Gerrits, H., Weisbeek, P.J. and Van Loon, L.C. (1998) A novel signaling pathway controlling induced systemic resistance in *Arabidopsis*. *The Plant Cell* 10, 1571–1580.

Reinhold-Hurek, B. and Hurek, T. (2011) Living inside plants: bacterial endophytes. *Current Opinion in Plant Biology* 14, 435–443.

Rosenblueth, M. and Martínez-Romero, E. (2006) Bacterial endophytes and their interaction with hosts. *Molecular Plant-Microbe Interactions* 19, 827–837.

Zamioudis, C. and Pieterse, C.M.J. (2012) Modulation of host immunity by beneficial microbes. *Molecular Plant-Microbe Interactions* 25, 139–150.

Zitouni, A. (1995) Les genres *Nocardiopsis* et *Saccharothrix (Actinomycetales)* dans les sols sahariens: taxonomie numérique, extraction, purification et caractérisation de quelques antibiotiques synthétisés. Thesis, Magister de Microbiologie, E.N.S. de Kouba, Algeria.

Zitouni, A., Lamari, L., Boudjella, H., Badji, B., Sabaou, N., Gaouar, A., Mathieu, F., Lebrihi, A. and Labeda, D.P. (2004) *Saccharothrix algeriensis* sp. nov., isolated from Saharan soil. *International Journal of Systematic and Evolutionary Microbiology* 54, 1377–1381.

Zitouni, A., Boudjella, H., Lamari, L., Badji, B., Mathieu, F., Lebrihi, A. and Sabaou N (2005) *Nocardiopsis* and *Saccharothrix* genera in Saharan soils in Algeria: isolation, biological activities and partial characterisation of antibiotics. *Research in Microbiology* 156, 984–993.

8 DOPE-FISH, a Tool to Visualize Beneficial Bacteria Systemically Controlling *Botrytis cinerea*?

S. Compant,[1,2]* S. Muzammil[1] and F. Mathieu[1]

[1]*LGC UMR 5503 (CNRS/INPT/UPS), Département Bioprocédés et Systèmes Microbiens, Université de Toulouse, Castanet-Tolosan, France;* [2]*AIT Austrian Institute of Technology GmbH, Health and Environment Department, Bioresources Unit, Tulln an der Donau, Austria*

Introduction

Plants can interact with a wide range of microorganisms. Some of these can have neutral or beneficial effects on their hosts, but others can lead to diseases that result in a reduction of plant growth and yield and, in some cases, this may also lead to the death of the plants. To reduce pathogen infection, various phytosanitary products are used. In grapevines, for instance, many chemical products are used to protect the crop against different phytopathogens, even though it is well known that the use of such products can lead to environmental and human health problems (Leroux, 2004), and other strategies can be employed to counteract pathogen attacks. One in particular is associated with the use of beneficial microorganisms, such as bacteria, which can have intimate associations with plants, including symbiosis (Compant *et al.*, 2005, 2010, 2013a). Symbiotic microorganisms can reduce pathogen infection either directly and/or indirectly by inducing systemic resistance (Lugtenberg and Kamilova, 2009). Some strains have shown their potential to do this when applied in the vineyard, where they can significantly reduce pathogenic infection (Compant, 2011; Compant *et al.*, 2013b). However, there is now a need to know where these microorganisms are localized following inoculation. This knowledge is important both from a microbial ecology viewpoint, and also in order to be able to respond to consumers' questions.

There are currently various techniques available for tracking microbes on and inside plants; for example, *gfp*, *Dsred* and *gus* reporter gene markers and their derivatives (Compant *et al.*, 2010), but the use of such methods implies that the microbe will have been genetically transformed before any application, and this is impossible in the field. Other techniques, such as quantitative PCR, can be used to detect a specific population of a microbe colonizing plant tissues following inoculation, but this will not allow the visualization of the microbes during their interaction with the plants. There are many other methods that can be used as well, and one of these is fluorescence *in situ* hybridization (FISH; Wagner *et al.*, 2003). FISH is a method that can allow localization and *in situ* visualization of microbes on and inside samples of plants taken from the field. One of the most frequently reported problems with FISH is a low

*Corresponding author: stephane.compant@ait.ac.at

signal intensity of the detected microbes in case of some environmental samples. This can be caused by a low cellular concentration of the target molecules (16S rRNA or 23S rRNA). Another problem corresponds to the low *in situ* accessibility of many regions of the 16S and 23S rRNA for singly labelled probes (Fushs *et al.*, 1998, 2001). catalysed reported deposition (CARD)-FISH has been used to counteract such effects, and there are other techniques that could be used too. In the vineyard, if simple FISH is used to track microbes, this will only allow the visualization of some microbes; others may be not visualized or a low signal may be obtained.

Stoecker *et al.* (2010) improved the FISH technique by creating the double labelling of oligonucleotide probes (DOPE)-FISH tool. It has been demonstrated that doubly labelled probes strongly increase the *in situ* accessibility of rRNA target sites and that this technique provides more flexibility for probe design (Stoecker *et al.*, 2010). Moreover the fluorescence signal is increased by two times more when DOPE-FISH is used.

Recently, we applied this technique to track a beneficial bacterium after application to grapevine (Compant *et al.*, 2013b). This microbe was *Saccharothrix algeriensis* NRRL B-24137, a beneficial actinobacterium that can reduce infection by *Botrytis cinerea* as well as by other plant pathogens (Muzammil *et al.*, 2012). This chapter presents an example of the results obtained.

Materials and Methods

Bacterial culture

S. algeriensis NRRL B-24137 was used throughout this work. This strain was grown at 30°C on International *Streptomyces* Project 2 (ISP 2) solid agar plated medium (pH 7.0) according to Muzammil *et al.* (2012). After 8 days' growth on the plate, the aerial mycelium and spores of strain NRRL B-24137 were harvested in phosphate-buffered saline (PBS) and the concentration was adjusted to 5×10^7 colony forming units (cfu) ml^{-1}.

Plant material

The grapevine plants used were *Vitis vinifera* L. cv. Cabernet Sauvignon clone 15 with 44-53 M (*V. riparia* × Malègue) as a rootstock. They were provided by Pépinières Colombie Vendries (Camparnaud, France). These plants were used because it has been previously described that the strain NRRL B-24137 could be beneficial when applied to them (Muzammil *et al.*, 2012). The plants were prepared according to Muzammil *et al.* (2012) and planted in twice autoclaved soil containing $\frac{1}{3}$ perlite, $\frac{1}{3}$ potting soil and $\frac{1}{3}$ sand. Plants were allowed to grow in a phytotronic growth chamber (16 h photoperiod, 20:25°C night:day, 70% relative humidity) and watered with sterile tap water.

Plant inoculation

One month after planting, the grapevine plants were delicately separated from their soil medium. Their root systems were dipped in a suspension of *S. algeriensis* NRRL

B-24137 bacteria for 3 min before the plants were replaced in soil-filled pots (as described above). The plants were then allowed to grow for 10 days before plant parts were sampled for microscopic analysis.

Probe specific for *Saccharothrix* spp., labelling and DOPE-FISH microscopy

A probe specific to *Saccharothrix* spp. was designed (see Compant *et al.*, 2013b) and then 5′ and 3′ labelled with a Dylight 488 fluorochrome (which fluoresces green at ultraviolet (UV) wavelengths). This fluorochrome was used to distinguish the *S. algeriensis* NRRL B-24137 strain from the normal root surface environment, which could be yellow autofluorescent both in control roots and in roots treated with strain NRRL B-24137.

To visualize the colonization of the rhizosphere/rhizoplane by inoculated strain NRRL B-24137, grapevine roots were cut into small parts 10 days post-inoculation. The samples were then fixed overnight at 4°C in a paraformaldehyde solution (4% in PBS, pH 7.2) in Eppendorf tubes. The samples were rinsed twice with PBS before being treated with 1 mg ml^{-1} lysozyme at 37°C for 15 min. Next, they were rinsed with PBS and dehydrated in an ethanol series. The DOPE-fluorescence *in situ* hybridization was then done as described by Compant *et al.* (2013b), using the labelled probe. After this treatment, the samples were kept in the dark for at least a day before examination with an epifluorescence microscope (BH2, Olympus, Japan) under a UV light. Photographs were taken with a TCC-3.3ICE-N camera (Tucsen, China).

Results and Discussion

After the application of strain NRRL B-24137 to the roots of grapevine plants, the strain with the specific probe created for some *Saccharothrix* spp, was detected. An example of this is illustrated in Fig. 8.1, which shows a hypha green-stained with Dylight 488 coupled with the specific probe. Because of the specificity of the probe, this hypha is known to correspond to the NRRL B-24137 strain of *Saccharothrix*. Spores, short filaments and long filaments of the beneficial strain were also detected colonizing the root surface of grapevine plants (Fig. 8.1A–H). The actinobacterial strain can be therefore detected on the root surface of grapevine plants, which is not surprising as they were inoculated with the beneficial bacterium under controlled conditions.

Although more work is needed to characterize the niches of colonization by strain NRRL B-24137, it can be concluded that the DOPE-FISH technique can be used to both detect and visualize specific microbes after inoculation (for more information, see Compant *et al.*, 2013b).

Although the normal single labelled FISH probe can also be used for this purpose, the use of DOPE-FISH provides more fluorescent signal and allows more flexibility in probe design (Stoecker *et al.*, 2010). As there is more and more research on uncommon bacteria that could act as beneficial bacteria, we suggest that this tool is used to track them once applied, as this kind of information can be of special interest from a microbial ecology viewpoint. It also allows visualization of the microbes following application without the requirement for transformed strains, which are of no use under field conditions.

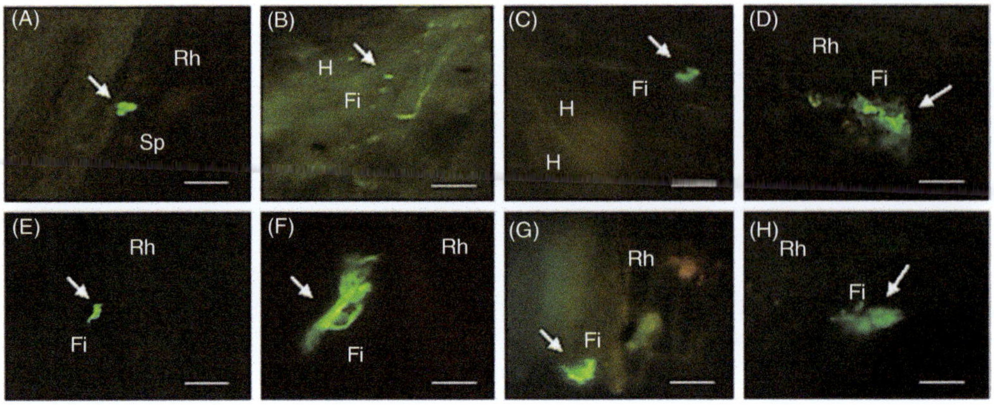

Fig. 8.1. Visualization of *Saccharothrix algeriensis* NRRL B-24137 on the rhizoplane of grapevine by the DOPE-FISH technique, showing the beneficial strain as spores (Sp) or filaments (Fi) on the rhizoplane (Rh) and or near root hairs (H). Scale bars: (A) 5 µm; (B, C) 20 µm; (D) 20 µm; (E) 25 µm; (F) 5 µm; (G) 20 µm; and (H) 15 µm. (A), (D) and (F) from Compant *et al.* (2013b) with permission.

We recently applied this technique to track several microbes thriving as endophytes inside the tissues of grapevine (unpublished), as well as by combining different fluorochromes that fluoresce at differing wavelengths. We further showed that this technique can be an adequate tool for monitoring plant–microbe interactions in general, and that it can replace single FISH to monitor plant–microbe associations. This is because single FISH does not always allow the tracking of microbes, whereas DOPE-FISH, owing to its double signal, can be suitable for this purpose. If FISH does not work well for some samples, it is possible to use the DOPE-FISH tool instead.

Acknowledgements

We are grateful to Dr Brian Murphy (Trinity College Dublin, Ireland) for English correction of the manuscript. This work was supported by a grant from the Higher Education Commission (HEC) Pakistan. Additional funding came from Centre National de la Recherche Scientifique, France (CNRS).

References

Compant, S. (2011) Editorial: Lutte biologique des agents pathogènes de la vigne avec des bactéries symbiotiques, une avancée pour nos vignobles? *Revue des Œnologues* 140, 7–8.

Compant, S., Duffy, B., Nowak, J., Clément, C. and Ait Barka, E. (2005) Use of plant growth-promoting bacteria for biocontrol of plant diseases: principles, mechanisms of action, and future prospects. *Applied and Environmental Microbiology* 71, 4951–4959.

Compant, S., Clément, C. and Sessitsch, A. (2010) Plant growth-promoting bacteria in the rhizo- and endosphere of plants: their role, colonization, mechanisms involved and prospects for utilization. *Soil Biology and Biochemistry* 42, 669–678.

Compant, S., Brader, G., Muzammil, S., Sessitsch, A., Lebrihi, A. and Mathieu, F. (2013a) Use of beneficial bacteria and their secondary metabolites to control grapevine pathogen diseases. *BioControl* 58, 435–455.
Compant, S., Muzammil, S., Lebrihi, A. and Mathieu, F. (2013b) Visualization of grapevine root colonization by the Saharan soil isolate *Saccharothrix algeriensis* NRRL B-24137 using DOPE-FISH microscopy. *Plant and Soil* 370, 583–591.
Fuchs, B.M., Wallner, G., Beisker, W., Schwippl, I., Ludwig, W. and Amann, R. (1998) Flow cytometric analysis of the *in situ* accessibility of *Escherichia coli* 16S rRNA for fluorescently labeled oligonucleotide probes. *Applied and Environmental Microbiology* 64, 4973–4982.
Fuchs, B.M., Syutsubo, K., Ludwig, W. and Amann, R. (2001) *In situ* accessibility of *Escherichia coli* 23S rRNA to fluorescently labeled oligonucleotide probes. *Applied and Environmental Microbiology* 67, 961–968.
Leroux, P. (2004) Chemical control of *Botrytis* and its resistance to chemical fungicides. In: Elad, Y., Williamson, B., Tudzynski, P. and Delen, N. (eds) *Botrytis: Biology, Pathology and Control*. Kluwer, Dordrecht, The Netherlands, pp. 195–222.
Lugtenberg, B. and Kamilova, F. (2009) Plant-growth-promoting rhizobacteria. *Annual Review of Microbiology* 63, 541–556.
Muzammil, S., Compant, S., Yu, Z., Mathieu, F. and Lebrihi, A. (2012) *Saccharothrix algeriensis* NRRL B-24137, a new endophyte with high potential to protect grapevine towards *Botrytis cinerea* in case of high temperature conditions. In: Faculté d'Oenologie de Bordeaux (ed.) *Oeno 2011: Actes de Colloques du 9e Symposium International d'Oenologie de Bordeaux*. Dunod, Paris, France.
Stoecker, K., Dorninger, C., Daims, H. and Wagner, M. (2010) Double labelling of oligonucleotide probes for fluorescence *in situ* hybridization (DOPE-FISH) improves signal intensity and increases rRNA accessibility. *Applied and Environmental Microbiology* 76, 922–926.
Wagner, M., Horn, M. and Daims, H. (2003) Fluorescence *in situ* hybridisation for the identification and characterisation of prokaryotes. *Current Opinion in Microbiology* 6, 302–309.

9 Endophytic Fungi in Asymptomatic *Vitis vinifera* L. and their Effects on *Plasmopara viticola*

S. Burruano,* V. Mondello and G. Conigliaro

Department Scienze Agrarie e Forestali, University of Palermo, Palermo, Italy

Introduction

Most plants, or probably all, live in symbiosis with several organisms or organism communities. Among these, fungal endophytes, which influence the number and diversity of other associated organisms (such as pathogenic bacteria or fungi, nematodes, phytophagous insects) can have a major impact on host plant ecology and fitness (Kobayashi and Palumbo, 2000; Omacini *et al.*, 2001; Ragazzi *et al.*, 2004; Schultz and Boyle, 2005; Reininger and Sieber, 2013; Suryanarayanan, 2013; Vázquez-de-aldana *et al.*, 2013; Aschehoug *et al.*, 2014). Originally, endophytic fungi were classified into two main groups – clavicipitaceous and non-clavicipitaceous – according to their phylogeny and life characteristics. Clavicipitaceous endophytes were reported as inhabiting various grasses, and non-clavicipitaceous endophytes as found in asymptomatic tissues of many land plants from different ecosystems (Sherwood and Carroll, 1974; Carroll, 1988; Stone *et al.*, 2004). However, a recent review suggests that non-clavicipitaceous endophytes include different fungi, and they have now been split into three distinct functional groups according to their life history features and their ecological roles (Rodriguez *et al.*, 2009).

Regardless of the group to which they belong, some endophytic fungi are able to affect the distribution of resources in the host, thereby inducing a reduction of host fitness. In contrast, others can provide ecological and agricultural benefits to plants (vegetative growth increase, adaptation to dryness, tolerance to biotic and abiotic stresses, induction of plant defensive mechanisms) through the secretion of secondary metabolites or hyperparasitism, or via competition for nutrients, light and space (Schulz and Boyle, 2005; Rodriguez *et al.*, 2009; Johnson *et al.*, 2013; May and Nelson, 2014). The use of the plant growth-promoting abilities of these fungi can be considered as an effective alternative to the use of pesticides in the control of plant pathogens (Graniti, 2002). With this in mind, we have searched for some beneficial endophytic fungi for grapevine plants.

*Corresponding author: santella.burruano@unipa.it

During a multi-year study on the life cycle of *Plasmopara viticola* to assess the typical epidemiology of downy mildew on grapevine (*Vitis vinifera* L.) in west Sicily (Italy), several leaves on which the downy mildew had collapsed were found and observed in a vineyard located in Marsala, in Trapani Province. An antagonistic fungus was isolated from these leaves and named A20; through its morphology, it was identified as a hyphomycete belonging to the genus *Acremonium*. The antagonism of A20 against the asexual and sexual spores of *P. viticola*, as well as endophytic nature of the genus *Acremonium* in *V. vinifera* were confirmed by further studies which have been carried out over the past decade.

In order to inform winegrowers about this new future possibility for the biocontrol of grapevine downy mildew, this chapter reports the results of investigations on the interaction between the endophytic fungus A20, the presence of endophytic fungi in the grapevine and downy mildew.

Materials and Methods

Endophytism of *Acremonium* in grapevine

The endophytic presence of the genus *Acremonium* in *V. vinifera* was first ascertained by isolation assays (Burruano *et al.*, 2008) made from three native Italian cultivars (Regina Bianca, Catarratto and Insolia) growing in an experimental field vineyard of the DEMETRA Department (agrarian and forest science) of the University of Palermo, in Sicily. Samples were collected monthly from epigeal organs (buds, petioles, leaves, shoots and seeds, when present) over 2 consecutive years. After harvesting, all samples were surface sterilized by shaking in 70% ethanol for 1 min, dipping into an aqueous sodium hypochlorite solution (5% v/v) for 2 min and shaking in 70% ethanol again for 1 min before rinsing in sterile distilled water. When they were dry, the organs were aseptically fragmented using sterile scalpels in a sterile laminar flow chamber, and then inoculated onto Petri dishes containing a suitable growth substrate (2% malt extract agar amended with 50 mg l^{-1} oxytetracycline hydrochloride) and kept at 22–24°C. All dishes were inspected daily with a stereomicroscope (Zeiss, Germany) to detect the occurrence of fungal colonies. Colonies that were morphologically similar to the A20 isolate were grown in pure culture and identified on the basis of morphological macroscopic and microscopic features according to descriptions by Barnett (1962), Von Arx (1974) and Domsch *et al.* (1980).

Next, the interaction between the endophytic hyphomycete strains isolated and *P. viticola* was tested. From January to June, leaves of cv. Insolia naturally colonized by *Acremonium* were harvested fortnightly and surface sterilized. Test leaves were inoculated with drops of an aqueous suspension of *P. viticola* sporangia (8×10^5 sporangia ml^{-1}) obtained by shaking heavily infected leaf pieces in water, after which they were exposed to natural light at 20 ± 1°C. After 2, 4 and 8 days of incubation, both inoculated and non-inoculated leaves were cut into discs (0.9 mm diameter) and submitted to a previously developed procedure of tissue clearing and specific staining for fungal chitin (according to Burruano *et al.*, 2003). All of the leaf discs were observed using differential interference contrast microscopy (DIC) and laser scanning microscopy (LSM) (Zeiss, Germany) (Burruano *et al.*, 2008).

Antagonism of the A20 isolate towards spores of *P. viticola*

The biocontrol activity of the A20 *Acremonium* strain was investigated on both the asexual and sexual spores of *P. viticola*. For the asexual spores, a preliminary test was carried out by dipping sporangiophores and sporangia from mildew-infected leaves into a suspension in distilled water of A20 germinated conidia (3×10^6 ml^{-1}). At the same time, other tests were done in which asexual spores were suspended in the following media: distilled water, different concentrations (25, 50 and 100%) of sterile potato dextrose broth (PDB), as well as different concentrations (25, 50 and 100%) of the culture filtrate (CF) of A20 gown in PDB.

After 2 h and 24 h under each treatment, sporangium vitality was ascertained by calculating the percentage sporangial germination (number of empty *P. viticola* sporangia out of the total sporangia present in each drop; Burruano *et al.*, 1999).

After chemical identification of some of the A20 secondary metabolites, previously named acremines A–F (Assante *et al.*, 2005), their antibiotic ability was tested using solutions of each dissolved in dimethyl sulfoxide (DMSO).

The vitality of *P. viticola* sporangia was also checked in the following other media (after Burruano *et al.*, 2008): DMSO in water, malt extract–peptone–glucose–broth (MPGB), corn (maize)-steep-broth (CSB), A20 CF in MPGB, A20 CF in CSB and CE from malt extract–peptone–glucose–agar (CE MPGA).

Antagonism of the A20 isolate to the pathogenic activity of *P. viticola*

CFs, crude extracts (CEs) and germinated conidia of the *Acremonium* isolate were assayed *in vitro* as inhibitors of the pathogenic activity of *P. viticola*. For these tests, leaves of the grapevine cv. Insolia, which had previously and repeatedly been tested by isolation assays to verify the absence of the endophyte, were placed in Petri dishes (9 cm diameter) containing fresh sporangia (9×10^5 ml^{-1}) suspended in various media: deionized sterile water, malt extract broth (MEB), DMSO in water, CF and CE from MEB culture and an aqueous suspension of germinating conidia (3×10^6 ml^{-1}) of the hyphomycete. After the incubation period (6–8 days from inoculation), the activity of each medium was evaluated on the basis of invasion percentage by *P. viticola* (Conigliaro *et al.*, 2008).

Antagonism of the A20 isolate to maturation of the sexual structures of *P. viticola*

The antagonism of the *Acremonium* endophyte to the maturation dynamics of *P. viticola* sexual structures was also studied over 2 years by nuclear staining and germination assays (Burruano *et al.*, 2007; Conigliaro *et al.*, 2010). At the end of October 2002 and 2004, *Acremonium*-free leaves of the grapevine cv. Catarratto that were strongly affected by downy mildew were collected and fragmented. In autumn 2002 and 2007, leaves with numerous 'mosaic spots' were also collected from cv. Insolia grapevines that had been naturally colonized by the endophyte. After each leaf sampling, the presence of and the nuclear stage of *P. viticola* spores in the mesophyll tissues were ascertained, the first by stereomicroscope, the second by staining with a specific fluorochrome for nuclear DNA (DAPI: 4,6-diamidino-2-phenylindole) (Burruano *et al.*, 2002). The naturally *Acremonium*-infected leaves of cv. Insolia with sexual structures of *P. viticola* were put to overwinter under the prevailing environmental conditions. The leaves of the *Acremonium*-free cv. Catarratto were also put to overwinter under the prevailing environmental conditions but, in addition, they were fortnightly irrigated with either distilled water, A20 CFs or A20 mycelial or conidial suspensions. From February to March, overwintered spores were put to germinate in water drops, and some were stained with DAPI (Burruano *et al.*, 2007; Conigliaro *et al.*, 2008).

Results

Acremonium endophytism in *V. vinifera*

The results from isolation assays for the endophytic presence of *Acremonium* strains over 2 years are presented in Table 9.1. They demonstrate the presence of *Acremonium* in all the epigeal grapevine organs analysed in three Italian cultivars (Regina Bianca, Catarratto and Insolia). During the 2-year study, the isolation percentages of the endophytic hyphomycete varied according to the weather and the grapevine cultivar and growth stage. In damp cooler periods, there was a higher spatial and temporal incidence of *Acremonium* in buds and leaves than in other organs tested. Hot temperatures and late growing stage seemed to hamper endophyte isolation. This trend was particularly evident in the cv. Catarratto organ isolation assays, which were negative for *Acremonium* from May until December (which is generally the end of the vegetative season in Sicilian grapevines.

The isolation surveys enabled the collection of many *Acremonium* isolates, above all from cv. Insolia. Half of the latter, which were similar to A20 in morphological features, were initially identified as belonging to the species *A. byssoides* (Burruano *et al.*, 2008). Recently, however, using PCR analysis and nucleotide sequence comparison, the previous species identification has been revaluated as *A. persicinum* (Lo Piccolo *et al.*, 2015).

The leaf clearing and staining procedure with observations by DIC and LSM microscopy permitted the visualization of the location of the *Acremonium* endophytes in both healthy leaves and leaves artificially infected by *P. viticola* (cv. Insolia). During winter samplings, latent hyphae of *Acremonium* (Fig. 9.1) were observed as red strings in the veins of healthy leaves. In spring samplings, natural infection by powdery mildew would have activated the endophytic hyphomycete, which then grew fast between the mesophyll cells and could reach and damage both epidermal cells and pathogen structures. *Acremonium* hyphae and conidia could be noted between epidermal cells and inside pathogenic forms (Fig. 9.1).

The endophytic continuity around hyphomycete development was visualized by LSM observations.

Table 9.1. Percentage of grapevine organs positive to the endophytic presence of *Acremonium* indicated by isolation assays from the epigeal organs of three Italian grapevine cultivars over 2 years.

Plant part by cultivar	Month											
	Jan.	Feb.	Mar.	Apr.	May	Jun.	Jul.	Aug.	Sep.	Oct.	Nov.	Dec.
Regina Bianca												
Shoot		11.1	20.5	0	13.5	9.7	13.4	0	0.3	0	0	0.3
Leaf		–	–	0	4.2	4.2	3.4	0	6.6	0	0	0
Bud		0	66.3	0	–	–	–	–	–	–	–	–
Petiole		–	–	0	4.8	4.8	0	0	0	4.2	4.5	13.8
Catarratto												
Shoot		27.7	17.3	0	24	0	0	0	0	0	0	0
Leaf		–	–	0	6.9	0	0	0	0	0	0	0
Bud		0	36.8	0	–	–	–	–	–	–	–	–
Petiole		–	–	0	0	0	0	0	0	0	0	0
Insolia												
Shoot	5.5	13.8	13.8	2.8	0	5.5	2.7	0	0	0	0	0
Leaf	66.6	–	0	0	0	8.3	5.5	0	5.5	0	0	0
Bud	0	33.3	0	22.2	0	36.1	–	–	–	–	0	0
Petiole	0	–	0	0	0	13.8	0	0	0	0	0	0
Seed	–	–	–	–	–	–	–	–		5.5	–	–

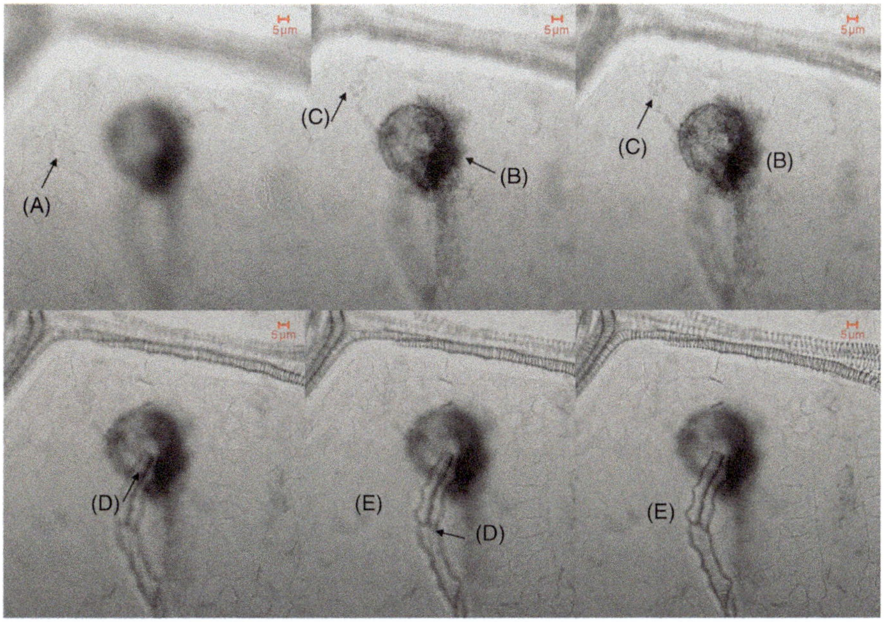

Fig. 9.1. A series of confocal images through the focal planes of foliar fragments showing the occurrence of *Acremonium* in non-inoculated (or apparently healthy) grapevine cv. Insolia: *Acremonium* conidia emerging from stoma (A); granular mass consisting of hyphal ribbons (B); phialide with conidia (C) of *Acremonium* endophyte; *Acremonium* hypha passing from the vein (D) and extending through the mesophyll cells (E). Observations by laser scanning microscopy (LSM). Scale bars = 5 μm.

Acremonium (A20) activity against *P. viticola* spores

In hyphomycete-free leaves of grapevine cv. Insolia artificially inoculated with a suspension of *P. viticola* sporangia and maintained under optimal humidity and temperatures, the emergence of uniform downy mildew (invasion by *P. viticola*) usually occurred in 4–5 days (Fig. 9.2A). In contrast, in leaves naturally colonized by the hyphomycete, the invasion by *P. viticola* was limited to the inoculum edge and development was delayed by 7–8 days (Fig. 9.2B).

DIC observations of discs from leaves collected in winter and cleared and stained for chitin allowed the visualization within 4–6 days of endophyte growth from the veins to *P. viticola* mycelium in leaves that had been inoculated with the pathogen (Fig. 9.3). On day 8 after inoculation, anomalous and rare sporangiophores emerged from the stomata that contained *Acremonium* structures (hyphae, phialides and conidia) (Fig. 9.3).

In grapevine leaves removed in spring, hyperparasitized asexual and sexual spores of *P. viticola* (Fig. 9.4) were already observed 2–5 and 7–8 days, respectively, after inoculation with the pathogen. Moreover, some host plant cells were deformed and red stained. LSM observations confirmed that the endophytic *Acremonium* strain can invade pathogenic structures.

In assays of the antagonism of the A20 *Acremonium* strain towards the asexual structures of *P. viticola*, strong degeneration of its sporangiospores and sporangia was ascertained after they were treated with an aqueous suspension of A20 (Burruano *et al.*,

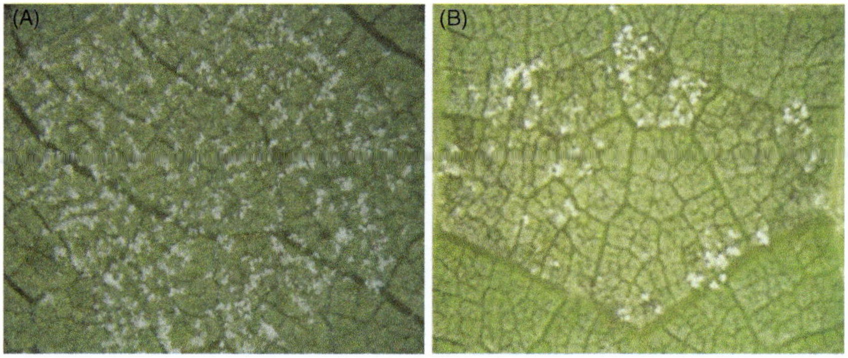

Fig. 9.2. (A) Emergence of white mildew on leaves of grapevine cv. Insolia free of the *Acremonium* endophyte and artificially inoculated with *Plasmopara viticola*. (B) Emergence of white mildew on leaves similarly treated but naturally colonized by the endophyte (magnification 2.5×).

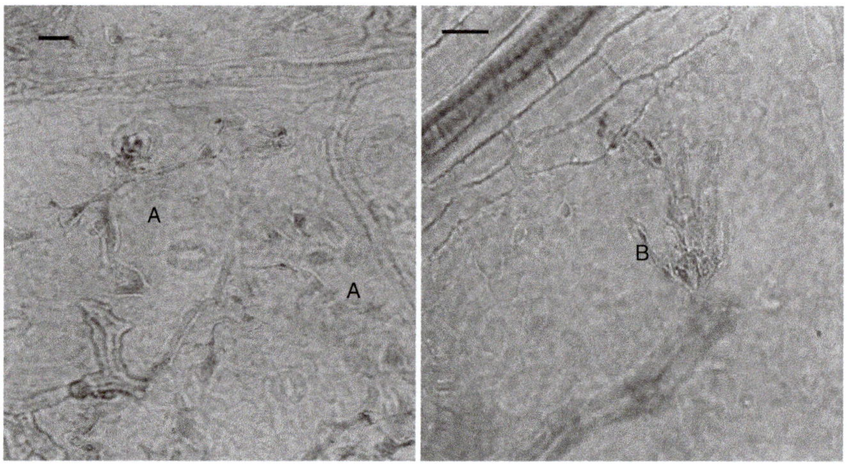

Fig. 9.3. *Plasmopara viticola* mycelium (A) and an anomalous sporangiophore structure affected by *Acremonium* endophyte (B) observed in discs from grapevine leaves naturally colonized by *Acremonium* collected in winter and then inoculated with *P. viticola*. Observations by differential interference contrast (DIC) microscopy. Scale bars = 10 μm.

1999). In particular, hyphomycete hyphae and conidia were observed bridling sporangiophores and collapsed sporangia (Fig. 9.5) 2 h after treatment.

When suspended in different media, *P. viticola* sporangia showed the highest germination percentage in water and the lowest in 100% CF of the A20 *Acremonium* strain grown in potato dextrose broth (PDB) (Table 9.2).

No degenerated sporangia were able to germinate or to release zoospores (propagules). Strong inhibition of the germination process was also observed in other CFs and CEs tested (Table 9.3).

Grapevine leaves inoculated with *P. viticola* sporangia suspended in deionized sterile water showed abundant (100% after 8 days) and homogeneous invasion by the

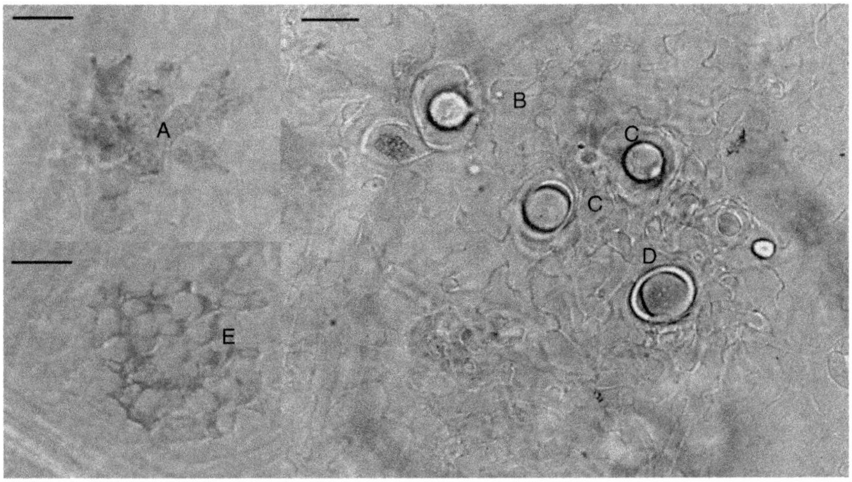

Fig. 9.4. Degenerated *Plasmopara viticola* sporangiophores (A) in grapevine leaves naturally colonized by *Acremonium* endophytic strain collected in spring and then inoculated with *P. viticola*; (B) oogonium; (C) oosphere; (D) oospore; (E) degenerated host epidermal cells. Observations by laser scanning microscopy (LSM). Scale bars = 20 µm.

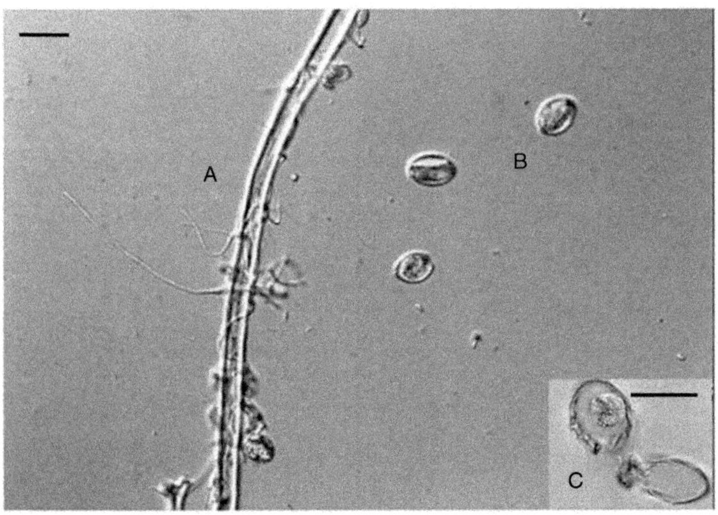

Fig. 9.5. *Plasmopara viticola* sporangiophores invaded by hyphae and conidia after treatment with a suspension of the *Acremonium* A20 strain (A); collapsed (B) and germinating *P. viticola* sporangia (C) after the same treatment. Observations by differential interference contrast (DIC) microscopy. Scale bars = 20 µm.

oomycete on the leaf lower surface (Fig. 9.6A and Table 9.4). In leaves inoculated with a sporangial suspension in MEB (Fig. 9.6B) and in DMSO in water (Fig. 9.6C), sporangiophores covered, respectively, about 83.3 and 91.6% of the lower surfaces after 8 days (Table 9.4). When the leaves were inoculated with sporangia suspended in CF and in CE from MEB (Figs 9.7A, B), after 6 and 8 days the pathogenic invasion was

Table 9.2. Germination percentages of *Plasmopara viticola* sporangia suspended in different media.

Medium[a]	Sampling[b]	
	After 2 h	After 24 h
Water	97.0aA	95.5aA
CF 25%	96.6aA	95.4aAB
PDB 25%	94.3abAB	93.9abAB
CF 50%	80.0bcAB	82.3bBC
PDB 50%	70.3cB	64.3cC
PDB 100%	45.0dC	38.2dD
CF 100%	1.0eD	2.2eE

[a]CF, culture filtrate of *Acremonium* A20 from potato dextrose broth; PDB, potato dextrose broth.
[b]Percentages followed by the same letter are not significantly different at $P \leq 0.05$ (lower case) or $P \leq 0.01$ (capitals).

Table 9.3. Germination percentages of *Plasmopara viticola* sporangia suspended in other different media.

Medium[a]	Germination (%)[b]
Distilled water	89.7a
DMSO	83.7b
MPGB	32.4c
CSB	36.6c
CF MPGB	0.9e
CF CSB	3.6d
CE MPGA	0.0f

[a]CE MPGA, crude extract from *Acremonium* A20 culture on malt extract–peptone–glucose–agar; CF CSB, crude filtrate from *Acremonium* A20 culture in corn (maize) steep broth; CF MPGB, crude filtrate from *Acremonium* A20 culture in malt extract–peptone–glucose–agar; CSB, corn (maize) steep broth; DMSO, dimethyl sulfoxide in water; MPGB, malt extract–peptone–glucose–broth.
[b]Percentages followed by the same letter are not significantly different at $P = 0.05$ (Tamhane's test).

Fig. 9.6. Grapevine leaves inoculated with *Plasmopara viticola* sporangia suspended in: deionized sterile water (A); malt extract broth (B); and dimethyl sulfoxide in water (C).

2.6 and 16.8%, respectively (Table 9.4). The lowest invasion percentage (0.5 and 1.1%) was observed in leaves inoculated with a suspension of *P. viticola* sporangia and with germinating A20 conidia (Fig. 9.7C and Table 9.4). Moreover, asexual spores of *P. viticola* were hyperparasitized by *Acremonium* hyphae.

Results from the study of the germination of *P. viticola* oospores in *Acremonium*-free but strongly downy mildew-affected leaves of grapevine cv. Catarratto that had been collected at the end of October 2002 and 2004, and then overwintered until February–March, showed that those that were already mononucleate in October (Fig. 9.8A, A′), and had overwintered under natural conditions with or without water irrigation, had germinated by the spring (Fig. 9.8B, B′). Treatment of the oospores during overwintering with a suspension of *Acremonium* spores or with an *Acremonium* CF caused a decrease in the relative germination percentages of the (mononucleate) oospores (Table 9.5). In particular, after treatment with a suspension of *Acremonium* spores, even if the oospores had sporadically germinated, they quickly collapsed and were invaded by A20 hyphae and conidia (Fig. 9.8C, C′). Following treatment with *Acremonium* CF during overwintering, many more spores, even if mononucleate at collection, quickly degenerated and never germinated (Fig. 9.8D, D′).

In leaves of grapevine cv. Insolia naturally colonized by *Acremonium* that were collected in autumn 2002 and 2007, the numbers of oospores of *P. viticola* decreased

Table 9.4. *Plasmopara viticola* invasion (% of leaf surface covered by downy mildew) 6 and 8 days after inoculation with sporangial suspensions in different media.

	% Leaf surface invaded after:	
Medium[a]	6 days	8 days
Distilled water	98.3	100.0
Water + DMSO	80.0	91.6
MEB[b]	73.3	83.3
CF MEB	2.6[b]	2.6[b]
CE MEB	16.8[b]	16.8[b]
Conidial suspension	0.5[b]	1.1[b]

[a]CE MEB, crude extract from *Acremonium* A20 culture in malt extract broth; CF MEB, culture filtrate from *Acremonium* A20 culture in malt extract broth; DMSO, dimethyl sulfoxide in water; MEB, malt extract broth.
[b]Statistically significant at $P \leq 0.01$.

Fig. 9.7. Grapevine leaves inoculated with *Plasmopara viticola* sporangia suspended in: culture filtrate (A); crude extract (B) of cultures of *Acremonium* A20 in malt extract broth; and a suspension of *Acremonium* germinating conidia (C).

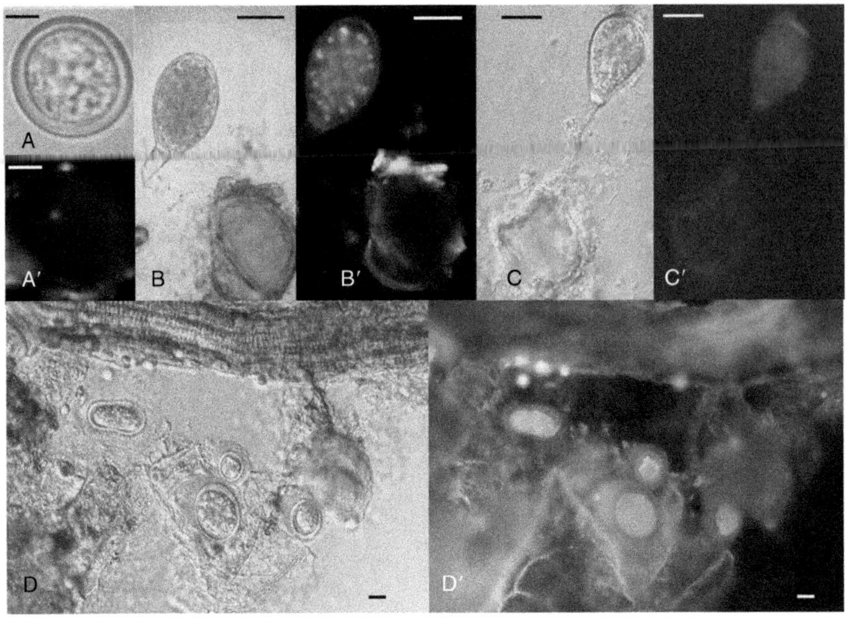

Fig. 9.8. Observations under transmitted (X) and fluorescent light (X′) of *Plasmopara viticola* structures (oospores and germinating oospores) in grapevine leaves collected in autumn and overwintered. Both healthy structures and those degenerated by hyphae and spores of the endophyte *Acremonium* are shown. A, A′, healthy and mononucleated mature oospores; B, B′, oospores overwintered under natural conditions with and without water irrigation and germinating normally; C, C′, degeneration in the germination of oospores treated with *Acremonium* spore suspension or culture filtrate during overwintering; D, D′, degenerated mature oospores (mononucleate at collection in the autumn) treated with *Acremonium* A20 during overwintering. Scale bars = 10 μm.

Table 9.5. Germination percentages of *Plasmopara viticola* oospores in leaves of grapevine cv. Catarratto collected free of *Acremonium* in October 2002 and 2004 and overwintered until February–March under different conditions.

Overwintering condition	Year 1[a]	Year 2
Natural weather	16.75bA	19.03bB
Natural weather + irrigation with water	14.11abA	22.13bB
Natural weather + irrigation with *Acremonium* spores	9.58aA	1.89aA
Natural weather + *Acremonium* culture filtrate	7.44aA	3.11aA

[a]Percentages followed by the same letter are not significantly different at $P \leq 0.05$ (lower case) or $P \leq 0.01$ (capitals).

during the winter–spring season of 2002–03 and were found to have still further reduced by the winter–spring season of 2007–08. Moreover, oospore germination was not observed in either 2002–03 or 2007–08 (Fig. 9.9).

Overall, the results described above demonstrate that *Acremonium* endophytism in grapevines was ascertained by isolation tests carried out on three different cultivars

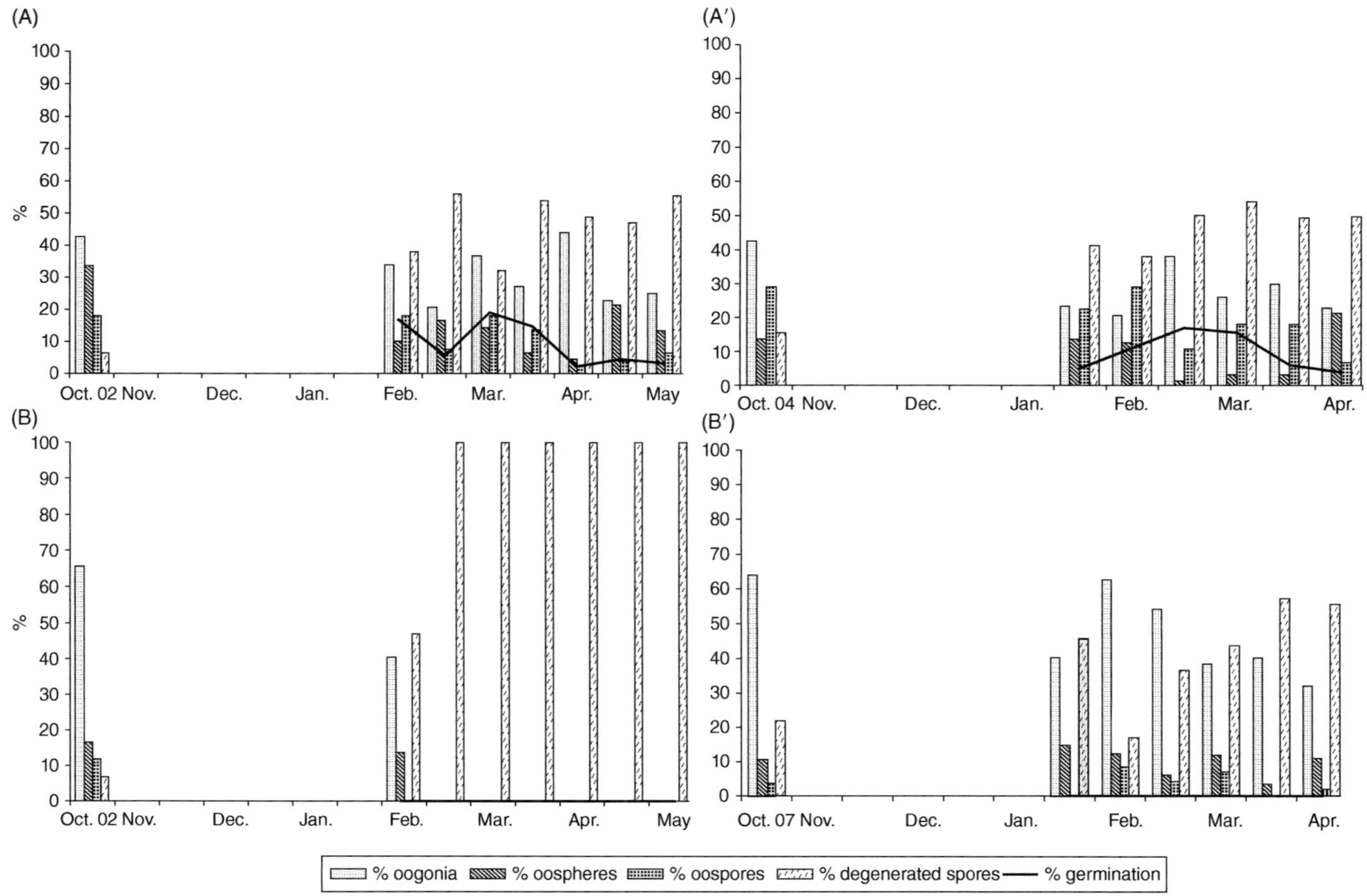

Fig. 9.9. Percentages of *Plasmopara viticola* sexual stages (oogonia, oospores and oospheres) and oospore germination observed from October to May 2002–03 and from October to April 2004–05 on leaves strongly affected by downy mildew collected from Acremonium-free grapevines cv. Catarratto (A, A′) in October 2002 and 2004 and overwintered. The same data are presented for leaves from cv. Insolia (B, B′) that had been naturally colonized by *Acremonium* endophytes and had been collected in October 2002 and 2007 and overwintered.

over 2 years. Histological assays that detected latent and active hyphomycete hyphae in asymptomatic and symptomatic grapevine leaves, respectively, have confirmed the interactions between the genus *Acremonium* and *V. vinifera*, as well as highlighting the parasitic capacity of the endophytic fungus against pathogenic structures. *Acremonium* antibiosis and parasitism towards *P. viticola* has also been shown by direct and indirect antagonism essays.

Discussion

Studies over the past decade seem to have identified some of the basic aspects of the interactions between *Acremonium* species, *V. vinifera* and *P. viticola*. In particular, in leaf samples from cv. Insolia grapevines, both the mutualistic nature of the symbiosis of the plant with the endophytic *Acremonium* strain A20 and the antagonistic symbiosis of strain A20 with *P. viticola* have been verified.

Investigations under the microscope have enabled visualization of how the A20 endophyte can change rapidly from a latent to an active state when stimulated by natural biotic factors (infection by powdery mildew and punctures by *Altica* sp. flea beetles) or abiotic influences (wounds caused by climatic or physiological changes), leading to the hypothesis that the endophyte has great plasticity. This hypothesis has been confirmed by the seasonal variability of A20 isolation from cv. Insolia grapevines and by histological assays of apparently healthy vine leaves artificially inoculated with *P. viticola*.

In every leaf fragment observed, A20 hyphae in the veins and their growth within the leaf parenchyma until sporulation and emergence were noted. The activation of A20 was also very rapid, but depended on grapevine vegetative growth and pathogenic factors, as well as its antagonistic activity against *P. viticola*. However, the greater bioactivity against *P. viticola* of A20 culture filtrates and crude extracts than that of the pure A20 compounds that have been isolated (Assante *et al.*, 2005), suggested that these acremines alone are not responsible for inhibiting pathogen germination.

In addition, the antagonism of the grapevine endophyte towards the host pathogen seems to include two mechanisms that follow in sequence: latent hyphomycete endophytes in the grapevine would be triggered by natural or artificial pathogenic infection to produce secondary metabolites toxic to the pathogen as well as related degradation products; by these means of antibiosis, the hyphomycete could parasitize the asexual and sexual forms of the pathogen. This sequence – antibiosis followed by parasitism, could strongly reduce the pathogen inoculum in the vineyard, thus enabling a containment of primary and secondary infections of downy mildew.

References

Aschehoug, E.T., Callaway, R.M., Newcomb, G., Nahant, T. and Chen, L. (2014) Fungal endophyte increases the allelopathy effects of an invasive forb. *Ecologies* 175, 285–291.

Assante, G., Delvalle, S., Malpezzi, L., Nasini, G., Burruano, S. and Torta, L. (2005) Acremines A–F, novel secondary metabolites produced by a strain of an endophytic *Acremonium*, isolated from sporangiophores of *Plasmopara viticola* in grapevine leaves. *Tetrahedron* 61, 7686–7692.

Barnett, H.L. (1962) *Illustrated Genera of Imperfect Fungi*. Burgess Publishing, Minneapolis, Minnesota.

Burruano, S., Conigliaro, G. and Torta, L. (1999) Prime osservazioni sull'antibiosi di *Acremonium* sp. contro *Plasmopora viticola* (Berck. et Curt) Berl.et de Toni. *Phytophoga* 9, 117–124.

Burruano, S., Conigliaro, G., Lo Piccolo, S. and Torta, L. (2002) Sull'interazione tra *Acremonium* sp., *Vitis vinifera* e *Plasmopara viticola*. In: Anselmi, N., Capretti, P., Cellerino, G.P., Franceschini, A., Granata, G., Luisi, N., Marras, F., Mazzaglia, A., Mutto Accordi, S., Ragazzi, A. and Vannini, A. (eds) *L'Endofitismo di Funghi e Batteri Patogeni in Piante Arboree e Arbustive. Atti del Convegno Nazionale, Sassari, Tempio Pausania, Italy, 19–21 Maggio*. Carlo Delfino, Sassari, Italy, pp. 103–112.

Burruano, S., Conigliaro, G., Lo Piccolo, S. and Torta, L. (2003) Detection of endophytic *Acremonium* sp. in healthy or naturally damaged leaves of *Vitis vinifera*. *Journal of Plant Pathology* 85, 318. [Abstract.]

Burruano, S., Conigliaro, G., Lo Piccolo, S. and Torta, L. (2007) Oospore di *Plasmopara viticola*: dinamica di maturazione e possibile antagonismo di *Acremonium byssoides*. *Micologia Italiana* 36(2), 53–58.

Burruano, S., Alfonzo, A., Lo Piccolo, S., Conigliaro, G., Mondello, V., Torta, L., Moretti, M. and Assante, G. (2008) Interaction between *Acremonium byssoides* and *Plasmopara viticola* in *Vitis vinifera*. *Phytopathologia Mediterranea* 47, 122–131.

Carroll, G. (1988) Fungal endophytes in stems and leaves: from latent pathogen to mutualistic symbiont. *Ecology* 69, 2–9.

Conigliaro, G., Lo Piccolo, S., Torta, L. and Burruano, S. (2008) *In semi-vivo* antagonism of *Acremonium byssoides* towards *Plasmopara viticola*. *IOBC-WPRS Bulletin* 36, 69–72.

Conigliaro, G., Ferraro, V., Martorana, A. and Burruano, S. (2010) *In vivo* antagonism of *Acremonium byssoides* endophyte in *Vitis vinifera, towards Plasmopara viticola. IOBC-WPRS Bulletin* 54(Part 2), 535–538.

Domsch, K.H., Gams, W. and Anderson, T.H. (1980) *Compendium of Soil Fungi*. Academic Press, London.

Graniti, A. (2002) L'endofitismo nei funghi: un adattamento ecologico o un modo di vita? In: Anselmi, N., Capretti, P., Cellerino, G.P., Franceschini, A., Granata, G., Luisi, N., Marras, F., Mazzaglia, A., Mutto Accordi, S., Ragazzi, A. and Vannini, A. (eds) L'Endofitismo di Funghi e Batteri Patogeni in Piante Arboree e Arbustive. *Atti del Convegno Nazionale, Sassari, Tempio Pausania, Italy, 19–21 Maggio*. Carlo Delfino, Sassari, Italy, pp. 11–22.

Johnson, L.J., De Bonth, A.C.M., Briggs, L.R., Caradus, J.R., Finch, S.C., Fleetwood, D.J., Fletcher, L.R., Hume, D.E., Johnson, R.D., Popay, A.J. *et al.* (2013) The exploitation of epichloae endophytes for agricultural benefit. *Fungal Diversity* 60, 171–188.

Kobayashi, D.Y. and Palumbo, J.D. (2000) Bacterial endophytes and their effects on plants and uses in agriculture. In: Bacon, C.W. and White, J.F. (eds) *Microbial Endophytes*. Marcel Dekker, New York, pp. 199–233.

Lo Piccolo, S., Alfonzo, A., Giambra, S., Conigliaro, G., Lopez-Llorca, L.V., and Burruano, S. (2015) Identification of *Acremonium* isolates from grapevine and evaluation of their antagonism towards *Plasmopara viticola*. *Annals of Microbiology.* DOI:10.1007/s13213-015-1082-5.

May, G. and Nelson, P. (2014) Defensive mutualisms: do microbial interactions within hosts drive the evolution of defensive traits? *Functional Ecology* 28, 356–363.

Omacini, M., Chaneton, E.J., Ghersa, C.M. and Müller, C.B. (2001) Symbiotic fungal endophytes control insect host–parasite interaction webs. *Nature* 409, 78–81.

Ragazzi, A., Moricca, S. and Dellavalle, I. (2004) Endophytism: knowns and unknowns of an age-old phenomenon. In: Ragazzi, A., Moricca, S. and Dellavalle, I. (eds) *Endophytism in Forest Trees*. Accademia Italiana Scienze Forestali, Firenze, Italy, pp. 15–32.

Reininger, V. and Sieber, T.N. (2013) Mitigation of antagonistic effects on plant growth due to root co-colonization by dark septate endophytes and ectomycorrhiza. *Environmental Microbiology Reports* 6, 892–898.

Rodriguez, R.J., White, J.F., Arnold, A.E. and Redman, E.S. (2009) Fungal endophytes: diversity and functional roles. *New Phytologist* 182, 314–330.

Schulz, B. and Boyle, C. (2005) The endophyte continuum. *Mycological Research* 109, 661–686.

Sherwood, M. and Carroll, G. (1974) Fungal succession on needles and young twigs of old-growth Douglas-fir. *Mycologia* 66, 499–506.

Stone, J.K., Polishook, J.D. Jr and White, J.F. (2004) Endophytic fungi. In: Mueller, G.M., Bills, G.F. and Foster, M.S. (eds) *Biodiversity of Fungi: Inventory and Monitoring Methods*. Elsevier Academic Press, Burlington, Massachusetts, San Diego, California and London, pp. 241–270.

Suryanarayanan, T.S. (2013) Endophyte research: going beyond isolation and metabolite documentation. *Fungal Ecology* 6, 561–568.

Vázquez-de-aldana, B.R., Zabalgogeazcoa, I., García-Ciudad, A. and García Criado, B. (2013) An *Epichloë* endophyte affects the competitive ability of *Festuca rubra* against other grassland species. *Plant and Soil* 362, 201–213.

Von Arx, J.A. (1974) *The Genera of Fungi Sporulating in Pure Culture*, 2nd edn. J. Cramer, Lehre, Germany.

10 Enhancing the Aggressiveness of the Powdery Mildew Mycoparasite *Ampelomyces quisqualis*

D. Angeli,[1,2]* S. Micheli,[1] M. Maurhofer[2] and I. Pertot[1]

[1]*Research and Innovation Centre, Fondazione Edmund Mach, San Michele all'Adige, Italy;* [2]*Institute of Integrative Biology, Swiss Federal Institute of Technology, Zurich, Switzerland*

Introduction

The mycoparasite *Ampelomyces quisqualis* is the most studied biocontrol agent of powdery mildews, and one strain isolated in Israel (CNCM I-807) has been commercialized in several countries (Falk *et al.*, 1995). *A. quisqualis* is a specific mycoparasite of *Erysiphales*, and has no impact on human health or the environment. Some inconsistency in the level of efficacy of *A. quisqualis* under field conditions has, however, been reported. In some experiments, a very good control of the powdery mildews of various crops has been achieved, but other trials have shown that the amount of biocontrol achieved was unsatisfactory (Sztejnberg, 1993; Angeli *et al.*, 2009a, 2012). There are a number of biotic and abiotic factors that can counteract the activity of *A. quisqualis* under certain environmental conditions (Caffi *et al.*, 2013). The research described in this chapter has focused on the identification and study of genetically different strains of *A. quisqualis* isolated from different fungal hosts and plants in order to verify whether there is specialization on a specific host or environment. Furthermore, we aimed at verifying if there is a different level of aggressiveness among *A. quisqualis* strains in order to select highly aggressive isolates to improve biocontrol efficacy.

Materials and Methods

A total of 16 *A. quisqualis* strains isolated from different crops, including grapevine, and associated with different powdery mildew species were included in this study (Table 10.1). Ten strains were obtained from different culture collections (Deutsche Sammlung von Mikroorganismen und Zellkulturen CDSM, DSMZ), Braunschweig,

*Corresponding author: dario.angeli@iasma.it

Table 10.1. Strains of *Ampelomyces quisqualis* from different host plants and from various culture collections (DSM; CABI; ATCC; CBS; L. Kiss, NKI Hungary[a] and D. Angeli, FEM[b]; see text for details) that were used in the aggressiveness experiments.

Strain designation	Geographic origin	Host plant
CNCM I-807[c]	Israel	Wild and ornamental plants
DSM4624	Germany	
CABI272851	Ecuador	
MYA3401[a]	UK	
MYA3391[a]	Hungary	Apple
MYA3394[a]	UK	
MYA3395[a]	UK	
DSM2222	Germany	Cucumber
CBS129	Canada	
CBS130	Canada	
ITA2[b]	Italy	Grapevine
ITA3[b]	Italy	
ITA4[b]	Italy	
ATCC245	USA	
ATCC249	USA	
ATCC250	USA	

[c]Commercial strain.

Germany; CAB International (CABI), Wallingford, UK; the American Type Culture Collection (ATCC), Manassas, Virginia, USA; and Centraalbureau voor Schimmelcultures (CBS), Utrecht, The Netherlands), three strains were obtained from the collection of L. Kiss, NKI Budapest (Plant Protection Institute of the Hungarian Academy of Sciences), Hungary and three strains from the collection of D. Angeli, Fondazione Edmund Mach (FEM), San Michele all'Adige, Italy.

The phylogenetic relationships among these strains were inferred using the sequence of internal transcribed spacer (ITS) rDNA regions (Angeli *et al.*, 2009b).

Powdery mildews from cucumber (*Podosphaera xanthii*) and strawberry (*P. aphanis*) were used as fungal hosts to evaluate the aggressiveness of the *A. quisqualis* strains under controlled conditions (18–20°C and relative humidity 70%), including the strains isolated from grapevine plants.

Two-week-old cucumber and strawberry plants were inoculated with the appropriate powdery mildew by spraying a suspension of distilled water containing 1×10^5 conidia ml^{-1}. Once the powdery mildew cultures were sporulating (7 days later), a suspension of distilled water containing 1×10^6 conidia ml^{-1} of *A. quisqualis* was sprayed on to the plants. Ten days after *A. quisqualis* infection the number of powdery mildew conidia on the leaf surfaces was measured. The conidia were counted using a Bürker chamber and results were expressed as the percentage of powdery mildew conidia found in comparison with untreated control.

Analysis of variance was applied on arcsin transformed data and significant differences among the treatments were separated by Tukey's test.

Results and Discussion

In the aggressiveness experiments against cucumber and strawberry powdery mildews, all the strains tested significantly reduced powdery mildew infection in comparison with the untreated control. Differences in aggressiveness were also observed among the strains (Fig. 10.1).

The CBS group, ITA group (grapevine isolates from Italy), DSM 4624 and CABI 272851 all greatly reduced the number of powdery mildew conidia on both cucumber and strawberry plants, whereas the commercial strain CNCM I-807, and DSM2222 and MYA 3401 gave significantly less reduction in the numbers of conidia. Very poor mildew reduction was obtained using the three ATCC and the MYA strains.

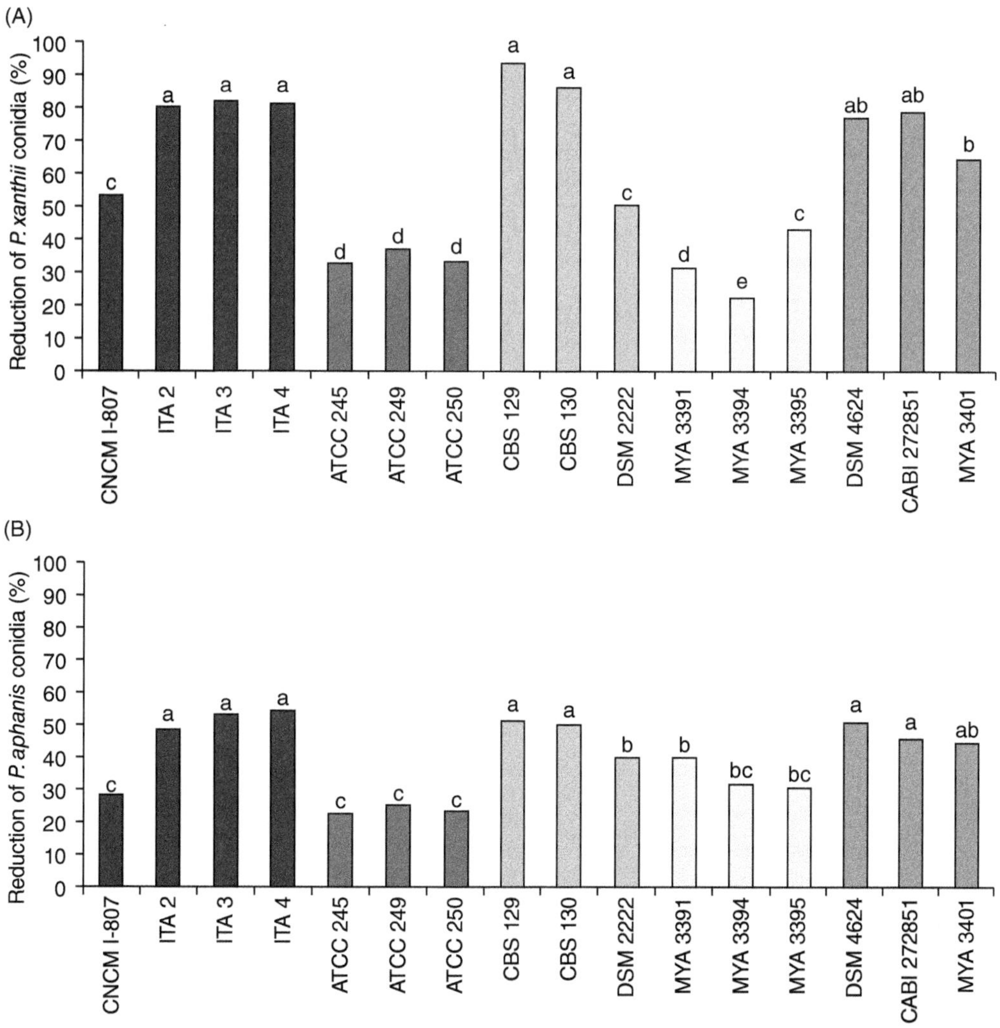

Fig. 10.1. Reduction in the number of powdery mildew conidia of *Podosphaera xanthii* (A) and *P. aphanis* (B) on leaves of cucumber (*P. xanthii*) or strawberry (*P. aphanis*) plants sprayed with different strains of *Ampelomyces quisqualis*. Columns with the same letters do not differ significantly ($P \leq 0.05$, Tukey's test).

Conclusions

Under controlled conditions, all the selected strains reduced the sporulation of powdery mildews but with a different level of reduction among strains. The susceptibility versus *A. quisqualis* varies among the different tested powdery mildews. Results suggest that there are not adaptations of *A. quisqualis* to powdery mildew species from the original host species as no differential interaction was observed, independently of their phylogenetic groups or geographical origin. Some of the tested strains, including the ones isolated from grapevine plants, show appropriate reduction of powdery mildews. They could be appropriate candidates for future commercial development in environmental conditions to control different powdery mildew species.

Acknowledgements

This research was financially supported by AMPELO, a research project funded by Autonomous Province of Trento, Italy.

References

Angeli, D., Pellegrini, E., Micheli, S., Ress, D., Maurhofer, M., Pertot, I. and Gessler, C. (2009a) Molecular characterization of *Ampelomyces* spp. isolates from different hosts and geographic origins and evaluation of their potential to control powdery mildew of cucumber. *IOBC-WPRS Bulletin* 43, 40–44.

Angeli, D., Pellegrini, E. and Pertot, I. (2009b) Occurrence of *Erysiphe necator* chasmothecia and their natural parasitization by *Ampelomyces quisqualis* in northern Italy (Trentino-Alto Adige Region). *Phytopathology* 99, 704–710.

Angeli, D., Puopolo, G., Maurhofer, M., Gessler, C. and Pertot, I. (2012) Is the mycoparasitic activity of *Ampelomyces quisqualis* biocontrol strains related to phylogeny and hydrolytic enzyme production? *Biological Control* 63, 348–358.

Caffi, T., Legler, S.E., Bugiani, R. and Rossi, V. (2013) Combining sanitation and disease modelling for control of grapevine powdery mildew. *European Journal of Plant Pathology* 135, 817–829.

Falk, S.P., Gadoury, D.M., Cortesi, P., Pearson, R.C. and Seem, R.C. (1995) Parasitism of *Uncinula necator* cleistothecia by the mycoparasite *Ampelomyces quisqualis. Phytopathology* 85, 794–800.

Sztejnberg, A. (1993) *Ampelomyces quisqualis* AQ10, CNCM I-807, for biological control of powdery mildew. US Patent No. 5190754. Available at: http://www.google.com/patents/US5190754 (accessed 23 September 2015).

11 Protection of Grapevines by *Pythium oligandrum* Strains Isolated from Bordeaux Vineyards Against Powdery Mildew

J. Gerbore,[1,2,3]* E. Bruez,[1,4] J. Vallance,[1,4] D. Grizard,[2] C. Regnault-Roger[3] and P. Rey[1,4]

[1]*Santé et Agroécologie du Vignoble (SAVE), INRA Bordeaux Sciences Agro, Villenave d'Ornon, France;* [2]*BIOVITIS, Saint Etienne de Chomeil, France;* [3]*Université de Pau et des Pays de l'Adour, Pau, France;* [4]*SAVE, Université de Bordeaux Bordeaux Sciences Agro, Villenave d'Ornon, France*

Introduction

The causal agent of powdery mildew, *Erysiphe necator*, is responsible for important damage to grapevines, which leads to economic losses worldwide (Gadoury *et al.*, 2012). To protect vineyards against this biotrophic ascomycete, winegrowers use fungicides very frequently (Compant *et al.*, 2013). However, due to the negative impact of these products on the environment and on human health, the development of alternative methods of plant protection, such as biological control, is required.

The oomycete *Pythium oligandrum*, which is known for its atypical relationship with plants whose roots it colonizes (Benhamou *et al.*, 2012; Gerbore *et al.*, 2014) and its ability to control phytopathogens through stimulation of plant defence systems and mycoparasitism, is an appropriate candidate for grapevine protection (Le Floch *et al.*, 2007; Rey *et al.*, 2008; Benhamou *et al.*, 2012; Horner *et al.*, 2012). *P. oligandrum* produces an extracellular glycoprotein named oligandrin and two cell wall proteins, POD1 and POD2, which have an elicitin-like protein biochemical signature. When applied to plants, these elicitin-like proteins induce resistance to plant pathogens (Picard *et al.*, 2000; Takenaka *et al.*, 2003; Masunaka *et al.*, 2010). For instance, grapevine root treatment with oligandrin induces systemic resistance against *Botrytis cinerea*, and can protect against the pathogen by as much as 75% (Mohamed *et al.*, 2007). In this study, the objectives were: (i) to molecularly characterize *P. oligandrum* strains isolated from vineyards of the Bordeaux region; (ii) to study the ability of *P. oligandrum* to induce systemic resistance against *E. necator* when the oospores of the biocontrol agent are

*Corresponding author: jonathan.gerbore@sabiovitis.fr

used; and (iii) to test whether *P. oligandrum* culture filtrate and mycelium extracts are able to induce resistance against powdery mildew.

Materials and Methods

P. oligandrum strains were isolated in 2010 from roots of grapevine cv. Cabernet Sauvignon planted in 12 Bordeaux vineyards. The strains originated from three types of soil (stony, sandy clay and silty clay) and were characterized at the molecular level by the inter-simple sequence repeat (ISSR) method. The repetitive oligonucleotide primer $(ACA)_5$ was used in a 30-μl reaction mixture composed of a 50-ng DNA template, 1.7 mM $MgCl_2$, 0.2 mM dNTPs, 0.2 μM of each primer (forward and reverse) and 1.7 U of Sylverstar Taq polymerase. A Mastercycler Gradient (Eppendorf AG 2233, France) was used for PCR reactions, and the following conditions were used: initial denaturation at 95°C for 3 min, 30 cycles of 1 min at 95°C, 45 s at 68°C and 1 min 45 s at 72°C. For each mixture, the amplification products were analysed by electrophoresis of an 8-μl aliquot on 2% agarose gel in 0.5 × Tris-borate-ethylenediamine tetraacetic acid (EDTA) buffer. A molecular weight ladder (GeneRuler™ 1000 bp DNA ladder, Fermentas, France) was used as the size standard.

Cabernet Sauvignon cuttings provided by the Chambre d'Agriculture de Gironde were grown in a greenhouse for the biocontrol assay in which *P. oligandrum* was used to control the development of the pathogen *E. necator*. One hundred cuttings were inoculated with 100 ml of three *P. oligandrum* strains (10^5 oospores ml^{-1}) at the root level. In the control treatment, a further 100 cuttings were inoculated with sterile distilled water. Nine leaves were sampled (the third from the top) per treatment and were collected at 0, 8 13, 26 and 36 days post-treatment with *P. oligandrum*. Leaf discs (24 mm diameter) were cut from disinfected leaves (10 min in 3% hypochlorite followed by rinsing in sterile distilled water) and placed, upper side up, in Petri dishes containing agar (20 g l^{-1}) and benzimidazole (30 mg l^{-1}). Inoculation with *E. necator* (strains B1 and B2) was done by blowing spores from 12–14-day-old sporulating leaves on to the upper surface of the leaf discs with an air pump. The density of deposited conidia was 700 spores cm^{-2}. The infected leaf discs were incubated in a growth chamber at 22°C. Disease was assessed 13 days after inoculation. For each of the nine replicates, the disc area covered by the fungus mycelium was assessed visually and the number of spores was counted using a particle counter (Coulter Counter Multisizer 3; Beckman Coulter, Paris-Nord, France).

The leaves (the third from the top) of Cabernet Sauvignon cuttings were also collected for another bioassay done in the laboratory. Leaf discs (12 replicates per treatment) were dipped in either *P. oligandrum* culture filtrate or crude *P. oligandrum* mycelium diluted in sterile distilled water before being challenged with *E. necator*. Control leaf discs were dipped in sterile distilled water. Strain B2 of *E. necator* was used for plant inoculation. The development of the pathogen was assessed after 13 days.

Results and Discussion

Genetic diversity of *P. oligandrum* strains isolated from Bordeaux vineyards

The DNA of a large number of *P. oligandrum* strains collected from 12 Bordeaux vineyards was successfully amplified with the primer $(ACA)_5$ and the ISSR banding patterns of each isolate were compared (Fig. 11.1). Distinct patterns were observed, with length of bands ranging from less than 500 bp to more than 1500 bp. Observation of the DNA bands allowed grouping into the same cluster of strains that displayed the same amplification pattern. Thus, a banding pattern designated 'A' was dominant. The cluster A was found in the three vineyard soils: sandy clay, stony and silty clay. It is characterized by three bands between 750 and 1000 bp, two bands around 400 bp and one band at 1500 bp. Two other banding patterns, designated 'B' and 'C', were found several times, each in a specific soil type. Cluster B, represented by three strains, was only isolated from sandy clay soil and cluster C was only detected in stony soils. The B and C clusters, unlike A, did not have the band at 750 bp; the B cluster, unlike C, is characterized by two bands at 1500 bp. Other patterns, at least four different patterns by soil type, were observed, but in each case, they were seen only once. The strains isolated from the vineyards were different from strains of *P. oligandrum* obtained from the Centraalbureau voor Schimmelcultures (CBS, Utrecht, The Netherlands) bank (1, 2, 3, 4 and 5). Finally, the ISSR analysis showed an important genetic diversity of *P. oligandrum* strains with a dominant group, the banding pattern A.

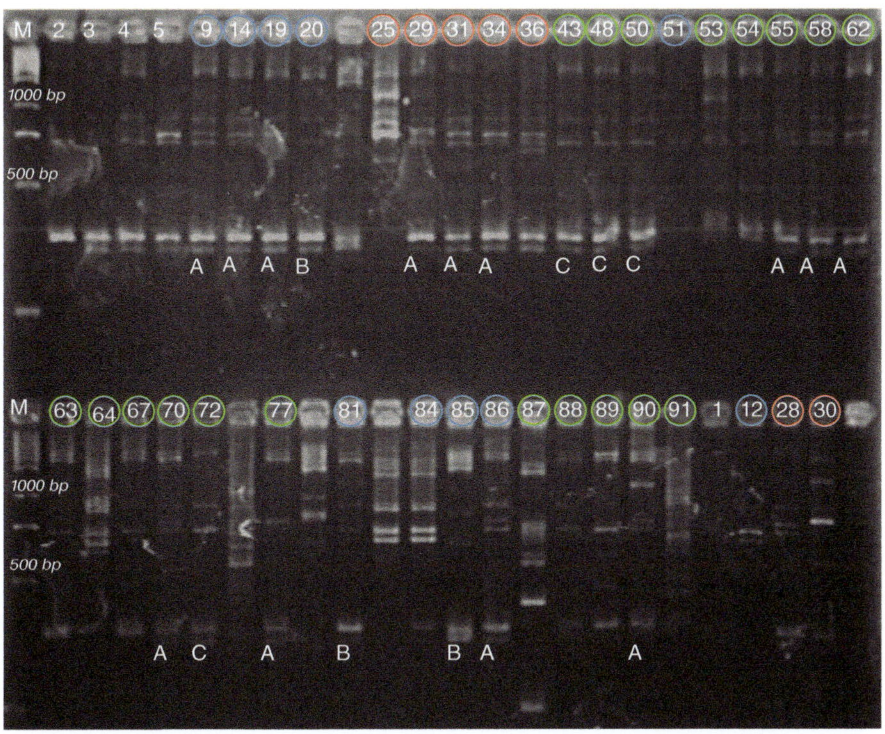

Fig. 11.1. DNA of *Pythium oligandrum* strains amplified with the ISSR primer $(ACA)_5$. The numbers on each lane refer to one strain. Strains from sandy clay, stony and silty clay soils are represented by numbers in blue, green and red circles, respectively. The numbers without circles are strains from the Centraalbureau voor Schimmelcultures (CBS, Utrecht) bank. Strains designated A, B and C show similar banding patterns. M is a 1000-bp ladder marker (GeneRuler™).

Control of powdery mildew after *P. oligandrum* root treatments

In this study, Cabernet Sauvignon cuttings being raised in a greenhouse were tested to determine whether oospores of the biocontrol agent *P. oligandrum* could protect grapevines against attack by *E. necator*. Leaf samples were collected on various days post-root treatment with *P. oligandrum*, and leaf discs cut from them were inoculated with *E. necator*. The aim was to determine if the oomycete could induce systemic resistance and, in the case of a positive result, to assess how long the protection period is. Table 11.1 shows the results obtained with *E. necator* strain B1 at 0, 8, 26 and 36 days post-infection. Leaf discs from the non-treated cuttings were more covered by *E. necator* mycelium at 0 and 8 days post-infection (the leaf coverage was of 98.0 and 95.6%, respectively), than at 26 and 36 days post-treatment, when leaf coverage was around 65%). However, although the leaves of cuttings inoculated with the oomycete 8 days after *P. oligandrum* root treatment were not able to stop or limit *E. necator* mycelium development, the data in Table 11.1 indicate that they were able to significantly reduce spore production by the pathogen. Thus, we hypothesize that *P. oligandrum* stimulates the plant defence reaction from the roots to the leaves, as described by Mohamed *et al.* (2007) against *B. cinerea* on grapevine. This mode of action is very different from the one displayed by *Ampelomyces quisqualis* against *E. necator* (Falk *et al.*, 1996), which mainly proceeds by mycoparasitism to destroy the pathogen. Nevertheless, the spread of powdery mildew cannot be stopped when the level of sporulation is high (Kiss, 2003). Angeli *et al.* (2013) reported that in a 2-year field experiment performed in five different Italian vineyards, the effectiveness of *A. quisqualis* was low

Table 11.1. Powdery mildew incidence (mycelial cover) and severity (number of spores produced) assessed on leaf discs sampled from Cabernet Sauvignon cuttings treated with *Pythium oligandrum* at the root level 0, 8, 26 and 36 days before leaf challenge with *Erysiphe necator* strain B1.

Days before challenge/ Treatment[a]	Incidence (%)[b]	Severity (no. spores)[b]
Day 0		
Treated	–	–
Non-treated	98.00 ± 1.11	1937 ± 349
Day 8		
Treated	85.00 ± 5.6b	713 ± 128a
Non-treated	95.56 ± 2.9b	1696 ± 172b
Day 26		
Treated	77.78 ± 6.2a	1462 ± 284a
Non treated	65.56 ± 4.1a	2020 ± 255a
Day 36		
Treated	60.56 ± 6.8a	1096 ± 208a
Non treated	64.38 ± 6.1a	1239 ± 223a

[a]Treated, *P. oligandrum* applied to roots; non-treated, sterile distilled water applied to roots.

[b]Assessment made 13 days after pathogen infection; mean values ± standard errors are indicated. According to analysis of variance (ANOVA), values followed by the same letter did not differ significantly ($P > 0.05$).

in the natural environment. It is possible to infer that a strategy of combining induced resistance conferred by one biocontrol agent (*P. oligandrum*) with the mycoparasitism of another (*A. quisqualis*) could provide an interesting control method to enhance the protection of plants under field conditions.

Figure 11.2 shows that the highest level of induced resistance to *E. necator* strain B2 was observed 13 days post-root inoculation with *P. oligandrum*. At this sampling time, *E. necator* leaf coverage and spore production was reduced by 34 and 64%, respectively, in *P. oligandrum*-treated cuttings compared with non-treated cuttings.

It is interesting to note that powdery mildews in general, and *E. necator* specifically, rely upon their ability to produce large amounts of conidia in a short time (Gadoury *et al.*, 2012) in order to spread the infection. As the results reported here show that *P. oligandrum* consistently reduces powdery mildew spores production, we can therefore hypothesize that under field conditions, it could slow down the dynamics of an epidemic by breaking down the cycle of asexual reproduction of the pathogen.

Control of powdery mildew after leaf treatments with *P. oligandrum* culture filtrate or mycelium extract

A bioassay was conducted in order to determine whether the induced resistance against *E. necator* reported above could be obtained after leaf disc treatments with culture filtrate or mycelium extract of *P. oligandrum*. Figure 11.3 shows that when leaf discs were treated with *P. oligandrum* culture filtrate or mycelium extract, disease incidence (mycelial cover) and severity (number of spores produced) were both reduced. The incidence of *E. necator* was estimated at 70% for the control but it was only 54 and 51%, respectively, in the culture filtrate and mycelium treatments. Disease severity was even further

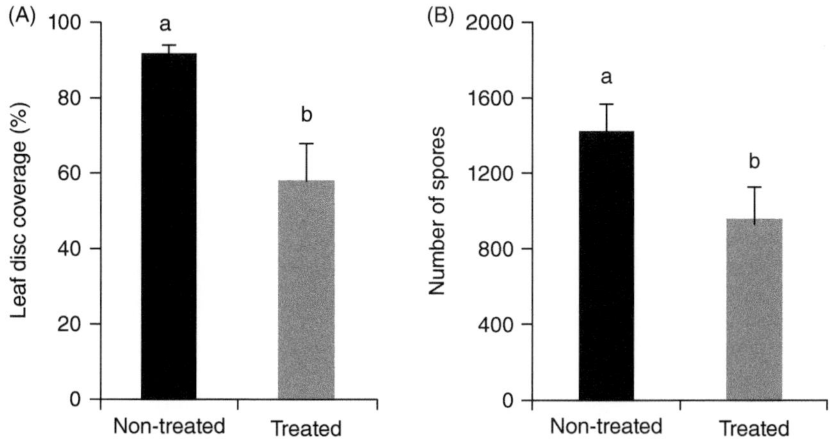

Fig. 11.2. Powdery mildew incidence (A) and severity (B) on leaves sampled from grapevine cv. Cabernet Sauvignon cuttings treated by inoculating with *Pythium oligandrum* at root level 13 days before the leaves were challenged with *Erisyphe necator* strain B2. Non-treated cuttings had sterile distilled water applied at root level and were similarly challenged with *Erisyphe necator*. Mean values are presented, and according to analysis of variance (ANOVA), values followed by the same letter did not differ significantly ($P > 0.05$).

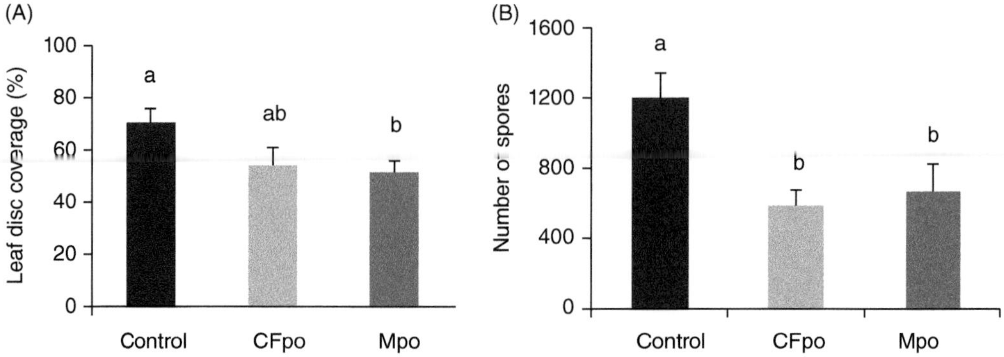

Fig. 11.3. Effects on powdery mildew incidence (A) and severity (B) of treatment of leaf discs from grapevine cv. Cabernet Sauvignon with *Pythium oligandrum* culture filtrate (CFpo) or mycelium extract (Mpo), followed by challenge with *Erisyphe necator* strain B2. Control discs were treated with distilled water only prior to challenge. Disease incidence and severity were assessed 13 days after infection. Mean values are presented, and according to analysis of variance (ANOVA), values followed by the same letter did not differ significantly ($P > 0.05$).

reduced, by 56 and 48%, respectively, in the culture filtrate and mycelium treatments. Interestingly, the differences between the culture filtrate and mycelium treatments were not statistically significant, either for disease incidence or severity. The effect of the culture filtrate may be due to the elicitin-like activity of the extracellular protein (oligandrin) that is produced by *P. oligandrum*, as Picard *et al.* (2000) have shown that the oomycete is able to produce high quantities of oligandrin in this kind of medium. Oligandrin has been reported to be able to stimulate plant defences in tomato, grapevine and tobacco (Picard *et al.*, 2000; Benhamou *et al.*, 2001; Mohamed *et al.*, 2007). Two other elicitin like proteins (POD1 and POD2) produced by *P. oligandrum* are also able to induce resistance in several crops (Takenaka *et al.*, 2003, 2006, 2011; Takenaka and Tamagake, 2009). In our experiment, these two cell-wall proteins are present in the mycelium extract and should thus induce resistance in grapevine leaves against *E. necator*.

Conclusions

Our work showed that there is a genetic polymorphism of *P. oligandrum* strains collected from various Bordeaux vineyards. One cluster (A) was predominant and present in stony, silty and sandy clay soils, but for most of the strains, no correlation was found between the soil type and the strain genotype. Three strains were selected and used as one inoculum that was applied to the roots of young vines. They appeared to induce disease resistance in the grapevine as powdery mildew incidence and severity were reduced in comparison with the control treatments. The protection seems to be strain dependent, because the highest protection was obtained when *E. necator* strain B2 rather than B1 was used. Protection was also obtained after culture filtrate or mycelium extract of *P. oligandrum* was applied to grapevine leaf discs. These findings highlight the potential of *P. oligandrum* for inducing resistance against powdery mildew when applied to the roots or leaves, and open new perspectives for its use to reduce the production of conidia by *E. necator*.

Acknowledgements

This research was supported by Biovitis SA, Saint-Etienne de Chomeil, France, and the French Association Nationale de la Recherche et de la Technologie.

References

Angeli, D., Masiero, C., Giovannini, O., Valentini, F. and Pertot, I. (2013) Reducing primary inoculum sources of grapevine powdery mildew by the hyperparasite *Ampelomyces quisqualis*. In: *Future IPM in Europe, 19–21 March 2013, PalaCongressi, Riva del Garda, Italy. Book of Abstracts: Biocontrol.* Fondazione Edmund Mach, Michele all'Adige and Land- und Forstwirtschaftliches Versuchszentrum Laimburg, Pfatten, Italy with PURE project (Innovative Crop Protection for Sustainable Agriculture), European Commission, p. 43. Available at: http://www.pure-ipm.eu/sites/default/files/content/files/BOOK%20OF%20ABSTRACTS_BIOCONTROL_0.pdf (accessed 16 September 2015).

Benhamou, N., Belanger, R.R., Rey, P. and Tirilly, Y. (2001) Oligandrin, the elicitin-like protein produced by the mycoparasite *Pythium oligandrum*, induces systemic resistance to *Fusarium* crown and root rot in tomato plants. *Plant Physiology and Biochemistry* 39, 681–696.

Benhamou, N., Le Floch, G., Vallance, J., Gerbore, J., Grizard, D. and Rey, P. (2012) *Pythium oligandrum*: an example of opportunistic success. *Microbiology* 158, 2679–2694.

Compant, S., Brader, G., Muzammil, S., Sessitsch, A., Lebrihi, A. and Mathieu, F. (2013) Use of beneficial bacteria and their secondary metabolites to control grapevine pathogen diseases. *Biocontrol* 58, 435–455.

Falk, S.P., Pearson, R.C., Gadoury, D.M., Seem, R.C. and Sztejnberg, A. (1996) *Fusarium proliferatum* as a biocontrol agent against grape downy mildew. *Phytopathology* 86, 1010–1017.

Gadoury, D.M., Wakefield, L.M., Cadle-Davidson, L., Dry, I.B. and Seem, R.C. (2012) Effects of prior vegetative growth, inoculum density, light, and mating on conidiation of *Erysiphe necator*. *Phytopathology* 102, 65–72.

Gerbore, J., Benhamou, N., Vallance, J., Le Floch, G., Grizard, D., Regnault-Roger, C. and Rey, P. (2014) Biological control of plant pathogens: advantages and limitations seen through the case study of *Pythium oligandrum*. *Environmental Science and Pollution Research* 21, 4847–4860.

Horner, N.R., Grenville-Briggs, L.J. and Van West, P. (2012) The oomycete *Pythium oligandrum* expresses putative effectors during mycoparasitism of *Phytophthora infestans* and is amenable to transformation. *Fungal Biology* 116, 24–41.

Kiss, L. (2003) A review of fungal antagonists of powdery mildews and their potential as biocontrol agents. *Pest Management Science* 59, 475–483.

Le Floch, G., Tambong, J., Vallance, J., Tirilly, Y., Levesque, A. and Rey, P. (2007) Rhizosphere persistence of three *Pythium oligandrum* strains in tomato soilless culture assessed by DNA macroarray and real-time PCR. *FEMS Microbiology Ecology* 61, 317–326.

Masunaka, A., Sekiguchi, H., Takahashi, H. and Takenaka, S. (2010) Distribution and expression of elicitin-like protein genes of the biocontrol agent *Pythium oligandrum*. *Journal of Phytopathology* 158, 417–426.

Mohamed, N., Lherminier, J., Farmer, M.J., Fromentin, J., Beno, N., Houot, V., Milat, M.L. and Blein, J.P. (2007) Defense responses in grapevine leaves against *Botrytis cinerea* induced by application of a *Pythium oligandrum* strain or its elicitin, oligandrin, to roots. *Phytopathology* 97, 611–620.

Picard, K., Ponchet, M., Blein, J.P., Rey, P., Tirilly, Y. and Benhamou, N. (2000) Oligandrin. A proteinaceous molecule produced by the mycoparasite *Pythium oligandrum* induces resistance to *Phytophthora parasitica* infection in tomato plants. *Plant Physiology* 124, 379–395.

Rey, P., Le Floch, G., Benhamou, N. and Tirilly, Y. (2008) *Pythium oligandrum* biocontrol: its relationships with fungi and plants. In: Ait Barka, E. and Clément, C. (eds) *Plant–Microbe Interactions*. Research Signpost, Trivandrum, Kerala, India, pp. 43–67.

Takenaka, S. and Tamagake, H. (2009) Foliar spray of a cell wall protein fraction from the biocontrol agent *Pythium oligandrum* induces defence-related genes and increases resistance against *Cercospora* leaf spot in sugar beet. *Journal of General Plant Pathology* 75, 340–348.

Takenaka, S., Nishio, Z. and Nakamura, Y. (2003) Induction of defense reactions in sugar beet and wheat by treatment with cell wall protein fractions from the mycoparasite *Pythium oligandrum*. *Phytopathology* 93, 1228–1232.

Takenaka, S., Nakamura, Y., Kono, T., Sekiguchi, H., Masunaka, A. and Takahashi, H. (2006) Novel elicitin-like proteins isolated from the cell wall of the biocontrol agent *Pythium oligandrum* induce defence-related genes in sugar beet. *Molecular Plant Pathology* 7, 325–339.

Takenaka, S., Yamaguchi, K., Masunaka, A., Hase, S., Inoue, T. and Takahashi, H. (2011) Implications of oligomeric forms of POD-1 and POD-2 proteins isolated from cell walls of the biocontrol agent *Pythium oligandrum* in relation to their ability to induce defense reactions in tomato. *Journal of Plant Physiology* 168, 1972–1979.

12 Bioactive Natural Products for Managing Downy Mildew Disease in Grapevine

M.T. Islam*

Department of Biotechnology, Bangabandhu Sheikh Mujibur Rahman Agricultural University, Gazipur, Bangladesh

Introduction

The peronosporomycetes (oomycetes or Oomycota in the old classification used elsewhere in this book) belong to the kingdom of Straminipila and are devastating pathogens of plants, animals, fishes and humans (Baldauf *et al.*, 2000; Dick, 2001; Islam and Tahara, 2001; Yoon *et al.*, 2002). These microorganisms are phylogenetically related to brown algae and diatoms and are thus insensitive to many fungicides. Several species of this class of microorganisms, such as *Plasmopara viticola* and *Phytophthora infestans*, are listed among the top ten economically most important plant pathogens, which cause multibillion dollar crop losses worldwide (Agrios, 1997; Gessler *et al.*, 2011). The downy mildew pathogen, *P. viticola*, which had been prevalent on wild species of *Vitis* in North America, was reported for the first time in Europe in 1878 and, since then, has posed a major threat to vineyards all over the world (Müller and Sleumer, 1934). Like other peronosporomycetes, *P. viticola* spreads by an extremely efficient cycle of asexual propagation (Kiefer *et al.*, 2002; Riemann *et al.*, 2002; Gessler *et al.*, 2011). The success of this obligate biotrophic pathogen can be attributed in part to the speed of its asexual differentiation to generate zoospores that target the stomata of grapevine leaves through water films. The biflagellate motile zoospores, which are released from airborne sporangia, are guided by host cues (not chemotaxis) to locate the stomata. They then encyst and germinate, forming germ tubes which penetrate the leaf surface through the stomata (Kiefer *et al.*, 2002; Islam and Von Tiedemann, 2008; Gessler *et al.*, 2011). Hence, the release of zoospores from the sporangia (zoosporogenesis), which then swim towards stomata, are critical stages for the disease cycles of this phytopathogen. Any disruption of zoosporogenesis or of motility of the zoospores markedly decreases the potential for pathogenesis (Judelson and Roberts, 2002; Judelson and Blanco, 2005; Islam *et al.*, 2011).

Although copper-containing synthetic pesticides are effective in controlling downy mildew in grapevine, the high cost of copper and the deleterious environmental consequences of these synthetic pesticides have suggested that their intensive application in viticulture should be restricted. The isolation and identification of new active molecules from non-host plants and microorganisms that selectively target the downy

*E-mail: tofazzalislam@yahoo.com

mildew pathogen without undesired impacts is possible (Abdalla *et al.*, 2011; Islam *et al.*, 2011; Islam and Von Tiedemann, 2011; Talontsi *et al.*, 2012a,b). Therefore, biological control of the downy mildew pathogen is considered to be an attractive alternative to the use of synthetic pesticides (Islam, 2010, 2011; Compant *et al.*, 2013; Islam and Hossain, 2013). Secondary metabolites or extracts from non-host plants and environmental microorganisms have been found to interfere with and/or inhibit the early developments of *P. viticola* on grapevine leaves (Abdalla *et al.*, 2011; Gessler *et al.*, 2011; Islam *et al.*, 2011; Islam and Von Tiedemann, 2011; Talontsi *et al.*, 2012a,b). The application of some of these novel metabolites has been shown to cause induced systemic resistance (ISR) in plants or to directly inhibit the growth of the pathogen (Hamiduzzaman *et al.*, 2005; Islam, 2011; Islam *et al.*, 2011). Numerous studies have reported on activation of ISR by inducers such as beta-aminobutyric acid (Hamiduzzaman *et al.*, 2005), oligochitosan (Aziz *et al.*, 2006; Yin *et al.*, 2010), laminarins (Aziz *et al.*, 2003; Allègre *et al.*, 2009), plant extracts (Godard *et al.*, 2009), secondary metabolites or compounds based upon them (Reuveni *et al.*, 2001) and biocontrol agents (Perazzolli *et al.*, 2008). Priming with resistance inducer has been identified as one of the most promising low-impact, low-cost strategies for use in modern disease management (Beckers and Conrath, 2007), and the mechanisms of such induced resistance, and its limitations and potential for disease control have extensively been discussed in a review by Gessler *et al.* (2011).

Several good reviews on the biocontrol of grapevine diseases have also recently been published including that by Gessler *et al.* (2011) noted above and one by Compant *et al.* (2013), but none of them so far has focused on bioactive natural products for managing downy mildew disease in grapevine. This chapter reviews updates on the effects of secondary metabolites from plants and of environmental microorganisms on zoosporogenesis, the motility of *P. viticola* zoospores and the germination of *P. viticola* cystospores. The efficacies of plant and microbial extracts or pure secondary metabolites on the suppression of downy mildew disease in grapevine, including the modes of their activities, are also discussed.

Bioassay Methods for Screening Antagonistic Microorganism and Plant Extracts

P. viticola asexually produces characteristic biflagellate (heterokont) motile spores, called zoospores from its lemon-shaped sporangia (Judelson and Blanco, 2005; Islam and Von Tiedemann, 2008), and it is these that infect grapevine plants. Several lines of evidence indicate that these zoospores locate their hosts guided by the specific host cues (specific signalling compounds or surface architecture) (Morris and Ward, 1992; Islam and Tahara, 2001; Islam, 2008). When the zoospores find potential infection sites on the hosts (such as stomata in the case of downy mildew), they cease their motility, attach to the host surface using their posterior flagella, and rapidly transform into round-shaped cystospores after shedding their flagella (Islam *et al.*, 2002a, 2003). Within 30–40 min, the cystospores germinate to form germ tubes which invade the plant tissues and initiate infection. This sequence is extremely rapid as it leads to infection within 30–40 min of the zoospores arriving at the host surface. The disruption of any of the asexual stages in the sequence eliminates the potential for pathogenesis (Islam *et al.*, 2005b, 2011). To assess the inhibitory activities of active molecules or extracts, two convenient methods have popularly been used (Hamiduzzaman *et al.*, 2005; Islam *et al.*, 2011): (i) the homogeneous solution method (Islam *et al.*, 2002b, 2005a, 2011); and (ii) detached leaf assay methods.

Homogenous solution assays

It was speculated that antimicrobial compounds can affect the release of zoospores from the sporangium (zoosporogenesis) of peronosporomycetes and inhibit motility of the released zoospores, the two critical stages in the disease cycle of these phytopathogens, and thereby decrease pathogenicity (Deora *et al.*, 2005; Islam, 2008; Islam *et al.*, 2011). In order to test the effects of crude extracts or pure metabolites from non-host plants, or the effects of antagonistic microorganisms, on these two key stages in the asexual life cycle of peronosporomycetes, the homogeneous solution assay has been used successfully (Islam *et al.*, 2002a,b, 2004, 2005b, 2011; Folman *et al.*, 2004). In this method, the culture filtrate or extract, or the pure metabolite, is directly added at the appropriate dose to an aqueous suspension of motile zoospores or sporangia taken from a tissue culture plate, and a certain period is allowed for direct interactions between the inhibitory substances and the pathogen (Islam and Von Tiedemann, 2011). The effect of antagonism, such as inhibition of zoospore release from the sporangia (zoosporogenesis), or inhibition of the motility, encystment or lysis of zoospores, can easily be recorded under a microscope (Islam *et al.*, 2005b, 2011). This has been established as a convenient and quantitative method of bioassay to test the effects of crude extracts and secondary metabolites on the asexual development of peronosporomycetes, including *P. viticola* (Islam and Tahara, 2001; Islam *et al.*, 2011).

However, it has been a challenging issue for researchers to find a convenient method to obtain samples of asexual life stages of *P. viticola* (such as sporangia, zoospores, cystospores and germinated cystospores) outside grapevine leaves, as they are recalcitrant to culture on laboratory culture media. Recently though, Islam and Von Tiedemann (2008) have developed convenient methods for the production of abundant sporangia on excised grapevine leaves (Fig. 12.1), which then release copious amounts of zoospores, cystospores and germinated cystospores in host-free sterilized water (Islam and Von Tiedemann, 2008; Islam *et al.*, 2011).

These methods have provided an opportunity to directly test the effect of living bacteria, crude extracts and pure secondary metabolites from non-host plants and microorganisms on the: (i) production of sporangia on grapevine leaves (Islam *et al.*, 2011); (ii) release of zoospores from sporangia (Islam and Von Tiedemann, 2011; Islam *et al.*, 2011); (iii) motility of zoospores (Abdalla *et al.*, 2011; Islam *et al.*, 2011; Zinad *et al.*, 2011; Talontsi *et al.*, 2012a,b); (iii) viability and lysis of zoospores (Abdalla *et al.*, 2011; Talontsi *et al.*, 2012a,b); and (iv) encystment and germination of cystospores (Islam and Von Tiedemann, 2008; Islam *et al.*, 2011).

The biological activities of several antibiotics, such as khatmiamycin from the terrestrial *Streptomyces* sp. ANK 313, staurosporine from *Streptomyces* sp. B5136 and 2,4-diacetylphloroglucinol from *Pseudomonas fluorescens*, have been successfully assessed by the homogenous solution method against the motility behaviour and viability of *P. viticola* zoospores (Abdalla *et al.*, 2011; Islam and Von Tiedemann, 2011; Islam *et al.*, 2011). It appears that this method is also convenient for bioassay-guided

Fig. 12.1. An artificially inoculated grapevine leaf fully covered with white powdery structures of sporangiophores with sporangia of *Plasmopara viticola* at day 6 of incubation in a growth chamber on an agar plate (15 cm i.d.) at 25°C and 95% relative humidity.

isolation and the purification of new active principles (compounds) from the crude plant extracts or culture filtrates of an antagonistic microorganism (Abdalla *et al.*, 2011; Islam *et al.*, 2011; Talontsi *et al.*, 2012a,b).

Detached leaf assays

The detached leaf assay has been used by several researchers as a suitable method for investigating the activity of crude extracts or pure secondary metabolites to the downy mildew pathogen (Hamiduzzaman *et al.*, 2005; Islam *et al.*, 2011). This is an assay that is performed on the basis of the interaction of pathogen, the bioactive compounds/extracts and the leaves of the host plant, which bears a resemblance to field conditions. In this method, freshly harvested young leaves are surface sterilized with 70% ethanol, rinsed with sterilized water and then placed on moistened papers or solid agar in a Petri dish. The leaves are then treated with the extract or pure compound and then inoculated with the pathogen, or co-inoculated with antagonistic bacteria and the pathogen, and incubated in a growth chamber. Leaves without extracts or antibiotic treatment serve as a control. Islam *et al.* (2011) described a leaf disc method in a recent study. In this study, leaf discs (1 cm in diameter) were punched from surface sterilized leaves and placed on 1.5% solidified water agar in each well of Nunc tissue culture multi-dishes. The leaf discs were sprayed with a suspension of freshly harvested *P. viticola* sporangia and a series of concentrations of staurosporine, a kinase inhibitor. The multi-dishes were incubated in a growth chamber and disease intensity visually assessed. In this study, staurosporine completely suppressed downy mildew disease in grapevine leaves at 2 µM (Fig. 12.2), although it displayed suppressed zoosporogenesis and impaired motility of *P. viticola* zoospores at concentrations as low as 5 and 2 nM, respectively.

Fig. 12.2. Suppression of sporangial growth of the downy mildew pathogen *Plasmopara viticola* on grapevine leaf discs by varying doses of staurosporine isolated from *Streptomyces* sp. B 5136 (adapted from Islam *et al.*, 2011). Staurosporine was suspended in aqueous dimethyl sulfoxide (DMSO, 1%) and an appropriate dose was sprayed onto leaf discs placed on 1.5% water agar 12 h before (pre-) inoculation with *P. viticola* sporangia (5×10^3 ml^{-1}). The inoculated leaf discs were incubated at 25°C in 95% relative humidity for 6 days. The application of 2 mM staurosporine completely suppressed downy mildew disease on the leaf disc.

Inhibitory Effects of Bioactive Secondary Metabolites from Plants and Microorganisms

The literature on the effects of natural products, including plant extracts, in controlling the grapevine downy mildew pathogen is scarce, most probably because of the complexity of its bioassay, because the pathogen is an obligate biotroph. However, interest in finding effective products of natural origin to control downy mildew in grapevine has recently increased (Dagostin *et al.*, 2011). This section reviews the inhibitory effects of crude plant extracts, living microorganisms and pure secondary metabolites on *P. viticola* that have been found.

Effects of plant extracts and secondary metabolites

Inula viscosa (L.) Aiton (syn. *Cupularia viscosa* G. et G., *Dittrichia viscosa* Grenter) (Asteraceae) is a perennial weed, which is native to the Mediterranean Basin. Oily paste extracts of *I. viscosa* leaves made using organic solvents (acetone and *n*-hexane) controlled downy mildew disease of grapevine caused by *P. viticola* in a detached leaf assay (Cohen *et al.*, 2006). Two major inhibitory compounds, each comprising 10.6% of the total paste weight, were identified as the sesquiterpene lactone, tomentosin and costic acid (Fig. 12.3). Other unidentified compounds were also found to be responsible for the higher inhibitory potency of the crude oily paste. The crude oily extract controlled 90% of disease in field-grown grapevine without any seasonal variation at a dose of 0.125% paste in water. Acetone extracts of *I. viscosa* have also shown inhibitory activity against late blight pathogen, *P. infestans* and many other fungi (Cohen *et al.*, 2002). Even though numerous secondary metabolites have been identified from the extracts of *I. viscosa*, the cytotoxic and genotoxic activities of crude extracts may restrict its commercial use in viticulture (Aşkin Cęlik and Aslantürk, 2010).

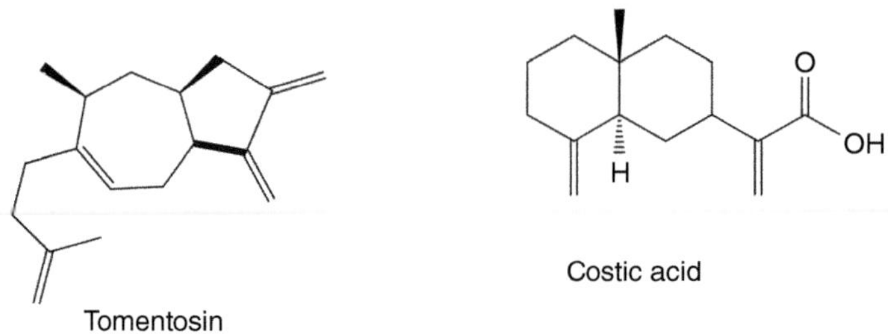

Fig. 12.3. Structures of the sesquiterpene lactone tomentosin and of costic acid isolated from leaves of the perennial weed *Inula viscosa* leaves, which suppress downy mildew disease in grapevine leaves.

A multi-year experiment searching for control strategies as alternatives to copper was carried out recently by Dagostin *et al.* (2011). Out of 112 different treatments, two plant extracts provided good control of *P. viticola* on both grapevine leaves and grape clusters, and six of them controlled infection on clusters only. Over a couple of years, Dagostin *et al.* (2011) concluded that *Yucca schidigera* and *Salvia officinalis* extracts, and *Trichoderma harzianum*, are promising candidates for organic viticulture. In another study, 58 plant extracts were tested. Of these, only extracts of *Chloris virgata*, *Dalbergia hupeana*, *Pinus massoniana*, *Paeonia suffruticosa* and *Robinia pseudoacacia* inhibited sporangial germination and controlled disease in plants effectively (Chen *et al.*, 2002). An extract of sage, *S. officinalis*, was found to suppress downy mildew disease under field conditions (Dagostin *et al.*, 2010). The spraying of extracts of a brown alga, *Ascophyllum nodosum*, also greatly reduced outbreaks of downy mildew disease (Lizzi *et al.*, 1998). Bioassay-guided isolation of the active principles from these plant extracts would lead to novel secondary metabolites useful for designing effective agrochemicals against the downy mildew pathogen.

Crude extracts of an African medicinal plant, *Endodesmia calophylloides*, displayed inhibitory activity against *P. viticola* zoospores (Talontsi *et al.*, 2012a). Xanthochymol and 1,5-dihydroxy-3-methoxy-xanthone were identified as the active compounds in the extracts (Fig. 12.4). *E. calophylloides* (Clusiaceae) is a small tree widely distributed in Nigeria, Cameroon, Gabon and Angola, which has been used for the treatment of eye-instillation and against filariae (Hutchinson *et al.*, 1994).

Inhibitory effects of antibiotics against downy mildew

A condition in which one or several secondary metabolites that are excreted by a microorganism have a harmful effect on other organisms is known as antibiosis and the compounds are referred to as antibiotics (Haas and Défago, 2005). Numerous environmental microorganisms (e.g. bacteria, fungi, etc.) including the endophytes of plants, are the reservoir of diverse classes of antibiotics. A good number of structurally diverse chemical compounds (antibiotics) have been identified from plants, bacteria and fungi as potential candidates for controlling *P. viticola* (Table 12.1).

Although the mechanisms of action of most these new compounds are unknown, they are structurally diverse and some of their actions are specific. Indeed, antibiosis

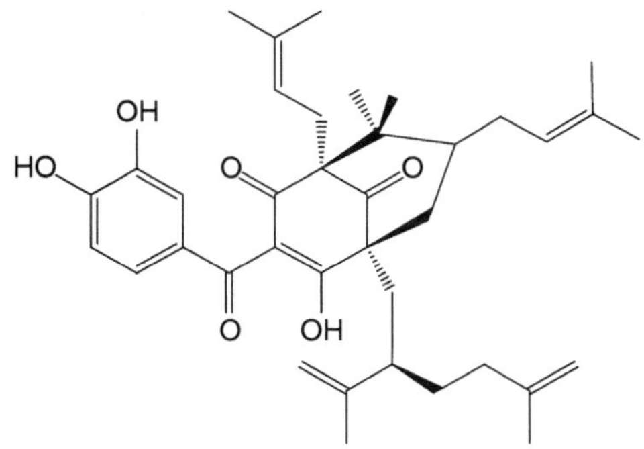

Xanthochymol

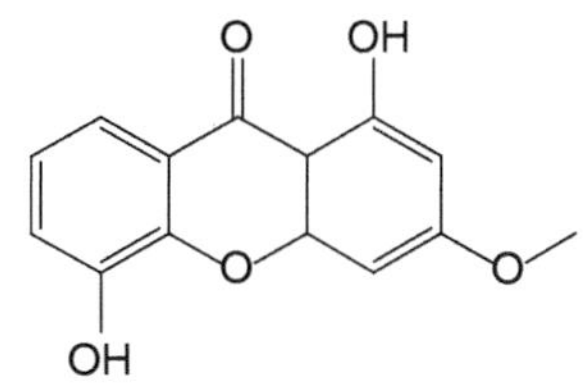

1,5-Dihydroxy-3-methoxyxanthone

Fig. 12.4. Structures of secondary metabolites isolated from the medicinal plant *Endodesmia calophylloides* that inhibit the motility of zoospores of *Plasmopara viticola*.

Table 12.1. Effect of secondary metabolites from environmental bacteria and actinomycetes on suppression of zoosporogenesis, inhibition of motility and lysis of zoospores of downy mildew pathogen *Plasmopara viticola*.

Compound	Source	IC_{50} (µM)	Activity	Reference
Staurosporine	*Streptomyces* sp. strain B5136	0.03	Suppresses zoosporogenesis	Islam *et al.*, 2011
	Streptomyces staurosporeus	0.01	Inhibits motility of zoospores	
7-Oxostaurosporine	*S. staurosporeus* ATCC 55006	0.05	Inhibits motility of zoospores	
	S. hygroscopicus ATCC 53730			
K-252a	*Nocardiopsis* sp.	0.25	Inhibits motility of zoospores	
Chelerythrine	*Chelidonium majus*	0.01	Inhibits motility of zoospores	
6,8-Dihydroxy-3-methylisocoumarin	*Streptomyces* sp. ANK302	20.00	Inhibits motility and lysis of zoospores	Zinad *et al.*, 2011
6,7,8-Trimethoxy-3-methylisocoumarin		90.00		
Khatmiamycin	*Streptomyces* sp. ANK313	30	Inhibits motility and lysis of zoospores	Abdalla *et al.*, 2011
Aloesaponarin II				
LL-C10037beta		90	Inhibits motility and lysis of zoospores	
		130	Inhibits motility of zoospores only	
2,4-Diacetylphloroglucinol (DAPG)	Synthetic (naturally produced by *Pseudomonas fluorescens*	70	Inhibits motility and lysis of zoospores	Islam and Von Tiedemann, 2011
2,4-Dipropylphloroglucinol (DPPG)	Synthetic	8		

is one of the most studied mechanisms of biological control of various peronosporomycete phytopathogens (Islam and Hossain, 2013). However, the literature on activity of antibiotics against the downy mildew pathogen is scarce, probably due to its recalcitrant behaviour in relation to culturing in culture media. The application of *Lysobacter capsici* AZ78 cells or its culture filtrates effectively controls both *P. infestans* and *P. viticola*, and *cyclo*(L-Pro-L-Tyr), which is produced by *L. capsici* AZ78, is toxic against the sporangia of both of these oomycetes (Puopolo *et al.*, 2014). These data support the potential of *Lysobacter* members for use in the control of plant pathogenic oomycetes and provide the basis for the development of new low-impact fungicides based on *cyclo*(L-Pro-L-Tyr).

Effects of phloroglucinol derivatives

The polyketide phenolic antibiotic 2,4-diacetylphloroglucinol (DAPG) is produced by different strains of *Pseudomonas fluorescens* that suppress the growth of several peronosporomycetes, including inhibitory activities against the zoosporogenesis and motility of *P. viticola* zoospores (Fenton *et al.*, 1992; Keel *et al.*, 1992; Shanahan *et al.*, 1992; Maurhofer *et al.*, 1995; Delany *et al.*, 2001; Iavicoli *et al.*, 2003; Bainton *et al.*, 2004; Islam and Fukushi, 2010; Islam and Von Tiedemann, 2011). Bioassays revealed that DAPG suppressed the release of zoospores from sporangia and impaired the motility of *P. viticola* zoospores in a dose- and time-dependent manner (Table 12.1) (Islam and Von Tiedemann, 2011). Interestingly, the zoospores whose release was halted by DAPG were lysed instead of progressing to encystment and germination.

In order to understand their structure–activity relationships, Islam and Von Tiedemann (2011) tested several derivatives of phloroglucinol, all structurally related to DAPG, to determine their effects on zoosporogenesis and motility behaviour in two peronosporomycetes, *P. viticola* and *Aphanomyces cochlioides*. These included phloroglucinol (PG), monoacetylphloroglucinol (MAPG), 2,4,6-triacetylphloroglucinol (TAPG) and 2,4-dipropylphloroglucinol (DPPG) (Fig. 12.5). According to the bioassay results, the activities of the compounds ranked in the order DPPG > TAPG > DAPG > MAPG > PG for both zoosporogenesis and zoospore motility inhibition. It appeared that: (i) the degree of substitution of hydrogen atoms in the benzene ring of PG by acyl groups (acetyl or propyl) increased bioactivity; and (ii) substitution with a larger aliphatic group (propyl) gave higher activity than that obtained with the shorter aliphatic group (acetyl). These findings help to elaborate an understanding of the structure–activity relationships of the polyketide antibiotic DAPG against peronosporomycete zoospores (De Souza *et al.*, 2003), but for the establishment of detailed structure–activity relationships, a further study is needed with a larger number of DAPG derivatives and zoospores from diverse species of the peronosporomycetes.

DAPG has shown a wide range of inhibitory activities, including antiviral, antibacterial, antifungal, anthelminthic and phytotoxic properties (Bainton *et al.*, 2004). The production of DAPG is considered to be one of the major determinants of disease suppression by *P. fluorescens* (Fujimoto *et al.*, 1995). The compound inhibits members of diverse genera of peronosporomycetes, such as *Pythium ultimum* (Fenton *et al.*, 1992; Shanahan *et al.*, 1992; Delany *et al.*, 2001), *A. cochlioides* (Islam and Fukushi, 2010) and *Peronospora parasitica* (Iavicoli *et al.*, 2003). Examples include: the inhibition of mycelial growth in *P. ultimum* and *A. cochlioides* through excessive

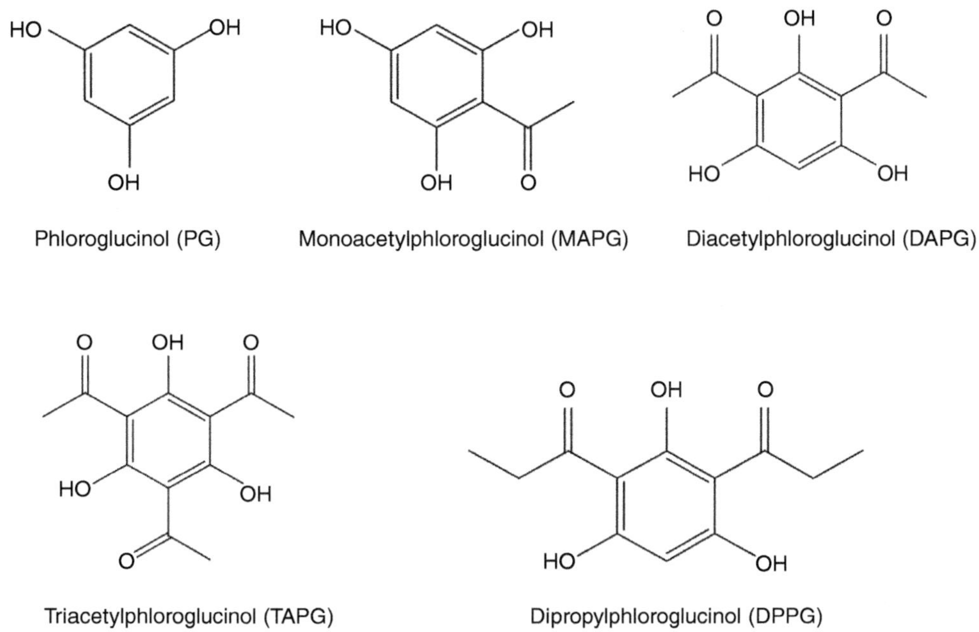

Fig. 12.5. Structures of phloroglucinol derivatives that impair motility and cause lysis of peronosporomycete zoospores.

branching and curling (Shanahan *et al.*, 1992; Islam and Fukushi, 2010); disruption of the organization of cytoskeletal filamentous actin in *A. cochlioides* hyphae (Islam and Fukushi, 2010); disorganization in the hyphal tips of *P. ultimum* var. *sporangiiferum*, including alterations (proliferation, retraction and disruption) of the plasma membrane, vacuolization and the disintegration of cell contents (De Souza *et al.*, 2003); inhibition of zoosporogenesis and zoospore motility in *A. cochlioides* and *P. viticola* (Islam and Von Tiedemann, 2011); and induction of systemic resistance in the host plants (Iavicoli *et al.*, 2003). DAPG has also recently been reported to inhibit mitochondrial function in yeast (Gleeson *et al.*, 2010). As cleavage of the nuclei and dramatic differentiation of the sporangia during zoosporogenesis require a supply of energy from the mitochondria, impairment of the mitochondrial function in the *P. viticola* sporangia and zoospores by DAPG might be associated with the suppression of zoospore release and inhibition of zoospore motility, respectively (Islam and Von Tiedemann, 2011). Although the mechanism is unknown, the motility inhibitory activity of DAPG against an osmotrophic amoeba, *Vahlkampfia* sp., has also been reported (Jousset *et al.*, 2006).

Motility inhibitory and lytic antibiotics from endophytic fungi

While in search of new and environmentally benign secondary metabolites that can regulate the motility and viability of *P. viticola* zoospores, Talontsi *et al.* (2012b) found that the crude extracts from cultures of the endophytic fungus *Cryptosporiopsis* sp. strain CAFT122-1, isolated from the healthy leaves, branches and stems of

Zanthoxylum leprieurii (Rutaceae), impaired zoospore motility of the grapevine downy mildew pathogen at 100 μg ml^{-1}. The active principles in the extracts were identified as cryptosporiopsin A, hydroxypropan-2′,3′-diol orsellinate, two cyclic pentapeptides and (–)-phyllostine. Cryptosporiopsin A and other compounds displayed motility inhibitory and lytic activities against zoospores of the grapevine downy mildew pathogen at 10–25 μg ml^{-1} (Table 12.2 and Fig. 12.6). Among the isolates, cryptosporiopsin A displayed the strongest motility inhibitory and lytic effects against zoospores at 5 μg ml^{-1}.

These compounds also exhibited potent inhibitory activity against the mycelial growth of two other peronosporomycete phytopathogens, *P. ultimum* and *A. cochlioides*, and a basidiomycete fungus *Rhizoctonia solani*. In addition, these novel metabolites exhibited modest cytotoxic activity against brine shrimp larvae (Talontsi *et al.*, 2012b). The synthesis of different analogues and derivatives of cryptosporiopsin A and their

Table 12.2. Inhibitory effects of secondary metabolites from fungi and plants against downy mildew pathogen *Plasmopara viticola*.

Compound	Source	IC_{50} (μM)	Activity	Reference
Cryptosporiopsin A	Endophytic fungus *Cryptosporiopsis* sp. from *Zanthoxylum leprieurii*	26	Inhibition of motility and lysis of zoospores	Talontsi *et al.*, 2012b
Hydroxypropan-2′, 3′-diol orsellinate		206		
cyclo(L-Ile-L-Leu-L-Leu-L-Leu-L-Leu)		17		
Excelsional	*Phomopsis* sp. CAFT69	14	Inhibition of motility	Talontsi *et al.*, 2012a
		28	Inhibition of motility and lysis of zoospores	
9-Hydoxyphomopsidin		27	Inhibition of motility	
		54	Inhibition of motility and lysis of zoospores	
Alternariol-5-*O*-methyl ether		34	Inhibition of motility and lysis of zoospores	
Diketopiperazines [*cyclo*(L-phenylalanine-*trans*-4-hydroxy-L-proline) *cyclo*(L-leucine-*trans*-4-hydroxy-L-proline), and *cyclo*(L-alanine-*trans*-4-hydroxy-L-proline)]	*Alternaria alternata*	300	Inhibits sporulation in grapevine leaves	Bobylev *et al.*, 1996
Xanthochymol	*Endodesmia calophylloides*	17	Inhibits motility and causes lysis of zoospores	Talontsi *et al.*, 2012a

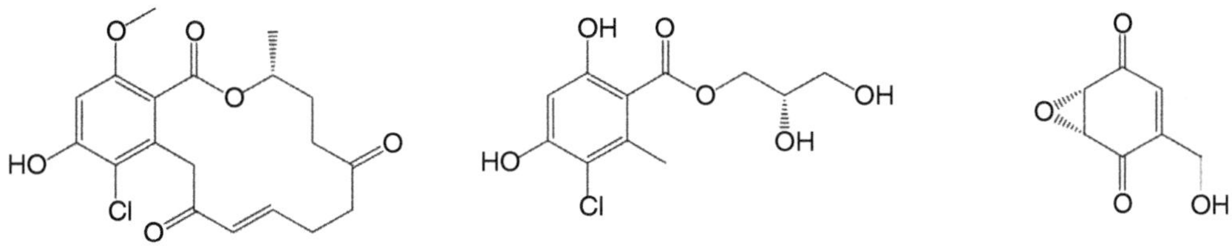

Fig 12.6. Secondary metabolites isolated from an endophytic fungus, *Cryptosporiopsis* sp. strain CAFT122-1, that showed motility inhibitory and lytic activities against zoospores of *Plasmopara viticola*.

bioassay would be helpful for establishing the structure–activity relationships involved. Further study on the mode of action of the motility inhibitory and zoosporicidal activities of the metabolites from *Cryptosporiopsis* sp. strain CAFT122-1 are needed so that their practical use as naturally occurring anti-peronosporomycetal agents can be considered.

In a similar study, Talontsi *et al.* (2012a) found that extracts of another endophytic fungus *Phomopsis* sp. CAFT69, isolated from *Endodesmia calophylloides*, markedly inhibited the motility of *P. viticola* zoospores. Using a series of chromatographic fractionations, they identified two new compounds, excelsional and 9-hydroxyphomopsidin, together with excelsione, phomopsidin, alternariol, alternariol-5-*O*-methyl ether, 5′-hydroxyalternariol and attenusin as active principles in the extracts (Fig. 12.7). Bioassays revealed that these compounds halted the motility and lytic activities of *P. viticola* zoospores at concentrations in the range of 1–10 μg ml^{-1} in a dose- and time-dependent manner.

The endophytic fungus *Acremonium byssoides* is a natural colonizer of grapevine plants. Culture filtrates and crude extracts of *A. byssoides* obtained from the grapevine cv. Insolia foliar strain inhibited *P. viticola* sporangia (Burruano *et al.*, 2008). Several novel secondary metabolites, namely acremines A–G (Fig. 12.8), have been discovered from a strain of *A. byssoides* which was isolated from the sporangiophores of the grapevine downy mildew pathogen (Assante *et al.*, 2005; Arnone *et al.*, 2008). Acremines are a class of compounds with a cyclohexane or aromatic component that has been replaced or cyclized by a phenyl moiety, and they inhibited sporangial germination more or less effectively depending on the acremine type (Assante *et al.*, 2005; Arnone *et al.*, 2008). However, the application of crude extracts or living fungus displayed higher efficiency than the pure compounds, indicating the possibility of the involvement of other factors, such as ISR in the host plants and/or the presence of other unidentified compounds in the extracts.

A new chomene, (*S*)-banchromene, and three known compounds, beauvericin, fusaproliferin and redicinin, isolated from a culture of the endophytic fungus *Fusarium* sp. strain CAMKT24b1 from *Piper guinense* (Piperaceae) showed moderate to high levels of motility-impairing activity against *P. viticola* zoospores at a concentration as low as 2.5 μg ml^{-1} (Tatong *et al.*, 2014). Among these compounds, beauvericin displayed the most powerful motility-impairing and lytic activities against the zoospores.

The application of suspensions of microconidia of *F. proliferatum* G6 reduced the production of *P. viticola* sporangia on grapevine leaves and thus protected against

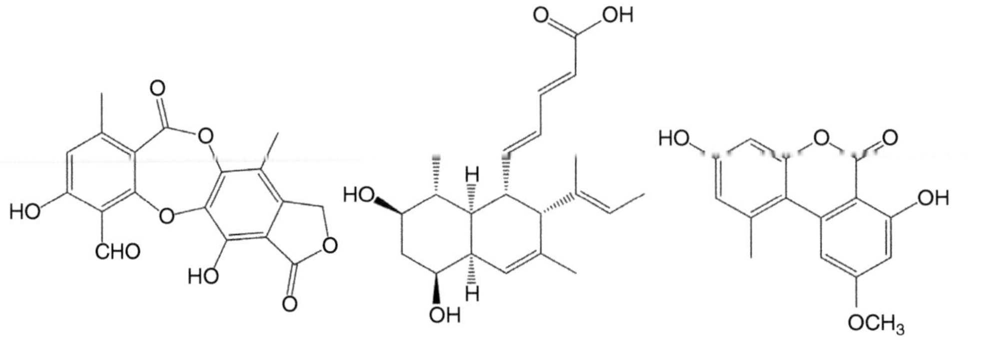

Excelsional 9-Hydroxyphomopsidin Alternariol-5-*O*-methyl ether

Fig. 12.7. Secondary metabolites isolated from an endophytic fungus, *Phomopsis* sp. CAFT69, that halted motility and had lytic activity against zoospores of *Plasmopara viticola*.

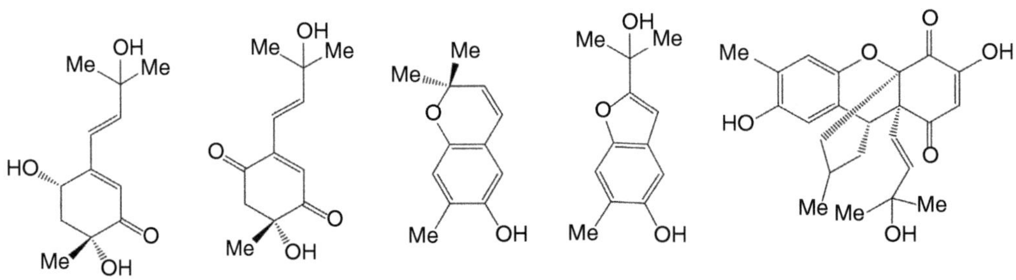

Acremine A Acremine B Acremine C Acremine D Acremine G

Fig. 12.8. Secondary metabolites isolated from an endophytic fungus, *Acremonium byssoides*, that inhibited sporangial germination of *Plasmopara viticola*.

downy mildew when applied post infection (Bakshi *et al.*, 2001). Furthermore, a hyphal preparation from UV-irradiated *F. proliferatum* G6, designated isolate 1505 (which did not produce microconidia) was even more effective in reducing the production of *P. viticola* sporangia. However, it should be noted that the production of fuminisin mycotoxins by *F. proliferatum* G6 probably limits its use as a biopesticide (Falk *et al.*, 1996).

Downy mildew protectant antibiotics from Gram-negative bacteria

Epothilone A is a macrolide with a 16-membered lactone ring that was isolated from the myxobacterium *Sorangium cellulosum* strain 90 and displays cytotoxic activities that include an inhibitory effect against *P. viticola* (Höfle *et al.*, 1996).

A composition of seratamolide and oocydin A isolated from another Gram-negative bacterium, *Serratia marcescens*, has been shown to protect plants from peronosporomycete phytopathogens including *P. viticola* (Fig. 12.9) (Strobel *et al.*, 2005).

Further studies on the isolation of more novel compounds and elucidation of their modes of action should lead to the development of new eco-friendly agrochemicals against the downy mildew pathogen *P. viticola*.

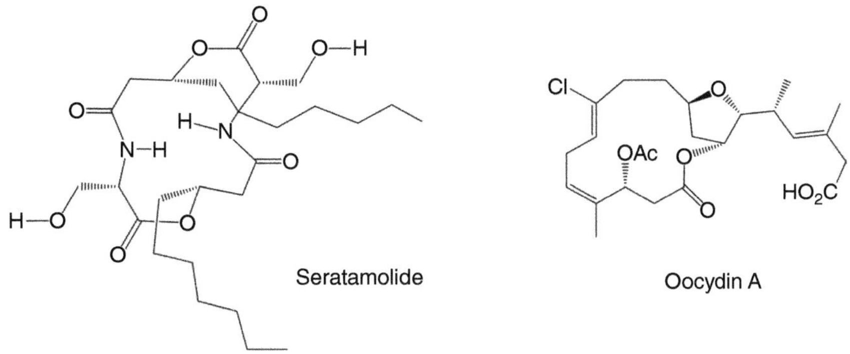

Fig. 12.9. Secondary metabolites isolated from the Gram-negative bacterium *Serratia marcescens* that protect plants from peronosporomycete phytopathogens.

Motility inhibitory and zoosporicidal antibiotics from Gram-positive *Streptomyces* bacteria

Streptomycetes spp. are known to produce diverse groups of interesting bioactive secondary metabolites. To find out about any bioactive compounds among these, a survey of the culture filtrates of more than 100 strains of marine and terrestrial *Streptomyces* spp. was tested against the motility of *P. viticola* zoospores (Islam, Von Tiedemann and Laatsch, unpublished). About 30% of the extracts displayed potent motility inhibitory and lytic activities against the biflagellate *P. viticola* zoospores at 200 μg ml^{-1}. Bioassay-guided chromatographic fractionations of some of the crude extracts resulted in the separation of several new bioactive secondary metabolites. The two subsections below describe the biological activities of these novel compounds.

Antibiotics from terrestrial Streptomyces *spp.*

During screening of a crude extract of the terrestrial *Streptomyces* sp. ANK313, it was shown to markedly inhibit the motility of *P. viticola* zoospores and caused subsequent lysis at 100 μg ml^{-1} (Abdalla *et al.*, 2011). Abdalla *et al.* isolated and characterized some motility inhibitory and lytic factors in the extracts by chromatographic fractionations. A new compound, khatmiamycin, together with five known metabolites, GTRI-02, 4-ethyl-5-methyl-heptanamide, aloesaponarin II, LL-C10037a and LL-C10037b, were characterized (Fig. 12.10) (Abdalla *et al.*, 2011). Khatmiamycin exhibited the highest motility inhibitory (100%) and lytic (83 ± 7%) activities against zoospores of *P. viticola* at 10 μg ml^{-1}.

Observations under the microscope revealed that the crude extract and the isolated metabolites showed zoosporicidal activity against *P. viticola* in an almost similar manner: initially, the zoospores were halted and then their cellular materials gradually fragmented and dispersed into the surrounding aqueous medium following bursting of the cell membranes. Reports of a similar inhibition of motility and subsequent lysis of zoospores of the sugarbeet damping-off pathogen *A. cochlioides* by various kinds of natural products, such as macrocyclic lactam antibiotics from a terrestrial bacterium, *Lysobacter* sp. SB-K88, have been published (Islam *et al.*, 2005a). Khatmiamycin also

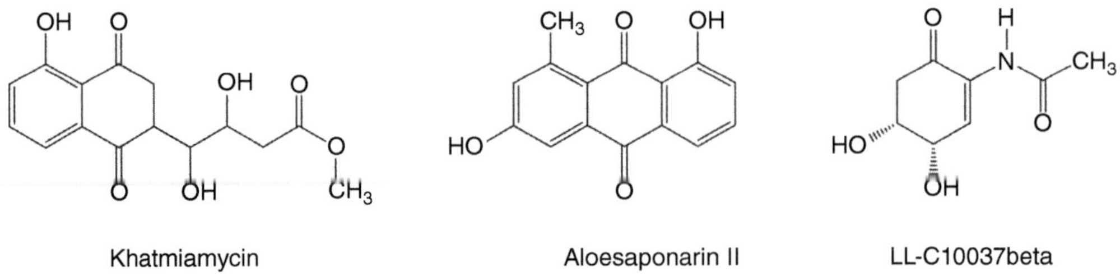

Fig. 12.10. Secondary metabolites isolated from the terrestrial *Streptomyces* sp. ANK313 that inhibit motility and cause lysis of *Plasmopara viticola* zoospores.

showed potent antibacterial activity against *Staphylococcus aureus* and *Streptomyces viridochromogenes* (Tü57) by causing inhibition zones of 11 and 14 mm diameter, respectively, at a dose of 40 µg per disc (Abdalla *et al*., 2011). This was the first report on motility inhibitory and lytic activities of metabolites from a terrestrial *Streptomyces* species against the zoospores of downy mildew pathogen *P. viticola*.

Crude extracts of another strain of *Streptomyces*, *Streptomyces* sp. ANK302, exhibited motility inhibitory activity against *P. viticola* zoospores at 100 µg ml^{-1} (Zinad *et al*., 2011). Chromatographic fractionations of crude extracts resulted in the isolation of four isocoumarins: 6,8-dihydroxy-3-methylisocoumarin, 6,7,8-trimethoxy-3-methylisocoumarin (Fig. 12.11), 6,8-dimethoxy-3-methylisocoumarin and 6,8-dihydroxy-7-methoxy-3-methylisocoumarin. Among these, 6,8-dihydroxy-3-methylisocoumarin displayed the strongest motility inhibitory and lytic activities against zoospores. Zoospores became paralysed (100%) and then immobilized (100%) within 60 min in the presence of 6,8-dihydroxy-3-methylisocoumarin. This compound also displayed antifungal activity against *Candida albicans*, causing an inhibition zone of 12 mm at 40 µg per disc (Zinad *et al*., 2011). The compound 6,7,8-trimethoxy-3-methylisocoumarin (25 µg ml^{-1}) also inhibited the motility of *P. viticola* zoospores in a dose- and time-dependent manner. Zoospores became paralysed (*c*.98%) and then immobilized (*c*.73%) within 60 min. This was the first report on the motility inhibitory and zoosporicidal effects of isocoumarins against an important peronosporomycete phytopathogen, which is insensitive to most synthetic chemical fungicides (Islam *et al*., 2002b).

The screening of more than 100 bioactive natural products from terrestrial (and marine) *Streptomyces* spp. showed that several antibiotics, such as macroterolide antibiotics (e.g. nonactin and dinactin), oligomycin F, the peptide mastoparan and antimycin A, were potent inhibitory and lytic factors against zoospores of *P. viticola* (Islam, Von Tiedemann and Laatsch, unpublished). Some of these natural compounds showed a potency almost equivalent to that of azoxystrobin in suppressing zoosporogenesis, and in inhibiting the motility of and causing lysis of *P. viticola* zoospores.

Inhibitory effects of indolocarbazole antibiotics from marine Streptomyces *spp.*

Marine microorganisms are known to produce diverse secondary metabolites with potential as inhibitors of specific enzymes or proteins in many signalling pathways (Sebolt-Leopold and English, 2006; Karaman *et al*., 2008). The indolocarbazole antibiotic staurosporine (Fig. 12.12) has been identified as a major active principle in the

6,8-Dihydroxy-3-methylisocoumarin 6,7,8-Trimethoxy-3-methylisocoumarin

Fig. 12.11. Structure of two potent zoosporicidal isocoumarins isolated from *Streptomyces* sp. ANK302.

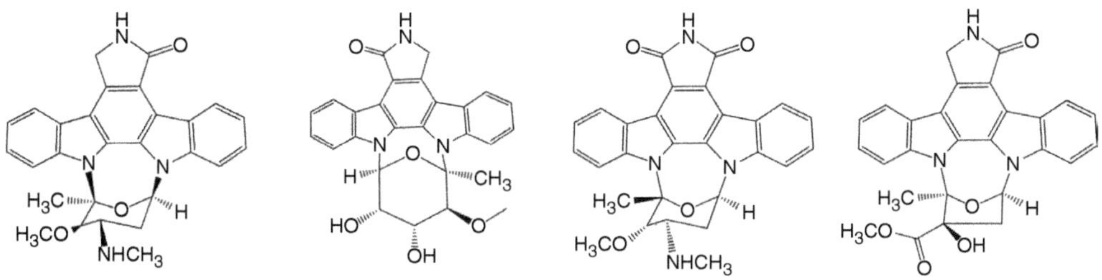

Staurosporine 4′-Demethylamino-4′,5′-dihydroxy-staurosporine 7-Oxo-staurosporine K-252a

Fig. 12.12. Structures of some indolocarbazole alkaloids isolated from *Streptomyces* spp. that impair the motility of peronosporomycete zoospores.

ethyl acetate extracts of the marine *Streptomyces* sp. strain B5136, and it rapidly impaired the motility of zoospores of the *P. viticola* (Islam *et al.*, 2011). In the presence of staurosporine (2 nM), the zoospores moved very slowly on their axes or spun in tight circles, instead of displaying straight swimming in a helical fashion. Compounds such as K-252a, K-252b, K-252c, *N*-formyl-staurosporine, arcyriaflavin A, *N*3-(α-L-rhamnopyranosyl)-staurosporine, 4′-demethylamino-4′,5′-dihydroxy-staurosporine and 7-oxistaurosporine (Fig. 12.12), which are structurally related to staurosporine, and have been isolated from other strains of *Streptomyces* spp., also impaired the motility of zoospores in a similar manner, but at varying doses (Islam *et al.*, 2011). Staurosporine is a known broad-spectrum inhibitor of protein kinases, including protein kinase C (PKC). Therefore, the involvement of the inhibition of protein kinases with the motility inhibitory activity of these compounds cannot be ruled out.

Biological Control of Grapevine Downy Mildew by Bioactive Natural Products

There are a few cases of crude extracts or pure secondary metabolites that have been proven to be effective as biocontrol agents against *P. viticola*. To evaluate the effect of the marine natural product staurosporine on downy mildew disease development on grapevine, varying doses (0, 0.002, 0.02, 0.2, 2, 5 and 10 μM) of staurosporine were applied to leaf discs of grapevine 12 h before and after inoculation with freshly harvested *P. viticola* sporangia. At 6 days post-inoculation (dpi), heavy sporulation of downy mildew was observed on the abaxial surface of the inoculated leaf discs of the control treatment (no staurosporine), while treatment of the leaf discs with staurosporine reduced the sporulation of *P. viticola* in a dose-dependent manner at 0.02 to 2 μM (Fig. 12.2). The application of staurosporine at 2.0 μM both before and after inoculation of the leaf discs completely suppressed *P. viticola* sporulation, although the application of 0.2 μM staurosporine before inoculation of the leaf discs reduced the sporulation of *P. viticola* significantly more than staurosporine treatment after pathogen inoculation. The treated leaf discs remained dark green when staurosporine was applied before inoculation with the pathogen (Islam *et al.*, 2011), whereas the former natural green colour of the heavily infected leaf discs turned to a dull to light yellow.

The endophytic fungus *Alternaria alternata* isolated from the leaves of grapevine completely inhibited the sporulation of *P. viticola* in grapevine leaves (Musetti *et al.*, 2006). The inhibitory compounds produced by *A. alternata* were isolated from a liquid culture broth of the fungus and characterized by thin layer chromatography followed by proton magnetic resonance (^{1}H NMR) spectroscopy as three diketopiperazines (DKPs): *cyclo*(L-phenylalanine-*trans*-4-hydroxy-L-proline), *cyclo*(L-leucine-*trans*-4-hydroxy-L-proline) and *cyclo*(L-alanine-*trans*-4-hydroxy-L-proline) (Fig. 12.13). When applied to both grapevine leaf discs and greenhouse plants at 0.3–2 mM concentrations, these compounds inhibited *P. viticola* sporulation. A mixture of the three diketopiperazines was very efficacious in limiting *P. viticola* sporulation. However, the DKPs displayed inhibitory activity against the pathogen only in the treated areas of the leaves, indicating poor translocation of DKPs in plants (Bobylev *et al.*, 1996). In contrast, the essential oil of lemon grass (*Cymbopogon citratus*) at a dose of 1.0–2.0 ml l^{-1} suppressed downy mildew disease in the field-grown grapevine cv. Isabel in Brazil (Maia *et al.*, 2014). The activity of chitinase in the leaves increased as a result of the essential oil treatments, whereas the activity of catalase decreased, suggesting that the essential oil acts as an elicitor in the grapevine.

Involvement of Protein Kinase C in Zoosporogenesis and Motility of Zoospores of *Plasmopara viticola*

Staurosporine, a broad-spectrum kinase inhibitor, suppressed zoosporogenesis and inhibited the motility of *P. viticola* zoospores (Islam *et al.*, 2011). To understand the role of any specific protein kinase in the maintenance of the flagellar motility of zoospores, Islam *et al.* (2011) tested 22 known kinase inhibitors. The protein kinase C

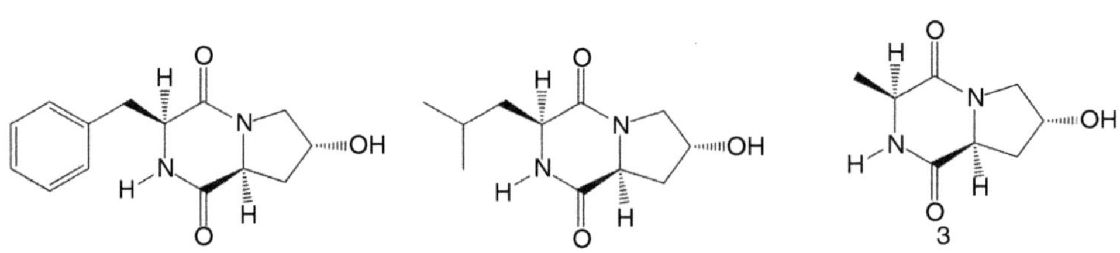

Cyclo(L-phenylalamine-*trans*-4-hydroxy-L-proline) *Cyclo*(L-leucine-*trans*-4-hydroxy-L-proline) *Cyclo*(L-alanine-*trans*-4-hydroxy-L-proline)

Fig. 12.13. Structure of some diketopiperazines isolated from an endophytic fungus *Alternaria alternata* that inhibit sporulation of *Plasmopara viticola* in grapevine leaves.

(PKC) inhibitor chelerythrine was the most potent in arresting the motility of zoospores at concentrations starting from 5 nM (Table 12.3).

Inhibitors that targeted kinase pathways other than PKC pathways did not show any practical activity in impairing zoospore motility. Both staurosporine (5 nM) and chelerythrine (10 nM) also inhibited the release of zoospores from the *P. viticola* sporangia in a dose-dependent manner (Fig. 12.14). In addition, staurosporine completely suppressed downy mildew disease in grapevine leaves at 2 μM, suggesting the potential of small-molecule PKC inhibitors for the control of peronosporomycete phytopathogens. This study, for the first time, discovered that PKC is a key signalling mediator associated with zoosporogenesis and the maintenance of flagellar motility in peronosporomycete zoospores (Islam *et al.*, 2011). Interestingly, this finding parallels earlier work on the role of PKC in the flagellar motility of mammalian sperm and the spermatozoa of aquatic vertebrates (Rotem *et al.*, 1990; White *et al.*, 2007). Because motility is critical for the life cycles and pathogenicity of pathogens, elucidation of the details of the signal transduction pathways involved might help us to design strategies for biorational management of the notorious peronosporomycete phytopathogens.

Table 12.3. Protein kinase inhibitors that suppress zoosporogenesis and inhibit the motility of *Plasmopara viticola* zoospores (adapted from Islam *et al.*, 2011).

Kinase or pathway targeted	Compound	Type of action or target	Inhibition of zoospore motility
Protein kinase C (PKC)	Chelerythrine	Inhibits PKC translocation	Total at 0.01 μM
	Gö6983	Diacylglycerol competitor	Total at 5 μM
	Calphostin C	Ca^{2+}-dependent PKC α1 and β1	Total at 30 μM
Protein tyrosine kinase (PTK) pathway	Tryphostin A47	Receptor type PTK	Total at 12 μM
	Herbimycin A	Non-receptor type PTK	Total at 50 μM
Broad spectrum	Staurosporine	ATP-site competitor	Total at 0.05 μM
	K-252a	ATP-site competitor	Total at 250 nM
	K-252b	ATP-site competitor	None
	K-252c	ATP-site competitor	Partial at ≥50 μM

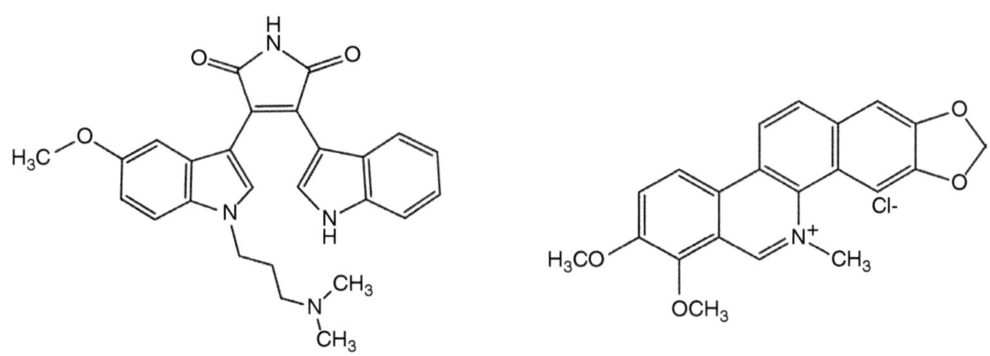

Gö6983

Chelerithrine chloride

Fig. 12.14. Structure of two selective inhibitors of protein kinase C (PKC) that inhibit the motility of *Plasmopara viticola* zoospores.

Elucidating the role of PKC in the swimming pattern and motility of zoospores will help to advance our understanding of the biology and pathogenicity of *P. viticola* and other peronosporomycete phytopathogens. Kinase inhibitors are a new class of therapeutics with the potential to inhibit multiple targets (Sebolt-Leopold and English, 2006). The selectivity of small-molecule kinase inhibitors has been assessed and described by Karaman *et al.* (2008). Staurosporine, a non-specific kinase inhibitor, not only inhibited zoosporogenesis and the motility of *P. viticola* zoospores, but also significantly suppressed downy mildew disease development in grapevine leaves (Islam *et al.*, 2011). Therefore, this naturally occurring low molecular weight protein kinase inhibitor might have high potential as a lead compound for designing effective agrochemicals against the economically important peronosporomycete phytopathogens.

PKC is known to be involved in many signal transduction pathways leading to a variety of cell responses, including motility, proliferation, differentiation and gene expression (Tamoki *et al.*, 1986; White *et al.*, 2007). One of the interesting findings of the Islam *et al.* (2011) study is the suppression of zoospore release from the sporangia (zoosporogenesis) by both staurosporine and the PKC inhibitor chelerythrine at nanomolar doses. They also showed that staurosporine completely suppressed the development of downy mildew disease on grapevine leaf discs artificially inoculated with freshly harvested sporangia of *P. viticola*. In earlier work, another staurosporine-producing isolate of *Streptomyces* was also found to inhibit the development of disease caused by *Phytophthora capsici* on peppers (Park *et al.*, 2006). Taken together, these results suggest that PKC might be an interesting target of control of the peronosporomycete phytopathogens.

Induction of Systemic Resistance to Grapevine Downy Mildew by Natural Products

The priming of plants with compounds eliciting ISR probably evolved to allow the plant to conserve energy under pathogen-free conditions (Walters and Heil, 2007) and has been identified as a promising low-impact and low-cost strategy for use in modern disease management (Beckers and Conrath, 2007). Elicitor-induced stomatal closure also showed protection against downy mildew disease in grapevine (Allègre *et al.*, 2009). Mechanisms of induced resistance against *P. viticola* in the grapevine include, as well as stomatal closure (Allègre *et al.*, 2009), the expression of defence genes (Hamiduzzaman *et al.*, 2005; Perazzolli *et al.*, 2011), increased enzymatic activity (Harm *et al.*, 2011), callose deposition (Hamiduzzaman *et al.*, 2005) and the accumulation of phytoalexins (Dercks and Creasy, 1989; Godard *et al.*, 2009).

Several compounds prime the plant to induce ISR against the downy mildew pathogen in grapevine plants. Boubakri *et al.* (2012) showed that thiamine (vitamin B_1) induces resistance in grapevine leaves against *P. viticola*. Mechanisms involved in the induction of resistance by thiamine include the generation of H_2O_2 in both grapevine suspension culture cells and plant leaves, the up-regulation of an array of defence-related genes and the induction of other defence responses at the subcellular level, such as callose deposition in stomatal cells, the accumulation of phenolic compounds and hypersensitive response (HR)-like cell death (Boubakri *et al.*, 2012). A non-protein amino acid, beta-aminobutyric acid (BABA) induced resistance in grapevine against downy mildew disease. The resistance developed in grapevine by the application of

BABA is likely to be associated with accumulation of pterostilbenes (phytoalexins) such as *trans*-piceide, *trans*-ε- and *trans*-δ-viniferin, and *trans*-pterostilbene (Slaughter *et al.*, 2008).

Not only priming with pure chemicals but also the application of biocontrol fungi induces systemic resistance in the grapevine against *P. viticola*. Perazzolli *et al.* (2008) observed that the application of *Trichoderma harzianum* T39 and benzothiadiazole (BHT) induced systemic resistance against *P. viticola*. Optimal disease suppression was observed when BHT was applied 24 h before pathogen inoculation (83% disease reduction) and *T. harzianum* was applied more than once, 48–72 h before inoculation (63% disease reduction). An aqueous extract of the dry mycelium of *Penicillium chrysogenum* (Pen) showed comparable control of downy mildew disease to that by copper-based pesticides, although the extract had no direct antiperonosporomycetal effect. Another efficient elicitor of defence responses in grapevine cells and plants is beta-1,3-glucan laminarin derived from the brown alga *Laminaria digitata* (Aziz *et al.*, 2003). Similarly, a sulfated laminarin has also been shown to elicit plant defence activities and protect grapevine against downy mildew (Trouvelot *et al.*, 2008). The mechanisms involved in laminarin-elicited defence reactions include calcium flux, alkalinization of the extracellular medium, an oxidative burst, activation of two-mitogen-activated protein kinases, the expression of ten defence-related genes, increases in chitinase and beta-1,3-glucanase activities and the production of phytoalexins, resveratrol and α-viniferin (Aziz *et al.*, 2003).

It is not clear whether the application of inducers for grapevine resistance to *P. viticola* can represent an effective alternative to copper-based pesticides for downy mildew control. However, inducers do seem reduce the susceptibility of plants to infection, improving the results obtained with other weaker biocontrol agents (Gessler *et al.*, 2011).

Conclusion and Future Perspectives

The obligate biotrophic pathogen *P. viticola* has remarkable biological features, including a cell wall made of β-1,3-glucan polymers and cellulose, the energy storage carbohydrate mycolaminarin, diploidy at the vegetative stage, motile zoospores and a sexual cycle, which make it difficult to control with common fungicides. The application of living bacteria or secondary metabolites or crude extracts of non-host plants and culture filtrates of fungi and bacteria has been found promising for the management of downy mildew disease in grapevine plant. A wide diversity of secondary metabolites from plants and microorganisms have been reported with inhibitory activity, and some have shown high promise *in vivo* of a biocontrol effect on the suppression of grapevine disease by different mechanisms, including antibiosis, ISR and the inhibition of PKC in the pathogen. This wealth of information has provided a firm foundation for the broader incorporation of these metabolites into sustainable strategies for the management of downy mildew diseases in the grapevine. Further research on the bioassay-guided isolation of more new secondary metabolites from endophytic and epiphytic microorganisms with potent inhibitory activity against *P. viticola* would help in the design of biorational strategies of downy mildew management in grapevine. Recent advances on the whole genome sequencing of the grapevine, several peronosporomycetes and biocontrol bacteria (Paulsen *et al.*, 2005) will provide the future basis for a comprehensive understanding of the interactions between biocontrol agents, the plant and the pathogen, and the development of improved strains with customized properties that will potentially function as effective biocontrol agents in low-input sustainable agriculture.

Acknowledgements

I am grateful to Prof. Andreas Von Tiedemann and Prof. Hartmut Laatsch of the University of Göttingen, Germany, for their kind support and collaboration during my postdoctoral research on bioactive compounds against the downy mildew pathogen *Plasmopara viticola* under a Georg Forster Fellowship of the Alexander von Humboldt Foundation. Sincere thanks are also due to the World Bank for funding during preparation of this book chapter under a sub-project CP # 2071 of Higher Education Quality Enhancement Project (HEQEP).

References

Abdalla, M.A., Win, S.Y., Islam, M.T., Von Tiedemann, A., Schüffler, A. and Laatsch, H. (2011) Khatmiamycin, a motility inhibitor and zoosporicide against the grapevine downy mildew pathogen *Plasmopara viticola* from *Streptomyces* sp. ANK313. *The Journal of Antibiotics* 64, 655–659.

Agrios, G.N. (1997) *Plant Pathology*. Academic Press, San Diego, California, pp. 1–952.

Allègre, M., Héloir, M.C., Trouvelot, S., Daire, X., Pugin, A., Wendehenne, D. and Adrian, M. (2009) Are grapevine stomata involved in the elicitor-induced protection against downy mildew? *Molecular Plant-Microbe Interactions* 22, 977–986.

Arnone, A., Nasini, G., Panzeri, W., Pava, O.V.D. and Malpezzi, I. (2008) Acremine G, dimeric metabolite from cultures of *Acremonium byssoides* A20. *Journal of Natural Products* 71, 146–149.

Assante, G., Dallavalle, S., Malpezz, L., Nasini, G., Burruano, S. and Torta, L. (2005) Acremines A–F, novel secondary metabolites produced by a strain of an endophytic *Acremonium byssoides*, isolated from sporangiophores of *Plasmopara viticola* in grapevine leaves. *Tetrahedron* 61, 7686–7687.

Aşkin Cęlik, T. and Aslantürk, O.S. (2010) Evaluation of cytotoxicity and genotoxicity of *Inula viscosa* leaf extracts with *Allium* test. *Journal of Biomedicine and Biotechnology*, Article ID 189252.

Aziz, A., Poinssot, B., Daire, X., Adrian, M., Bezier, A., Lambert, B., Joubert, J.M. and Pugin, A. (2003) Laminarin elicits defense responses in grapevine and induces protection against *Botrytis cinerea* and *Plasmopara viticola. Molecular Plant-Microbe Interactions* 16, 1118–1128.

Aziz, A., Trotel-Aziz, P., Dhuicq, I., Jeandet, P., Couderchet, M. and Vernet, G. (2006) Chitosan oligomers and copper sulfate induce grapevine defense reactions and resistance to gray mold and downy mildew. *Phytopathology* 96, 1188–1194.

Bainton, N.J., Lynch, J.M., Naseby, D. and Way, J.A. (2004) Survival and ecological fitness of *Pseudomonas fluorescens* genetically engineered with dual biocontrol mechanisms. *Microbial Ecology* 48, 349–57.

Bakshi, S., Sztejnberg, A. and Yarden, O. (2001) Isolation and characterization of a cold-tolerant strain of *Fusarium proliferatum,* a biocontrol agent of grape downy mildew. *Phytopathology* 91, 1062–1068.

Baldauf, S.L., Roger, A.J., Wenk-Siefert, I. and Doolittle, W.F. (2000) A kingdom-level phylogeny of eukaryotes based on combined protein data. *Science* 290, 972–977.

Beckers, G.J. and Conrath, M.U. (2007) Priming for stress resistance: from the lab to the field. *Current Opinion in Plant Biology* 10, 425–431.

Bobylev, M.M., Bobylev, L.L. and Strobel, G.A. (1996) Synthesis and bioactivities of analogs of maculysin, a host-specific phytoalexin produced by *Alternaria alternata* on spotted knapweed (*Centaurea maculosa*). *Journal of Agricultural and Food Chemistry* 44, 3960–3964.

Boubakri, H., Wahab, M.A., Chong, J., Bertsch, C., Mliki, A. and Soustre-Gacougnolle, I. (2012) Thiamine induced resistance to *Plasmopara viticola* in grapevine and elicited host-defense responses, including HR like cell death. *Plant Physiology and Biochemistry* 57, 120–133.

Burruano, S., Alfonzo, A., Piccolo, S.I., Conigliaro, G., Mondello, V., Torta, L., Moretti, M. and Assante, G. (2008) Interaction between *Acremonium byssoides* and *Plasmopara viticola* in *Vitis vinifera*. *Phytopathologia Mediterranea* 47, 122–131.

Chen, J., Dai, G., Gu, Z. and Miao, Y. (2002) Inhibition effect of 58 plant extracts against grape downy mildew (*Plasmopara viticola*). *Natural Product Research and Development* 14, 9–13.

Cohen, Y., Baider, A., Ben-Daniel, B. and Ben-Daniel, Y. (2002) Fungicidal preparations from *Inula viscosa*. *Plant Protection Science* 38, 629–630.

Cohen, Y., Wang, W.Q., Ben-Daniel, B.H. and Ben-Daniel, Y. (2006) Extracts of *Inula viscosa* control downy mildew of grapes caused by *Plasmopara viticola*. *Phytopathology* 96, 417–424.

Compant, S., Brader, G., Muzammil, S., Sessitsch, A., Lebrihi, A. and Mathieu, F. (2013) Use of beneficial bacteria and their secondary metabolites to control grapevine pathogen diseases. *BioControl* 58, 435–455.

Dagostin, S., Formolo, T., Giovannini, O., Pertot, I. and Schmitt, A. (2010) *Salvia officinalis* extract can protect grapevine against *Plasmopara viticola*. *Plant Disease* 94, 575–580.

Dagostin, S., Schärer, H., Pertot, I. and Tamm, L. (2011) Are there alternatives to copper for controlling grapevine downy mildew in organic viticulture? *Crop Protection* 30, 776–788.

De Souza, J.T., Arnould, C., Deulvot, C., Lemanceau, P., Gianinazzi-Pearson, V. and Raaijmakers, J.M. (2003) Effect of 2,4-diacetylphloroglucinol on *Pythium*: cellular responses and variation in sensitivity among propagules and species. *Phytopathology* 93, 966–975.

Delany, I.R., Walsh, U.F., Ross, I., Fenton, A.M., Corkery, D.M. and Gara, F.O. (2001) Enhancing the biocontrol efficacy of *Pseudomonas fluorescens* F113 by altering the regulation and production of 2,4-diacetylphloroglucinol. *Plant and Soil* 232, 195–205.

Deora, A., Hashidoko, Y., Islam, M.T. and Tahara, S. (2005) Antagonistic rhizoplane bacteria induce diverse morphological alterations in peronosporomycete hyphae during in vitro interaction. *European Journal of Plant Pathology* 112, 311–322.

Dercks, W. and Creasy, L.L. (1989) Influence of fosetyl-Al on phytoalexin accumulation in the *Plasmopara viticola*–grapevine interaction. *Physiological and Molecular Plant Pathology* 34, 203–213.

Dick, M.W. (2001) The peronosporomycetes. In: McLaughlin, D.J., McLaughlin, E.G. and Lemke, P.A. (eds) *The Mycota VII, Part A, Systemics and Evolution*. Springer, Berlin and Heidelberg, Germany, pp. 39–72.

Falk, S.P., Pearson, R.C., Gadoury, D.M., Seem, R.C. and Sztejnberg, A. (1996) *Fusarium proliferatum* as a biocontrol agent against grape downy mildew. *Phytopathology* 8, 1010–1017.

Fenton, A.M., Stephens, P.M., Crowley, J., O'Callaghan, M. and O'Gara, F. (1992) Exploitation of genes involved in 2,4-diacetylphloroglucinol biosynthesis to confer a new biocontrol capability to a *Pseudomonas* strain. *Applied and Environmental Microbiology* 58, 3873–3878.

Folman, L.B., De Klein, M.J.E.M., Postma, J. and Van Veen, J.A. (2004) Production of antifungal compounds by *Lysobacter enzymogenes* isolate 3.1 T8 under different conditions in relation to its efficacy as a biocontrol agent of *Pythium aphanidermatum* in cucumber. *BioControl* 25, 1121–1221.

Fujimoto, D.K., Weller, D.M. and Thomashow, L.S. (1995) Role of secondary metabolites in root disease suppression. In: Inderjit, K.M.M. and Dakshini, F.A.E. (eds) *Allelopathy Organisms, Processes, and Applications*. American Chemical Society, Washington, DC, pp. 330–347.

Gessler, C., Pertot, I. and Perazzolli, M. (2011) *Plasmopara viticola*: a review of knowledge on downy mildew of grapevine and effective disease management. *Phytopathologia Mediterranea* 50, 3–44.

Gleeson, O., O'Gara, F. and Morrissey, J.P. (2010) The *Pseudomonas fluorescens* secondary metabolite 2,4-diacetylphloroglucinol impairs mitochondrial function in *Saccharomyces cerevisae*. *Antonie Van Leeuwenhoek* 97, 261–273.

Godard, S., Slacanin, I., Viret, O. and Gindro, K. (2009) Induction of defence mechanisms in grapevine leaves by emodin- and anthraquinone-rich plant extracts and their conferred resistance to downy mildew. *Plant Physiology and Biochemistry* 47, 827–837.

Haas, D. and Défago, G. (2005) Biological control of soilborne pathogens by fluorescent *pseudomonads*. *Nature Reviews Microbiology* 3, 307–319.

Hamiduzzaman, M.M., Jakab, G., Barnavon, L., Neuhaus, J.M. and Mauch-Mani, B. (2005) β-Aminobutyric acid-induced resistance against downy mildew in grapevine acts through the potentiation of callose formation and jasmonic acid signaling. *Molecular Plant-Microbe Interactions* 18, 819–829.

Harm, A., Kassemeyer, H.-H., Seibicke, T. and Regner, F. (2011) Evaluation of chemical and natural resistance inducers against downy mildew (*Plasmopara viticola*) in grapevine *American Journal of Enology and Viticulture* 62, 184–192.

Höfle, G., Bedrof, N., Steinmetz, I.H., Shomburg, D., Gerth, K. and Reichenbach, H. (1996) Epothilone A and B – novel 16-membered macrolides with cytotoxic activity: isolation, crystal structure, and conformation in solution. *Angewandte Chemie International Edition* 35, 1567–1569.

Hutchinson, L.L., Dalziel, J.M. and Hepper, F.N. (ed.) (1994) *Flora of West Tropical Africa*, 2nd edn, reprint. Crown Agents for the Colonies, London.

Iavicoli, A., Boutet, E., Buchala, A. and Metraux, J.P. (2003) Induced systemic resistance in *Arabidopsis thaliana* in response to root inoculation with *Pseudomonas fluorescens*. *Molecular Plant-Microbe Interactions* 16, 851–858.

Islam, M.T. (2008) Disruption of ultrastructure and cytoskeleton network is involved with biocontrol of damping-off pathogen *Aphanomyces cochlioides* by *Lysobacter* sp. SB-K88. *Biological Control* 46, 312–321.

Islam, M.T. (2010) Mode of antagonism of a biocontrol bacterium *Lysobacter* sp. SB-K88 toward a damping-off pathogen *Aphanomyces cochlioides*. *World Journal of Microbiology and Biotechnology* 26, 629–637.

Islam, M.T. (2011) Potentials for biological control of plant diseases by *Lysobacter* spp., with special reference to strain SB-K88. In: Maheshwari, D.K. (ed.) *Bacteria in Agrobiology: Plant Growth Responses*. Springer, Berlin, pp. 335–363.

Islam, M.T. and Fukushi, Y. (2010) Growth inhibition and excessive branching in *Aphanomyces cochlioides* induced by 2,4-diacetylphloroglucinol is linked to disruption of filamentous actin cytoskeleton in the hyphae. *World Journal of Microbiology and Biotechnology* 26, 1163–1170.

Islam, M.T. and Hossain, M.M. (2013) Biological control of peronosporomycete phytopathogens by antagonistic bacteria. In: Maheshwari, D.K. (ed.) *Bacteria in Agrobiology: Plant Disease Management*. Springer, Berlin and Heidelberg, Germany, pp. 167–218.

Islam, M.T. and Tahara, S. (2001) Chemotaxis of fungal zoospores, with special reference to *Aphanomyces cochlioides*. *Bioscience, Biotechnology, and Biochemistry* 65, 1933–1948.

Islam, M.T. and Von Tiedemann, A. (2008) Zoosporogenesis and differentiation of grapevine downy mildew pathogen *Plasmopara viticola* in host-free system. *Phytopathology* 98, S72.

Islam, M.T. and Von Tiedemann, A. (2011) 2,4-Diacetylphloroglucinol suppresses zoosporogenesis and impairs motility of the peronosporomycete zoospores. *World Journal of Microbiology and Biotechnology* 27, 2071–2079.

Islam, M.T., Ito, T. and Tahara, S. (2002a) Microscopic studies on the attachment and differentiation of zoospores of the phytopathogenic fungus *Aphanomyces cochlioides*. *Journal of General Plant Pathology* 68, 111–117.
Islam, M.T., Ito, T., Sakasai, M. and Tahara, S. (2002b) Zoosporicidal activity of polyflavonoid tannin identified in *Lannea coromandelica* stem bark against phytopathogenic oomycete *Aphanomyces cochlioides*. *Journal of Agricultural and Food Chemistry* 50, 6697–6703.
Islam, M.T., Ito, T. and Tahara, S. (2003) Host-specific plant signal and G-protein activator, mastoparan, trigger differentiation of zoospores of the phytopathogenic oomycete *Aphanomyces cochlioides*. *Plant and Soil* 255, 131–142.
Islam, M.T., Hashidoko, Y., Ito, T. and Tahara, S. (2004) Interruption of the homing sequence of phytopathogenic *Aphanomyces cochlioides* zoospores by secondary metabolites from nonhost *Amaranthus gangeticus*. *Journal of Pesticide Science* 29, 6–14.
Islam, M.T., Hashidoko, Y., Deora, A., Ito, T. and Tahara, S. (2005a) Suppression of damping-off disease in host plants by the rhizoplane bacterium *Lysobacter* sp. strain SB-K88 is linked to plant colonization and antibiosis against soilborne peronosporomycetes. *Applied and Environmental Microbiology* 71, 3776–3786.
Islam, M.T., Hashidoko, Y., Deora, A., Ito, T. and Tahara, S. (2005b) *Lysobacter–Aphanomyces Interactions: An Ecological Role for Biocontrol of Damping-Off Disease*. Nippon Nogei Kagakkai Taikai Koen Yoshishu 2005a: 221.
Islam, M.T., Von Tiedemann, A. and Laatsch, H. (2011) Protein kinase C is likely to be involved in zoosporogenesis and maintenance of flagellar motility in the peronosporomycete zoospores. *Molecular Plant-Microbe Interactions* 24, 938–947.
Jousset, A., Lara, E., Wall, L.G. and Valverde, C. (2006) Secondary metabolites help biocontrol strain *Pseudomonas fluorescens* CHA0 to escape protozoan grazing. *Applied and Environmental Microbiology* 72, 7083–7090.
Judelson, H.S. and Blanco, F.A. (2005) The spores of *Phytophthora*: weapons of the plant destroyer. *Nature Reviews Microbiology* 3, 47–58.
Judelson, H.S. and Roberts, S. (2002) Novel protein kinase induced during sporangial cleavage in the oomycete *Phytophthora infestans*. *Eukaryotic Cell* 1, 687–695.
Karaman, M.W., Herrgard, S., Treiber, D.K., Gallant, P., Atteridge, C.E., Campbell, B.T., Chan, K.W., Ciceri, P., Davis, M.I., Edeen, P.T. *et al.* (2008) A quantitative analysis of kinase inhibitor selectivity. *Nature Biotechnology* 26, 127–132.
Keel, C., Schnider, U., Maurhofer, M., Viossard, C., Laville, J., Burger, U., Wirthner, P., Haas, D. and Defago, G. (1992) Suppression of root diseases by *Pseudomonas fluorescens* CHA0: importance of the bacterial secondary metabolite 2,4-diacetylphloroglucinol. *Molecular Plant-Microbe Interactions* 5, 4–13.
Kiefer, B., Riemann, M., Kassemeyer, C.B.H. and Nick, P. (2002) The host guides morphogenesis and stomatal targeting in the grapevine pathogen *Plasmopara viticola*. *Planta* 215, 387–393.
Lizzi, Y., Coulomb, C., Polian, C., Coulomb, P.J. and Coulomb, P.O. (1998) L'algue face au mildiou: quel avenir? Des résultats de laboratoire très encourageants. *Phytoma* 508, 29–30.
Maia, A.J., Oliveira, J.S.B., Schwan-Estrada, K.R.F., Faria, C.M.R., Batista, A.F., Costa, W.F. and Batista, B.N. (2014) The control of isariopsis leaf spot and downy mildew in grapevine cv. Isabel with the essential oil of lemon grass and the activity of defensive enzymes in response to the essential oil. *Crop Protection* 63, 57–67.
Maurhofer, M., Keel, C., Haas, D. and Defago, G. (1995) Influence of plant species on disease suppression by *Pseudomonas fluorescens* CHA0 with enhanced antibiotic production. *Plant Pathology* 44, 44–50.
Morris, P.F. and Ward, E.W.B. (1992) Chemoattraction of zoospores of the phytopathogen *Phytophthora sojae* by isoflavones. *Physiological and Molecular Plant Pathology* 40, 17–22.
Müller, K. and Sleumer, H. (1934) Biologische Untersuchungen uber die Peronosporakrankheit des Weistockes. *Landwirtschaftliches Jahrbuch* 79, 509–576.

Musetti, R., Vecchione, A., Stringher, L., Borselli, S., Zulini, L., Marzani, C., D'Ambrosio, M., Sanità di Toppi, L. and Pertot, I. (2006) Inhibition of sporulation and ultrastructural alterations of grapevine downy mildew by the endophytic fungus *Alternaria alternata*. *Phytopathology* 96, 689–698.

Park, S.K.R., George, R., Cai, Y., Chang, H.Y., Krantz, D.E., Friggi-Grelin, F., Birman, S. and Hirsh, J. (2006) Cell-type-specific limitation on *in vivo* serotonin storage following ectopic expression of the *Drosophila* serotonin transporter, dSERT. *Journal of Neurobiology* 66, 452–462.

Paulsen, I.T., Press, C., Ravel, J., Kobayashi, D.Y., Myers, G.S.A., Mavrodi, D.V., Debpy, R.M., Seshadri, R., Ren, Q., Madupu, R. *et al.* (2005) Complete genome sequence of the plant commensal *Pseudomonas fluorescens* Pf-5, insights into the biological control of plant disease. *Nature Biotechnology* 23, 873–878.

Perazzolli, M., Dagostin, S., Ferrari, A., Elad, Y. and Pertot, I. (2008) Induction of systemic resistance against *Plasmopara viticola* in grapevine by *Trichoderma harzianum* T39 and benzothiadiazole. *Biological Control* 47, 228–234.

Perazzolli, M., Roatti, B., Bozza, E. and Pertot, I. (2011) *Trichoderma harzianum* T39 induces resistance against downy mildew by priming for defence without costs for grapevine. *Biological Control* 58, 74–82.

Puopolo, G., Cimmino, A., Palmieri, M.C., Giovannini, O., Evidente, A. and Pertot, I. (2014) *Lysobacter capsici* AZ78 produces cyclo(L-Pro-L-Tyr), a 2,5-diketopiperazine with toxic activity against sporangia of *Phytophthora infestans* and *Plasmopara viticola*. *Journal of Applied Microbiology* 117, 1168–1180.

Reuveni, M. (2001) Activity of trifloxystrobin against powdery and downy mildew diseases of grapevines. *Canadian Journal of Plant Pathology* 23, 52–59.

Riemann, M., Buche, C., Kassemmeyer, H.-H. and Nick, P. (2002) Cytoskeletal responses during early development of the downy mildew of grapevine (*Plasmopara viticola*). *Protoplasma* 219, 13–22.

Rotem, R., Paz, G.F., Homonnai, Z.T., Kalina, M. and Naor, Z. (1990) Protein kinase C is present in human sperm: possible role in flagellar motility. *Proceedings of the National Academy of Sciences of the United States of America* 87, 7305–7308.

Shanahan, P., O'Sullivan, D.J., Simpson, P., Glennon, J.D. and O'Gara, F. (1992) Isolation of 2,4-diacetylphloroglucinol from a fluorescent pseudomonad and investigation of physiological parameters influencing its production. *Applied and Environmental Microbiology* 58, 353–358.

Sebolt-Leopold, J.S. and English, J.M. (2006) Mechanisms of drug inhibition of signaling molecules. *Nature* 441, 457–462.

Slaughter, A.R., Hamiduzzaman, M.M., Gindro, K., Neuhaus, J.M. and Mauch-Mani, B. (2008) Beta-aminobutyric acid-induced resistance in grapevine against downy mildew: involvement of pterostilbene (Special Issue: The downy mildews – genetics, molecular biology and control). *European Journal of Plant Pathology* 122, 185–195.

Strobel, G.A., Morrison, S.L. and Cassella, M. (2005) Protecting plants from oomycete pathogens by treatment with compositions containing serratamolide and oocydin A from *Serratia marcescens*. US Patent US006926892B2. Available at: https://patentimages.storage.googleapis.com/pdfs/US6926892.pdf (accessed 25 September 2015).

Talontsi, F.M., Islam, M.T., Facey, P., Douanla-Meli, C., Von Tiedemann, A. and Laatsch, H. (2012a) Depsidones and other constituents from *Phomopsis* sp. CAFT69 and its host plant *Endodesmia calophylloides* with potent inhibitory effect on motility of zoospores of grapevine pathogen *Plasmopara viticola*. *Phytochemistry Letters* 5, 657–664.

Talontsi, F.M., Facey, P., Tatong, M.D.K., Islam, M.T., Frauendorf, H., Draeger, S., Von Tiedemann, A. and Laatsch, H. (2012b) Zoosporicidal metabolites from an endophytic fungus *Cryptosporiopsis* sp. of *Zanthoxylum leprieurii*. *Phytochemistry* 83, 87–94.

Tamoki, T., Nomoto, H., Takahashi, I., Kato, Y., Morimoto, M. and Tomita, F. (1986) Staurosporine, a potent inhibitor of phospholipid calcium dependent protein kinase. *Biochemical and Biophysical Research Communications* 135, 397–402.

Tatong, M.D.K., Talontsi, F.M., Rahim, H.M.D.A., Islam, M.T., Oswald, R.B. and Laatsch, H. (2014) Banchromene and other secondary metabolites from the endophytic fungus *Fusarium* sp. obtained from *Piper guineense* inhibit the motility of phytopathogenic *Plasmopara viticola* zoospores. *Tetrahedron Letters* 55, 4057–4061.

Trouvelot, S., Varnier, A.L., Allègre, M., Mercier, L., Baillieul, F., Arnould, C., Gianinazzi-Pearson, V., Klarzynski, O., Joubert, J.M., Pugin, A. and Daire, X. (2008) A beta-1,3 glucan sulfate induces resistance in grapevine against *Plasmopara viticola* through priming of defense responses, including HR-like cell death. *Molecular Plant-Microbe Interactions* 21, 232–243.

Walters, D.R. and Heil, M. (2007) Costs and trade-offs associated with induced resistance. *Physiological and Molecular Plant Pathology* 71, 3–17.

White, D., De Lamirande, E. and Gagnon, C. (2007) Protein kinase C is an important signaling mediator associated with motility of intact sea urchin spermatozoa. *Journal of Experimental Biology* 210, 4053–4064.

Yin, H., Zhao, X. and Du, Y. (2010) Oligochitosan; a plant disease vaccine – a review. *Carbohydrate Polymers* 82, 1–8.

Yoon, H.S., Hackett, J.D., Pinto, G. and Bhattacharya, D. (2002) The single, ancient origin of chromist plastids. *Proceedings of the National Academy of Sciences of the United States of America* 99, 15507–15512.

Zinad, D.S., Shaabana, K.A., Abdalla, M.A., Islam, M.T., Schuffler, A. and Laatsch, H. (2011) Bioactive isocoumarins from a terrestrial *Streptomyces* sp. ANK302. *Natural Product Communications* 6, 45–48.

13 Biological Control of Grape Canker Pathogens

L. Costadone* and W.D. Gubler*

Department of Plant Pathology, University of California, Davis, USA

Introduction

Grapes (and raisins) represented the second most important agricultural commodity in the state of California in 2012, and was only narrowly third in 2013 (USDA Californian Agricultural Statistics Service, 2013). Fungal trunk diseases of grapevines are known to occur in most viticulture regions of the world, including California, and are responsible for significant economic losses (Úrbez-Torres *et al.*, 2006). Fungal species in the family Botryosphaeriaceae are the causal agents of canker diseases and dieback on many woody perennial hosts worldwide (Van Niekerk *et al.*, 2004). In California, recent research has shown that nine different species of Botryosphaeriaceae can infect grapevines (Úrbez-Torres and Gubler, 2009). Symptoms of grapevine cankers include dead spurs and cordons, and eventual vine death, due to canker development in the vascular tissue (Úrbez-Torres and Gubler, 2009). Wedge-shaped cankers in cross section are diagnostic of the disease, but cankers may appear in other shapes as well owing to fungal growth in the xylem of infected wood. Usually, no foliar symptoms are associated with the disease because of the sudden death of cordons and spur positions.

Canker diseases reduce vineyard productivity by killing spur positions. Production costs are therefore increased as a result of the pruning of cankered wood from the vine that is then necessary after the infection, and the cultural and chemical preventive measures that are required to control the disease. In California, the overall loss in net income for wine grapes caused by *Eutypa* dieback and canker and *Botryosphaeria* canker was estimated to be over US$260 million p.a. (Siebert, 2001). With large investments required for vineyard establishment and maintenance, the development of effective management strategies is then essential for the economic sustainability of vineyards.

Pruning wounds on cordons and spurs and mechanical injuries represent the main entry points of the fungi associated with canker diseases. Management of the disease is mostly achieved by fungicide treatments of wounds, such as large pruning cuts or mechanical damage, during the dormant period. Pruning wounds may remain susceptible to infection for up to 6 weeks after pruning, and chemical treatments do not protect the wounds for the entire period of their susceptibility (Úrbez-Torres and Gubler, 2010). Furthermore, the available commercial products registered that provide are persistent protection of pruning wounds are still limited, so the long-term protection of these wounds is still a challenge because the products that are available have only limited persistence. Biological agents capable of colonizing pruning wounds may provide better long-term protection than fungicides (John *et al.*, 2004).

*Corresponding authors: lcostadone@ucdavis.edu; wdgubler@ucdavis.edu

Promising results for the biological control of grapevine dieback caused by *Eutypa lata* have been obtained using antagonistic or competitive strains of fungi and bacteria (John *et al.*, 2005a,b; Rolshausen and Gubler, 2005). However, there is little information on the relative effectiveness of epiphytic and/or antagonistic fungi in reducing wound susceptibility to infection by species of Botryosphaeriaceae.

Interest in new biological control agents to prevent the infection of pruning wounds by canker pathogens has increased in recent years, partly due to the ineffectiveness of fungicides (as mentioned above) and partly due to the need for better protection. Applications of microorganisms that are competitive or antagonistic to canker-causing agents, or of compounds produced by these organisms, could provide additional opportunities of managing these canker diseases. The use of epiphytic and endophytic bacteria and fungi has shown promising results for the control of several pathogens of cacao (Mejìa *et al.*, 2008) and *E. lata* (Schmidt *et al.*, 2001). Furthermore, the successful biological control of *E. lata* on grapevines using *Fusarium lateritium*, *Cladosporium herbarum* and *Bacillus subtilis* has been demonstrated under field conditions (Ferreira *et al.*, 1990; Munkvold and Marois, 1993). In these studies, the same level of protection as benomyl was demonstrated in controlling infection by *E. lata*.

Despite the promising results for the biological control of grapevine dieback caused by *E. lata* that have been achieved using antagonistic strains of fungi and bacteria, there is little information on the relative effectiveness of epiphytic and or/antagonistic fungi in reducing wound susceptibility to infection by Botryosphaeriaceae. Preliminary *in vitro* results show that epiphytic microorganisms restrict the growth of pathogens causing grapevine canker disease, highlighting their status as a new source of biological control agents. This research should provide alternative control methods for disease control strategies for both organic and conventional vineyards. Therefore, the objectives of this study were: (i) to isolate and identify microorganisms that occur naturally on grapevine pruning wounds; and (ii) to evaluate *in vitro* and *in planta* the ability of these microorganisms to prevent infection by grapevine canker pathogens.

Materials and Methods

Isolation and identification of potential biocontrol organisms

Grapevine-associated microorganisms were recovered by placing Petri dishes containing 4% potato dextrose agar (PDA) amended with tetracycline (0.01%) directly in contact with freshly pruned healthy grapevine spurs obtained from various vineyards in California. Plates were then incubated at 22°C until colonies appeared. Individual colonies were transferred to fresh 4% PDA plates. Isolates were also collected from pruning wounds randomly collected on 1-year-old canes of *Vitis vinifera* L. 'Red globe' in Davis, California. The epiphytic microorganisms were isolated directly from the wound surface of each spur, and the endophytic microorganisms from the wood about 1 cm below the wound surface. The potential antagonistic activity of the fungal and bacterial epiphytes was first screened *in vitro* by means of dual culture assays against members of the Botryosphaeriaceae and also against *E. lata*. Among all the microorganisms tested, the isolates that showed a strong inhibitory effect against both Botryosphaeriaceae isolates and *E. lata* were selected for further evaluations.

The most effective isolates were identified using both morphological characteristics and molecular techniques.

Dual culture assay

The antagonistic potential of the isolated bacterial and fungal species against species of the family Botryosphaeriaceae and against *E. lata* was evaluated *in vitro* on artificial media using the dual inoculation technique mentioned above (Dass and Teyegaga, 1996). Pure cultures of each fungal pathogen and of the bacterial or fungal isolates were grown together on the same PDA Petri dish. For the bacterial isolates, a 3-mm diameter mycelial plug from a 10-day-old culture of a member of the Botryosphaeriaceae or *E. lata* was placed at the centre of the Petri plate and a bacterial streak was made along the periphery of the PDA plates. For the fungal isolates, a 3-mm diameter mycelial plug of the potential antagonist was placed 4 cm from a mycelial plug of a member of the Botryosphaeriaceae or *E. lata* on a PDA plate. The interaction between the possible antagonists and the mycelium of the member of the Botryosphaeriaceae or *E. lata* was assessed by measuring the inhibition of the radial growth of the pathogen towards the possible antagonist. The plates were incubated at room temperature (25 ± 2°C) and the radial growth of the mycelium was measured daily for evidence of growth inhibition. Furthermore, random plating of PDA plugs taken from the zone of interaction between the antagonistic organism and the fungal pathogen was also done to assess the viability of the pathogen. The experiment was done twice with five replicates using a completely randomized design.

Production of volatile antibiotics

The production of antifungal volatile compounds was assessed using sandwich agar plates. Both the lid and the bottom of Petri dishes were filled with PDA. Antagonistic bacterial streaks or fungal mycelium plugs were allowed to colonize the PDA surface of the Petri plate for 48 h. Mycelial plugs from 10-day-old cultures of members of the Botryosphaeriaceae or *E. lata* were placed at the centre of the bottom of the Petri dish. The experiment was conducted with five replications and repeated three times. The plates were incubated at room temperature (25 ± 2°C). Radial mycelial growth of the Botryosphaeriaceae or of *E. lata* was recorded daily in order to assess the percentage inhibition of growth of the fungal pathogens in the presence/absence of the possible antagonistic isolate.

Spore germination test

An experiment was conducted to determine the effect of antagonistic bacteria on spore germination. A spore suspension (2×10^5, 100 µl) of a member of the Botryosphaeriaceae was prepared and combined with a bacterial suspension (10^7 colony forming units (cfu) ml^{-1}, 100 µl) in a 150-ml sterile tube and incubated at room temperature (25 ± 2°C). For the control plate, sterilized water was added instead of the bacterial suspension. After an incubation time of 30 min, 100 µl of the mixture was spread over the PDA plates with a sterile bent glass rod. The plates were then incubated for 24 h at 25°C. Spore germination was assessed with a compound microscope by observing 50 randomly selected conidia. A spore was considered germinated when the germ tube length was half of the length of the spore. The experiment was repeated twice with five replicates.

Inoculation of autoclaved cane segments

Pathogen–antagonist interactions were also evaluated following the protocol described by Rolshausen and Gubler (2005). French square bottles (250 ml) (Fisher Scientific, Pittsburgh, Pennsylvania) filled with 20 ml PDA were autoclaved and laid flat on their sides for cooling. Members of the Botryosphaeriaceae or *E. lata* were grown on PDA Petri plates for 10 days before the experiment. A 4-mm diameter mycelial plug obtained from the margin of each culture was placed in the centre of each French square bottle and allowed to grow for 5 days. Blocks of grape wood were dipped for 1 min into a suspension of fungal or bacterial spores or a cell suspension of the test antagonist isolates and of *Cladosporium* sp. (a known fungal biocontrol agent). An experimental paste – now commercialized as VitiSeal™ (which contains US National Organic Program (NOP) ingredients), other experimental products (unnamed bacteria) and cinnamon oil were also tested on the wood blocks. Non-treated control wood blocks were dipped into water only. Standard fungicides such as Rally (myclobutanil, a triazole fungicide), Topsin M (thiophanate methyl) and Mettle (tetraconazole) were used as control reference chemical dips. The wood blocks were then placed on glass sterile bent glass rods resting on the agar in the bottles. The bottles were incubated for 4 weeks at room temperature (20 ± 2°C) to allow the fungi to colonize the agar. The fungicidal activity of the biocontrol isolates was also evaluated by allowing the test antagonist microorganisms to colonize the wood chips for 7 days before placing them in contact with the fungal pathogen. For each fungal isolate tested, each treatment was replicated three times. Treatments were evaluated as the percentage of wood block surface colonized by the fungal pathogen. After 60 days, the blocks were removed from the bottles and re-isolation of the fungal pathogens from the inner wood was done by plating segments of the inner wood on to PDA.

Field studies

Antagonistic organisms selected based on the results obtained *in vitro* were also evaluated under field conditions. Experiments were conducted in Cabernet Sauvignon and Chardonnay grapevines located in Napa and Lodi, California, respectively. The grapevines had been trained to bilateral cordons with typically 10–12 spurs per cordon. Individual spur positions served as replications. Five spurs were used per treatment. Spur positions for each grapevine were pruned to three buds and the treatments were sprayed to runoff or painted immediately after pruning. Different potential biocontrol agents were tested; these included Biocontrol I (Bp1), Bp1 culture filtrate and *Cladosporium* sp. The treatments also included inoculated (non-treated) controls, experimental products, B-lock (boric acid and latex paint), Rally, Topsin M and a combination of Rally and Topsin M as standard chemical control methods. After the treatments had dried, pruning wounds were artificially inoculated with a spore suspension of a member of the Botryosphaeriaceae or *E. lata*.

Members of the Botryosphaeriaceae isolates for inoculation were grown in PDA Petri plates for 10 days. Spore suspensions of each were prepared by pouring sterile water onto each plate and scraping the agar surface with a flat-headed spatula to harvest the pycnidial spores. Water containing the conidia was filtered through four layers of cheesecloth in order to remove mycelial and agar fragments. Using a haematocytometer, spore suspensions were adjusted to a final density of approximately 1×10^5 conidia ml^{-1}. A spore suspension of *E. lata* was prepared for inoculation by directly collecting ascospores from discharging perithecia that were found in a vineyard located at the University of California Davis Campus. The concentration was adjusted to approximately 1×10^5 ascospores. The inoculation of spur wounds was done with 100 μl of the spore suspensions directly on to pruning wounds at day 0 and then at 7-day intervals for 4 weeks following the treatments with antagonistic organisms. Treated spurs were allowed to stand for 6 months and were then harvested to determine the rate of infection. Untreated controls were reserved to evaluate the natural level of infection. The field trial was established in a randomized block design.

Results

Fifty isolates of fungi and bacteria have been collected from grapevines of different ages and cultivars growing in California. They were all recovered from pruning wounds and were identified based on morphological characteristics and PCR. The preliminary screening for antagonistic bacterial isolates was done on the basis of radial growth inhibition zones and the inhibition of mycelial growth and spore germination of the test fungi, members of the Botryosphaeriaceae and *E. lata*. Among the bacterial strains tested *in vitro*, one isolate (Biocontrol I, also designated Bp1) showed strong inhibitory effects. The mycelial growth of both *E. lata* and members of the Botryosphaeriaceae was significantly reduced ($P < 0.05$) by isolate Biocontrol I after 7 days of incubation, when the inhibition of *E. lata* mycelial growth was 81%. PDA plugs from the interaction zone were re-cultured on fresh PDA and significantly reduced mycelial growth occurred. There was no sporulation of the fungal pathogen. Microscopic observation of the hyphae of *Lasiodiplodia theobromae* (a member of the Botryosphaeriaceae) colonized by Biocontrol I (Fig. 13.1B) showed malformation and swellings. The hyphal tips of the fungus from the interaction zone appeared malformed and the hyphae were thickened and had more vacuoles compared with the control. In contrast, the hyphal walls from the control were smooth with no swellings or vacuolation (Fig. 13.1A). Encouraging results were obtained by evaluating the activity of diffusible antifungal substances in sandwich PDA plates prepared by filling both the lid and the bottom of Petri plates with PDA. Biocontrol I was found to produce both diffusible and volatile antibiotics, which showed conspicuous differences from the *L. theobromae* control in mycelial growth inhibition.

The preliminary results of an *in vitro* evaluation of suspensions of two of the isolates used to treat grapevine wood blocks placed into French square bottles in which members of the Botryosphaeriaceae had been grown are displayed in Figs 13.2 and 13.3. The first rating was conducted 7 days after the insertion of the treated wood blocks into the bottles. The treatments shown are fungicide and/or the isolate suspensions,

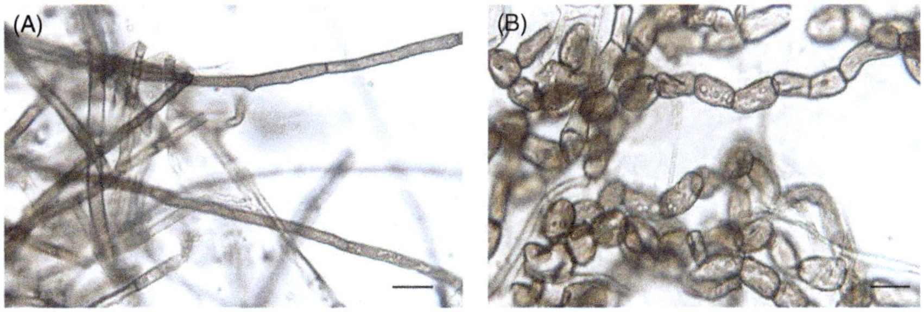

Fig. 13.1. (A) Normal hyphal morphology of *Lasiodiplodia theobromae* (A). (B) Thickened and swelled hyphae as affected by the Bp1 isolate from grapevine pruning wounds. Scale bars = 10 μm.

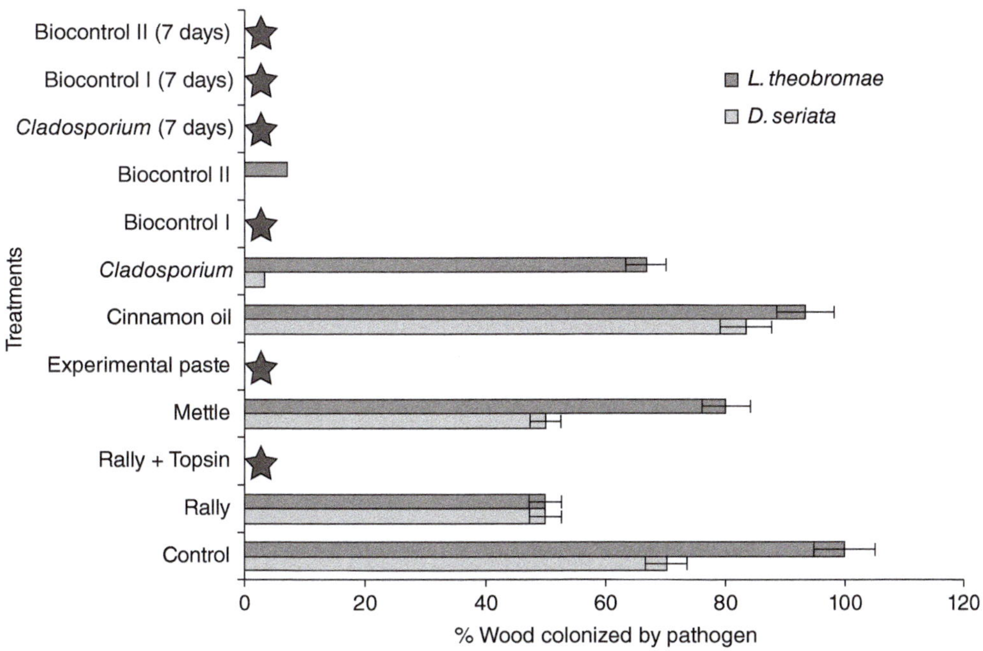

Fig. 13.2. *In vitro* evaluation of treatments to prevent the Botryosphaeriaceae *Lasiodiplodia theobromae* and *Diplodia seriata* (grapevine canker pathogens) from colonizing grapevine wood blocks over a period of 7 days. Treatment efficacy was determined based on the percentage recovery of pathogens from the wood blocks. Biocontrols I and II are isolates from fresh grapevine pruning wounds applied to grapevine wood blocks immediately before the assay; in Biocontrols I and II (7 days), the isolates were applied 7 days before the assay. The experimental paste is an organic preparation now commercialized as VitiSeal™.
★ Indicates 0% wood colonization.

along with various other biocontrol preparations (Fig. 13.2). At the end of 7 days, the isolates Biocontrol I and II and *Cladosporium* sp. showed the same amount of wood protection as an experimental paste and the fungicide mixture Rally + Topsin M used as a standard chemical control mixture. Subsequent ratings were made at further

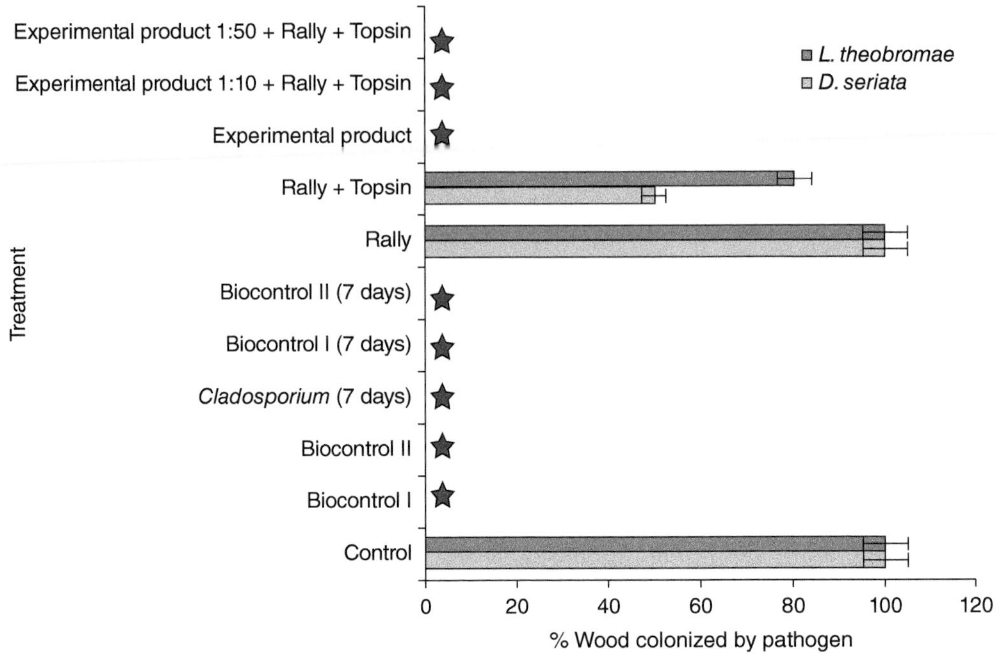

Fig. 13.3. *In vitro* evaluation of treatments to prevent the Botryosphaeriaceae *Lasiodiplodia theobromae* and *Diplodia seriata* (grapevine canker pathogens) from colonizing grapevine wood blocks over a period of 60 days. Treatment efficacy was determined based on the percentage recovery of pathogens from the wood blocks. Biocontrols I and II are isolates from fresh grapevine pruning wounds applied to grapevine wood blocks immediately before the assay; in Biocontrols I and II (7 days), the isolates were applied 7 days before the assay; in *Cladosporium* (7 days) the biocontrol fungus was also applied 7 days before the assay. The experimental products were unnamed bacteria. ★ Indicates 0% wood colonization.

7-day intervals, and overall, the results showed that fungal growth on wood blocks was prevented by Biocontrol I, Biocontrol II and an experimental product mixture with Rally and Topsin (Fig. 13.3). After 60 days of incubation, the pathogen was not recovered from wood blocks when Biocontrol I (Bp1), Biocontrol II and *Cladosporium* had been allowed to colonize the wood surface for 7 days prior to their insertion into the inoculated bottles (Fig. 13.3).

The most effective biocontrol isolates and experimental products, based on the *in vitro* laboratory observations were then evaluated under field conditions.

Natural infection and artificial inoculation of control grapevines by members of the Botryosphaeriaceae and *E. lata* was 100% successful for the field experiments conducted in Lodi (Figs 13.4 and 13.5). When pruning wounds were treated with the Biocontrol I bacterial suspension, the percentage of successful (fungal) pathogen recovery was 33% when inoculation was done immediately after treatment. Based on the percentage of pathogen re-isolation, the culture filtrate and Biocontrol I bacterial suspension provided 100% protection when the inoculation events occurred 3 weeks after they were applied (Figs 13.4 and 13.5).

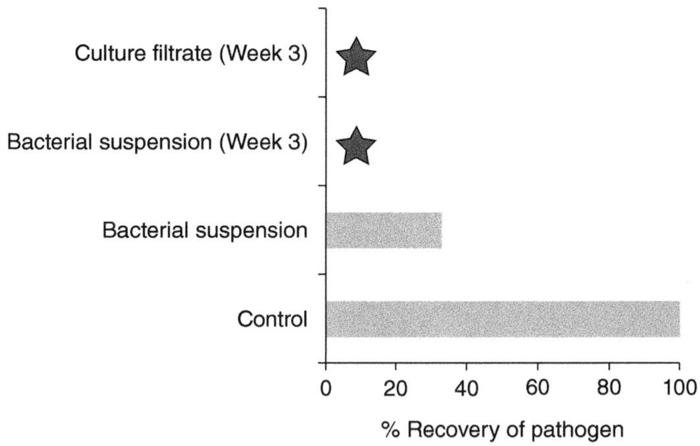

Fig. 13.4. Recovery of *Eutypa lata* from grapevine pruning wounds 6 months after treatment with a suspension of the bacterial isolate Biocontrol I (see text) or its culture filtrate, followed by inoculation with *E. lata*. Week 3 indicates that inoculation with *E. lata* was done 3 weeks after treatment with the bacterial suspension/culture filtrate. Field study at Lodi, California. ★ Indicates 0% pathogen recovery.

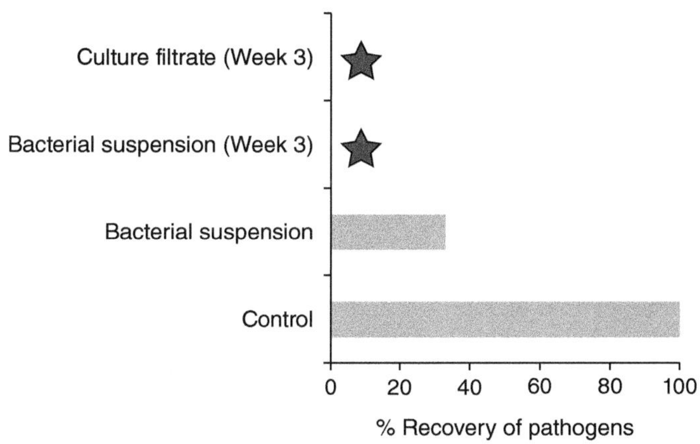

Fig. 13.5. Recovery of members of the Botryosphaeriaceae from grapevine pruning wounds 6 months after treatment with a suspension of the bacterial isolate Biocontrol I (see text) or its culture filtrate, followed by inoculation with members of the Botryosphaeriaceae. Week 3 indicates that inoculation was done 3 weeks after treatment with the bacterial suspension/filtrate. Field study at Lodi, California. ★ Indicates 0% pathogen recovery.

Natural and artificial inoculation of control grapevines by *E. lata* and members of the Botryosphaeriaceae was lower in the field trials conducted in Napa than in those in Lodi, averaging around 80% (Figs 13.6 and 13.7), and no pathogens were recovered from the pruning wounds treated with the Biocontrol I bacterial suspension. Visual observations also indicated that there was no phytotoxicity on the buds and leaves treated with the Biocontrol I suspension.

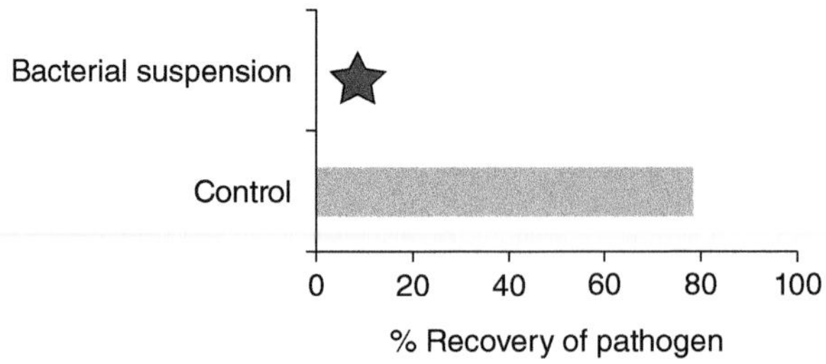

Fig. 13.6. Recovery of *Eutypa lata* from grapevine pruning wounds 6 months after treatment with a suspension of the bacterial isolate Biocontrol I (see text), followed by inoculation with *E. lata*. Field study at Napa, California. ★ Indicates 0% pathogen recovery.

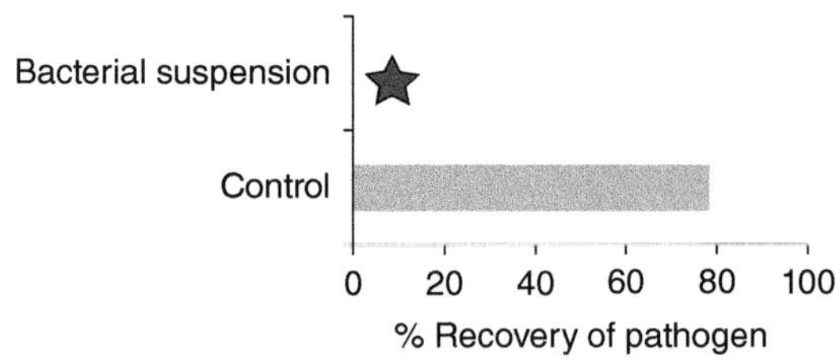

Figure 13.7. Recovery of members of the Botryosphaeriaceae from grapevine pruning wounds 6 months after treatment with a suspension of the bacterial isolate Biocontrol I (see text), followed by inoculation with members of the Botryosphaeriaceae. Field study at Napa, California. ★ Indicates 0% pathogen recovery.

Discussion

The results of this study showed the potential of using the isolate Biocontrol I as an antagonistic organism to control grapevine canker diseases. They indicate that isolate Biocontrol I produced antibiotic substances *in vitro* that inhibited mycelial growth and spore germination of some Botryosphaeriaceae and of *E. lata*. The observed hyphal malformations were probably caused by the toxic effect of antifungal compounds that interfere with the normal mycelial growth. However, further investigations are being conducted in order to clarify the nature of these compounds.

Based on the *in vitro* screening results, the Biocontrol I isolate offered optimal protection against some members of the Botryosphaeriaceae when it was able to completely colonize the surface of grapevine wood blocks before the wood made contact with the pathogen. In this assay, treatment with the Biocontrol I and II isolates and *Cladosporium* conferred the same level of protection as the fungicide mixture Rally + Topsin. Furthermore, results obtained 60 days after treatment indicated that Biocontrols I and II and *Cladosporium* provided 100% protection of the wood against the fungal colonization and were able to function on the wood surface for longer than the fungicide mixture used as a reference treatment.

Preliminary field results also confirmed the potential of Biocontrol I to be used as an effective biocontrol agent against grapevine canker pathogens.

References

Dass, C. and Teyegaga, A. (1996) Growth suppression of some wood-decay and other fungi by *Bacillus subtilis*. *Australian Journal of Botany* 44, 705–712.

Ferreira, J.H.S., Matthee, F.N. and Thomas, A.C. (1990) Biological control of *Eutypa lata* in grapevine by an antagonistic strain of *Bacillus subtilis*. *Phytopathology* 81, 283–287.

John, S., Scott, E.S., Wicks, T.J. and Hunt, J.S. (2004) Interactions between *E. lata* and *Trichoderma harzianum*. *Phytopathologia Mediterranea* 43, 95–104.

John, S., Wicks, T.J., Hunt, J.S., Lorimer, M.F., Oakey, H. and Scott, E.S. (2005a) Protection of grapevine pruning wounds from infection by *Eutypa lata* using *Trichoderma harzianum* and *Fusarium lateritium*. *Australasian Plant Pathology* 34, 569–575.

John, S., Scott, E.S., Wicks, T.J. and Hunt, J.S. (2005b) Potential of *Trichoderma harzianum* and *Fusarium lateritium* to protect grapevines from *Eutypa* dieback. *Australian and New Zealand Grapegrower and Winemaker* No. 497 (Extra, Annual Technical Issue), 18–21.

Mejìa, L.C., Rojas, E.I., Maynard, Z., Van Bael, S., Arnold, A.E., Hebbar, P., Samuels, G.J., Robbins, E. and Herre, E.A. (2008) Endophytic fungi as biocontrol agents of *Theobroma cacao* pathogens. *Biological Control* 46, 4–11.

Munkvold, G.P. and Marois, J.J. (1993) Efficacy of natural epiphytes and colonizers of grapevine pruning wounds for biological control of *Eutypa* dieback. *Phytopathology* 83, 624–629.

Rolshausen, P.E. and Gubler, W.D. (2005) Use of boron for the control of *Eutypa* dieback of grapevine. *Plant Disease* 89, 734–738.

Schmidt, C.S., Lorenz, D. and Wolf, G.A. (2001) Biological control of the grapevine dieback fungus *Eutypa lata* I: screening of bacterial antagonists. *Journal of Phytopathology* 149, 427–435.

Siebert, J.B. (2001) *Eutypa*: the economic toll on vineyards. *Wines and Vines* April 2001, 50–56.

Úrbez-Torres, J.R. and Gubler, W.D. (2009) Pathogenicity of *Botryosphaeriaceae* species isolated from grapevine cankers in California. *Plant Disease* 93, 584–592.

Úrbez-Torres, J.R. and Gubler, W.D. (2010) Susceptibility of grapevine pruning wounds to infection by *Lasiodiplodia theobromae* and *Neofusicoccum parvum*. *Plant Pathology* 69, 261–270.

Úrbez-Torres, J.R., Leavitt, G.M., Voegel, T. and Gubler, W.D. (2006) Identification and distribution of *Botryosphaeria* species associated with grapevine cankers in California. *Plant Disease* 90, 1490–1503.

USDA National Agricultural Statistics Service (2015) *California Agricultural Statistics: 2013 Annual Bulletin*. Pacific Regional Field Office, Sacramento, California. Available at: http://www.nass.usda.gov/Statistics_by_State/California/Publications/California_Ag_Statistics/Reports/2013cas-all.pdf (accessed 28 September 2015).

Van Niekerk, J.M., Crous, P.W., Groenewald, J.Z., Fourie, P.H. and Halleen, F. (2004) DNA phylogeny, morphology and pathogenicity of *Botryosphaeria* species on grapevines. *Mycologia* 96, 781–798.

14 Esquive® WP Limits Development of Grapevine Trunk Diseases and Safeguards the Production Potential of Vineyards

E. Mounier,[1*] F. Boulisset,[1] F. Cortes,[1] M. Cadiou,[1] P. Dubournet[2] and E. Pajot[1*]

[1]*Agrauxine, Groupe Lesaffre, Beaucouze, France;* [2]*Bayer SAS, Lyon, France*

Introduction

Grapevine trunk diseases (GTD) such as *Eutypa* dieback, esca decline of vines and black dead arm (BDA), which are widely believed to be caused by multiple fungal pathogens acting in concert, although the causes may well be more complex, cause great damage to vineyards. Since the banning of two key fungicides (sodium arsenite and carbendazim/flusilazole) in 2001, no chemical products have been available to growers to reduce GTD. However, the vine industry is today facing a significant resurgence of GTD (see Viguès *et al.*, 2007). According to a study carried out by the Observatoire National des Maladies du Bois (French National Grapevine Trunk Disease Observatory) in 18 French vine-growing regions and 29 varieties, the rates of GTD increased from 1.6% in 2003 to 4.79% in 2007. The study has also demonstrated the susceptibility of different varieties to esca/GTD and to *Eutypa* dieback, as well as a peak of prevalence in the group of plants aged 12–18 years (Doublet, 2013). A situation report in 2012 gave similar findings in that: (i) for *Eutypa* dieback, an increase in prevalence was noted in Trebbiano (Ugni Blanc) plots in the Charente region, with disease expression rates reaching 15%, although there was a slightly lesser increase in Chenin and Cabernet Franc varieties in the Loire Valley; (ii) for esca/GTD, almost half of the observed varieties showed prevalence rates nearing or exceeding 5%, and losses of productive surface area in vineyards (missing, dead or replanted areas) varied from 4 to 20% depending on the location of the vineyards, with an average of 13% (Doublet, 2013).

In view of the spread of GTD in France and in wider Europe, research teams have been working hard to increase knowledge of the pathogens responsible for GTD in plants. The fight against esca and BDA can only be undertaken once the modes of development and growth of the microorganisms that are responsible in the vine wood,

*Corresponding authors: emmanuelle.mounier@agrauxine.fr; emmanuel.pajot@agrauxine.fr

and the mechanisms of dissemination of the diseases in vineyards, are fully understood. Lecomte *et al.* (2011) stated that disease expression is the result of a negative combination of abiotic and biotic factors. Results from the University of Bordeaux have shown that esca decline of vines is characterized by a series of three types of symptoms: internal necrosis of the wood, and vascular disorders and foliar symptoms (caused by the necrosis) that vary widely in appearance and gravity (Lecomte *et al.*, 2008). For GTD, the periods of greatest sensitivity of plants to *Botryophaeria* spp. have been identified by Larignon *et al.* (2013). An improved understanding of GTD will facilitate decisions on the management of these diseases and when to use various approaches to control in the future.

In the absence of any conventional (chemical) treatment for GTD, preventive measures must be implemented as a first line of defence in order to limit the spread of these diseases and curtail their consequences. While indispensable, such measures alone will be insufficient as the goal is also to provide an acceptable level of protection in the vineyards, so the development of environmentally friendly, safe and sustainably efficient means of control is essential. In an attempt to make such products available, the French company Agrauxine has developed a biocontrol product called Esquive® WP based on the fungus *Trichoderma atroviride* I-1237. This product is presently the only anti-GTD product that is registered for use in France against *Eutypa* dieback and esca/BDA.

Because of its numerous modes of action and its proven efficacy against GTD-causing pathogens, as well as its ability to grow and multiply rapidly, *T. atroviride* strain I-1237 is of particular interest as a biocontrol agent. It has several known modes of action: it acts as a competitor for space and nutrients, which allows it to rapidly colonize pruning wounds; it secretes volatile and non-volatile compounds that inhibit pathogen growth; and it acts by mycoparasitizing and destroying the disease-causing pathogens.

Laboratory studies by several different national and international research institutes have focused on *Trichoderma* spp. and/or on strain I-1237 in order to better characterize and understand its mode of action against GTD. A study in 2005 showed that *T. atroviride* consumes toxins secreted by *Eutypa lata* and *Phaeoacremonium aleophilum*, two of the fungi that have been implicated in causing *Eutypa* dieback and esca decline, respectively (Christen *et al.*, 2005). These toxins are known to disturb photosynthesis, among other effects (Philippe *et al.*, 1992; Deswarte *et al*, 1996), and also to cause cavitation of plant vessels, which can lead to high levels of hydric stress, thus indirectly causing appearance of the typical symptoms of GTD. Laboratory tests have also shown that strain I-1237 can grow and spread 30 to 95% faster than the main GTD pathogens. Moreover, confrontation tests between strain I-1237 and the main GTD pathogens have shown that *T. atroviride* I-1237 is highly antagonistic to *Botryosphaeria* spp., *P. aleophilum*, *E. lata* and *Phaeomoniella chlamydospora*. Finally, research at the University of California at Davis has shown that *T. atroviride* I-1237 is highly antagonistic to *E. lata* and *Botryosphaeria*. Indeed, *in vitro* tests have shown that it prevents the penetration of pathogenic fungi into shoot cross-sections. Together, these studies have demonstrated conclusively that strain I-1237 is promising for the control of GTD.

Initial results from field trials of Esquive® WP conducted by various French and international research institutes appear to validate the reports from the research summarized above that indicate the promise of *T. atroviride* I-1237. In this chapter, we present data from two of these trials.

Materials and Methods

Trial sites

Two distinct trial networks in the field (i.e. in vineyards) were used for this study on the rating of the efficacy of Esquive® WP. The first network (Trial Network 1), consisting of 21 sites (but one did not yield results), was set up in 2007 (Fig. 14.1A). The second network (Trial Network 2) was established in 2012 with 17 sites (Fig. 14.1B).

The test sites are located in all of the main French vine-growing regions. Moreover, a wide range of varieties, rootstocks, ages, densities and locations of vine plants was used (see Tables 14.1 and 14.2). Researchers of the company Agrauxine, in partnership with French GTD specialists from L'Institut Français de la Vigne et du Vin (IFV), Institut National de la Recherche Agronomique (INRA) and Direction Générale de l'Alimentation (DGAL), have set up, monitored and analysed the results.

The method used in 2011 for the trials had several modifications of the protocol used in 2010 (Mounier *et al.*, 2011): (i) an increase in the average numbers of subjects per group, which, on each trial site, rose from 266 in 2010 to 784 in 2011; (ii) the use of plant by plant mapping; and (iii) methods of statistical analysis in line with the characteristics of the new protocol.

The location sites had already been used in correlation studies of the sensitivity of vines to *Eutypa* dieback, esca decline and BDA (Grosman, 2008). All of the plants on any particular site were of the same variety, had the same rootstock, the same clone and the same geographic origin. Soils and other agricultural characteristics were identical for the Esquive® WP-treated group and the control group. All grower activity (e.g. treatment applications, grass sowing) was noted.

Treatment

Treatment was carried out during the winter, when the vine was dormant. Esquive® WP was applied during the 2 weeks following vine pruning with a sprayer or with the equipment commonly used by the vine-grower (field sprayer, trailer sprayer, and so on). The product was also applied at the registered dose (4 kg ha^{-1}), mixed with water. Formulations were sprayed on to all plant surfaces, with special attention to pruning wounds, until a good soak was achieved. The control plot received no treatment.

Assessments

Esca/BDA ratings on test sites of both of the trial networks were conducted during August and September, when symptoms were most clearly observable. The analysis of plant health was undertaken by visual observation. Healthy, replanted, dead and missing plants, as well as those showing foliar symptoms, were precisely noted and plotted on the site map. For those plants showing symptoms of GTD, three severity criteria were applied: foliar symptoms, partial red–yellowish pigmentation of leaves (one branch showing symptoms), total red–yellowish pigmentation

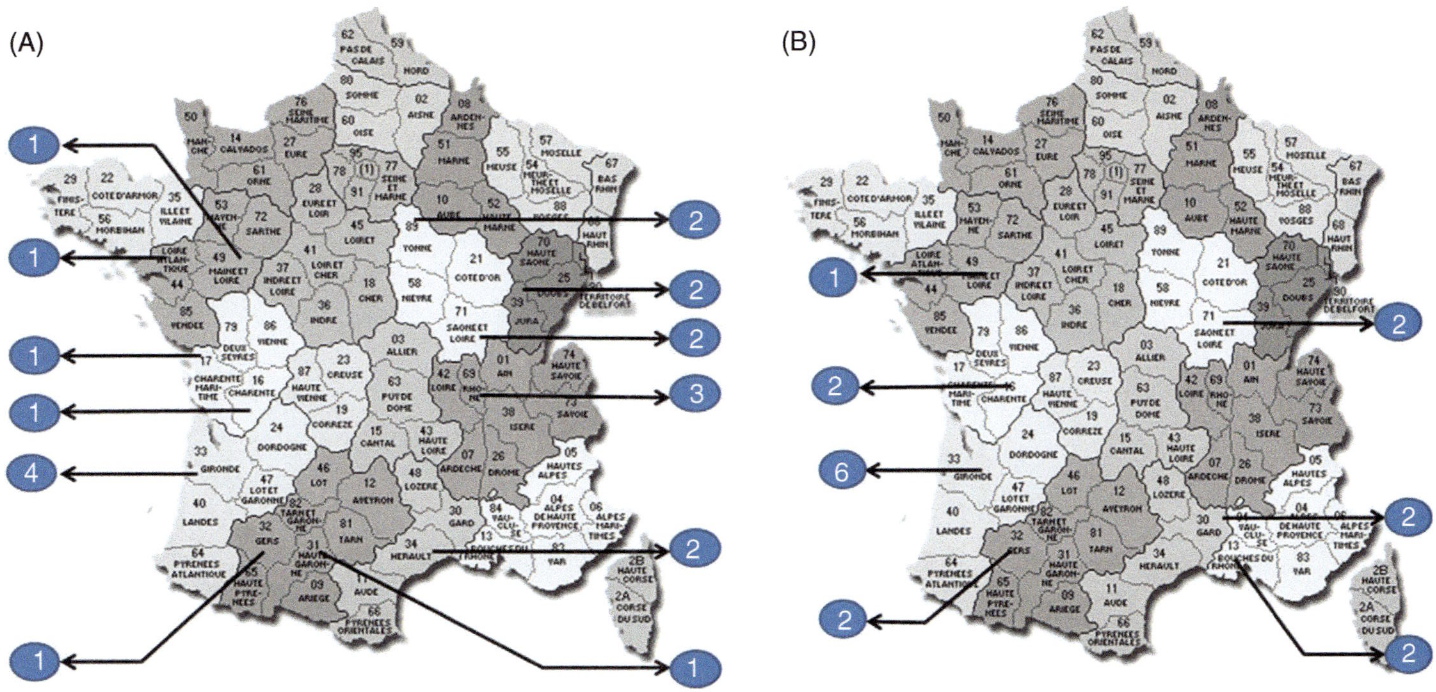

Fig. 14.1. Map showing the distribution of (A) Trial Network 1 and (B) Trial Network 2 experimental test sites in France. Blue numbers are the numbers of trials per region. Note that six of the sites in Trial Network 2 were not included in the analyses that are reported elsewhere as Esquive® WP had not been properly applied.

Table 14.1. Agricultural characteristics of Trial Network 1, set up in 2007 with partners.

Partners[a]	French Département	Variety	Density (plants ha^{-1})	Vine age at start of trials	Trial start date
AGX	34	Sauvignon Blanc	4000	13	2013
IFV	31	Cabernet Sauvignon	4500	10	2010
CA Jura	39	Trousseau	5000	6	2012
CA Charente Maritime	17	Ugni Blanc	4330	2	2011
CA Jura	39	Savagnin	5000	4	2012
CA Charente-Maritime[b]	16	Sauvignon Blanc	3200	2	2011
IFV	33	Cabernet Franc	5556	18	2011
AGX	34	Sauvignon Blanc	4000	12	2013
AGX	71	Chardonnay	8000	7	2010
AGX	69	Chardonnay	7800	11	2010
AGX	89	Chardonnay	6250	22	2009
AGX	69	Gamay	8330	9	2010
AGX	33	Cabernet Sauvignon	7200	11	2009
AGX	69	Gamay	9600	8	2010
CA Nantes	44	Melon de Bourgogne	6500	9	2010
AGX	33	Cabernet Sauvignon	7200	12	2009
AGX	89	Chardonnay	6250	28	2009
AGX	49	Sauvignon	5405	18	2011
IFV	33	Cabernet Franc	6500	15	2007
IFV	32	Sauvignon	3600	3	2010
AGX	71	Chardonnay	8000	17	2010

[a]AGX, Agrauxine; CA Charente-Maritime, La Chambre d'Agriculture de Charente-Maritime; CA Jura, La Chambre d'Agriculture de Jura; CA Nantes, La Chambre d'Agriculture de Nantes; IFV, L'Institut Français de la Vigne et du Vin.
[b]No results available.

Table 14.2. Agricultural characteristics of Trial Network 2, set up in 2012.

French Département	Variety	Density (plants ha^{-1})	Vine age at start of trials
71	Chardonnay	8500	25
71	Chardonnay	8500	25
49	Cabernet Sauvignon	5000	16
16	Ugni Blanc	2604	18
16	Ugni Blanc	2604	8
33	Sauvignon Blanc	5722	14
33	Sauvignon Blanc	5722	4
33	Cabernet Sauvignon	6666	10–11
33	Cabernet Sauvignon	6666	4
33	Cabernet Franc	10,162	3
33	Cabernet Franc	6666	11

Fig. 14.2. Esca/BDA symptoms in the vineyard. (A) Partial interveinal red–yellowish pigmentation of leaves; (B) total interveinal red–yellowish pigmentation of leaves; and (C) the apoplectic form of the disease. From Bayer and Agrauxine.

of leaves (all branches showing symptoms) and the (acute) apoplectic form of the disease, in which one or several arms or the whole plant may wilt, with leaf drop, shrivelling and (sometimes) the drying and, finally, death of the vine (Fig. 14.2). On the sites of both Trial Network 1 and Trial Network 2, this plant-by-plant mapping process allowed researchers to monitor plant health and productivity year by year for each site, and to monitor the year-to-year expression of symptoms on each plant.

Statistical analysis

Within each trial, statistical analysis was done using the chi squared (χ^2)-test. To determine the efficacy of Esquive® WP, three degrees of probability were used: (*) $P < 0.1$; (**) $P < 0.05$; (***) $P < 0.01$.

Results

Esquive® WP reduces foliar symptoms of esca/BDA

The efficacy of Esquive® WP against esca and BDA was first evaluated in terms of the three severity criteria of the disease complex (partial or total red–yellowish pigmentation of leaves and the apoplectic form of the disease) as described before. Figures 14.3 and 14.4 present, for each site in both trial networks, the percentage of plants showing the foliar symptoms of esca/BDA in the control group (untreated) and the Esquive® WP-treated group. In each graph (bar plot), the results from the 20 trials from Network Trial 1 and the 11 trials from Network Trial 2 are presented in increasing order of treatment effect.

Out of the 31 trials from which data are shown, eight showed a significant reduction in the number of plants showing foliar symptoms of esca/BDA, and 12 others showed a favourable (but not significant) response to treatment. In the eight trials conducted on young vine plants (less than 8 years of age), no meaningful parasite pressure was observed on the plots.

On those test sites demonstrating efficacy of Esquive® WP against GTD, an average reduction of 30% in the foliar symptoms of esca/BDA was observed.

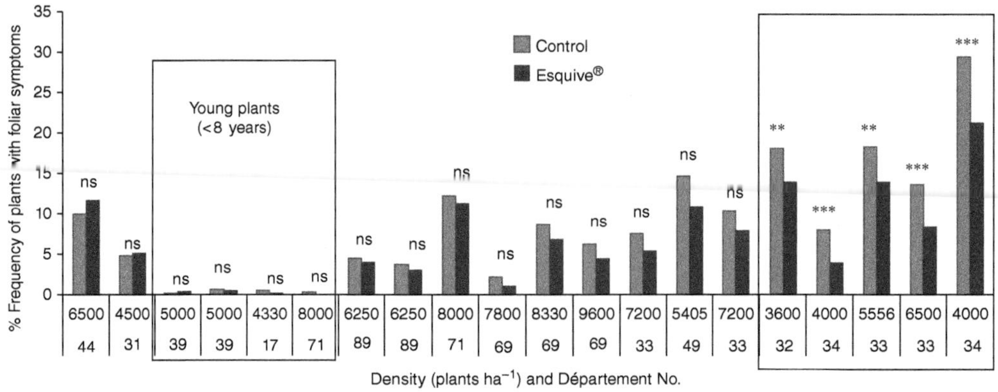

Fig. 14.3. Efficacy of Esquive® WP against esca/BDA in French vineyards in 2014: results from Trial Network 1. A statistical analysis using the χ^2 test gave the results as significant at: $P < 0.1$ (*), $P < 0.05$ (**) or $P < 0.01$ (***), or as not significant (ns). Box to left indicates young vine plants (<8 years old at start of trial in 2010–2012); box to right indicates significant disease reduction in treated vines.

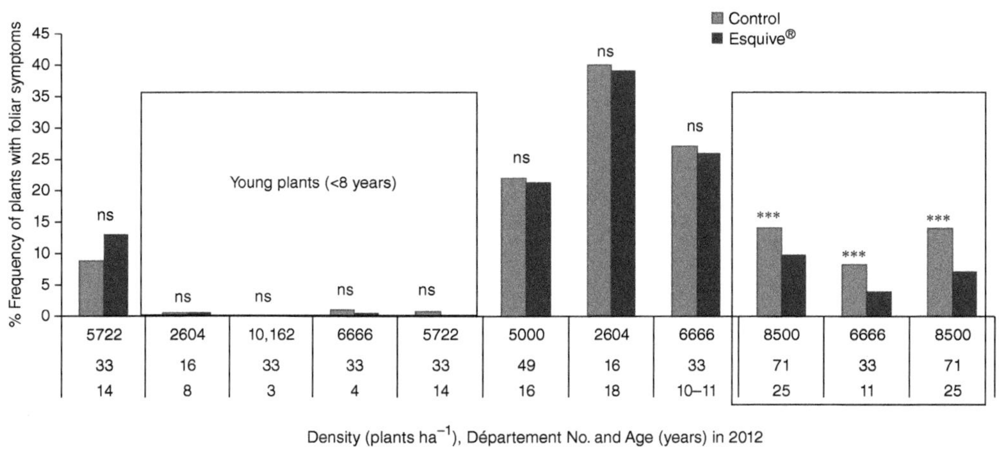

Fig. 14.4. Efficacy of Esquive® WP against esca/BDA in French vineyards in 2014: results from Trial Network 2. A statistical analysis using the χ^2 test gave the results as significant at: $P < 0.1$ (*), $P < 0.05$ (**) and $P < 0.01$ (***), or as not significant (ns). Box to left indicates young vine plants (<8 years old at start of trial in 2012); box to right indicates significant disease reduction in treated vines.

Esquive® WP helps to keep plants producing and more healthy

The effect of Esquive® WP on the overall health of plants was also analysed. An assessment of 'unproductive surface area' in each trial within each Network was made by calculating the totals of replanted, missing and dead plants, and adding these to the numbers of plants showing severe symptoms of *Eutypa* dieback and esca/BDA. Figure 14.5 shows the number of plants still producing as a result of treatment with Esquive® WP in Trial Network 1. These results were obtained by calculating the difference between the number of unproductive plants in the untreated control group and

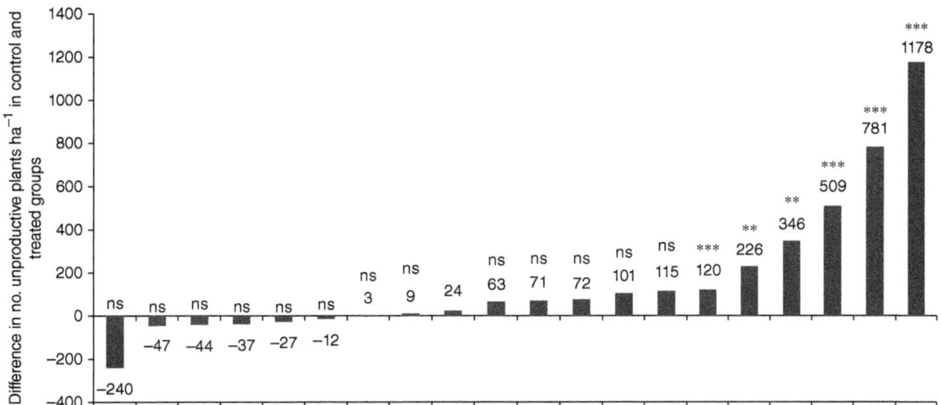

No. of treatments	5	3	5	3	4	4	6	4	3	5	6	6	8	5	3	5	5	6	5	5
Age at planting	7	13	9	6	18	18	11	2	4	10	28	22	15	3	12	8	11	12	17	9
Plant density ha^{-1}	8000	4000	6500	5000	5556	5405	7200	4330	5000	4500	6250	6250	6500	3600	4000	9600	7800	7200	8000	8330
Département No.	71	34	44	39	33	49	33	17	39	31	89	89	33	32	34	69	69	33	71	69

Fig. 14.5. Effect of Esquive® WP treatment on keeping vine plants in production in French vineyards in 2014: results from Trial Network 1. Unproductive plants are total numbers of replanted, missing and dead plants + numbers of plants with severe *Eutypa* dieback and esca/BDA. A statistical analysis using the χ^2 test gave the results as significant at: $P < 0.1$ (*), $P < 0.05$ (**) and $P < 0.01$ (***), or as not significant (ns).

in the Esquive® WP-treated group, expressed per hectare. The 20 trials (sites) are presented on the graph in increasing order of treatment effect. On each and every site, a χ^2 statistical analysis was performed. The number of plants still producing following the application of Esquive® WP was significant for six of the trials.

Esquive® WP kept plants in production in six trials, indicating that it prevents healthy plants, and those presenting symptoms of slow-developing forms of esca/BDA, from moving into apoplexy and, finally, death. It also significantly reduced the levels of infestation in five trials (data not shown). Of 20 trials, ten showed conclusive and beneficial results, while four had no symptoms owing to the young ages of the vine plants. Esquive® WP was effective in 60% of trials. In 2014, an average number of 525 plants ha^{-1} was observed to stay productive in the Esquive® WP-treated group compared with the untreated control group.

Plants suffering from slow-developing forms of esca/BDA in 1 year may, in the following year, either maintain their health status (remaining in a state of partial interveinal red–yellowish leaf pigmentation), lose all symptoms and appear to be fully healthy plants, evolve towards apoplexy, or even die, go missing or be replaced.

Data from Trial Network 2 indicated that the application of *T. atroviride* strain I-1237 (i.e. Esquive® WP) over 2 consecutive years yielded interesting and significant results on the progress of esca/BDA in treated plants. Application of the χ^2 test affords a way to analyse the changes in plants that presented slow-developing forms of esca/BDA in the years 2012 to 2014 when they were treated with Esquive® WP over that period. Just four trials from Trial Network 2 were chosen for this analysis, as in the others, the vine plants were either too young (<8 years old) or showed too few foliar symptoms for a meaningful analysis. The results are presented in Fig. 14.6, which shows the percentage frequency of different stages of development of esca/BDA after

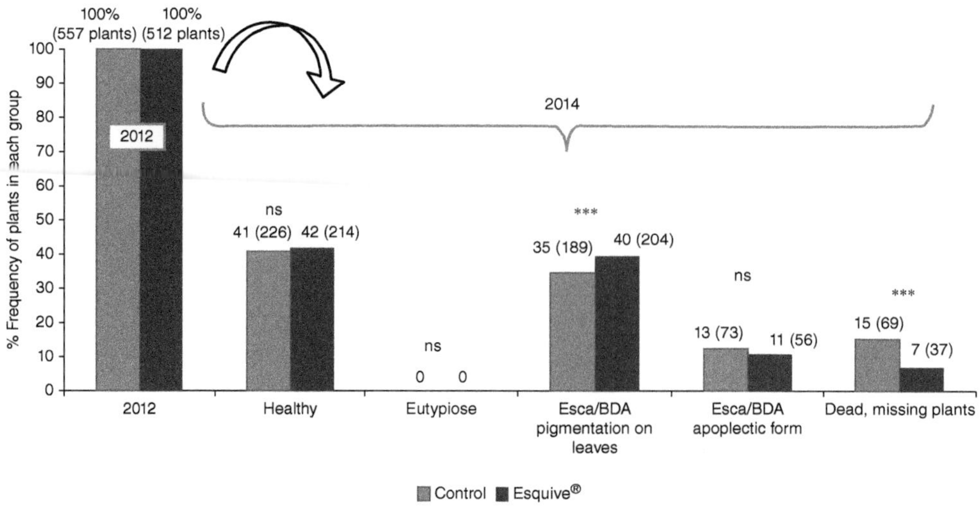

Fig. 14.6. Changes in the status of grapevine plants showing the slow-developing form of esca/BDA from 2012 to 2014 with treatment with Esquive® WP over that period. Control plants were not so treated. Numbers in parentheses indicates the actual numbers of plants in each group. A statistical analysis using the χ^2 test gave the results as significant at: $P < 0.1$ (*), $P < 0.05$ (**) and $P < 0.01$ (***), or as not significant (ns).

2 years of application of Esquive® WP. In order to show the changes of the plants in 2014 better, the number or plants with slow-developing forms (partial interveinal red–yellowish pigmentation of leaf forms) of esca/BDA in 2012 was taken as 100% of the untreated and Esquive® WP-treated group members.

Two significant differences are apparent between the plants:

- More plants presenting slow-developing forms of esca/BDA (forms with partial interveinal red–yellowish pigmentation of the leaves) in 2012 had evolved towards insensitivity (i.e. were dead, missing or replaced) in 2014 in the untreated control group than in the Esquive® WP-treated group.
- More plants in the Esquive® WP-treated group remained as slow-developing forms of Esca/BDA in 2014 than in the untreated control group.

Esquive® WP tends, therefore, to keep plants infested by esca and BDA in their slow-evolving forms by preventing them from dying and thus becoming unproductive in the following years.

In the long run then, the application of Esquive® WP maintains production potential in vineyards. Monitoring of plants over the next few years will allow us to confirm this analysis.

Discussion and Conclusion

The organization of two large-scale Trial Networks has provided significant quantities of data concerning GTD and its development, and the action of Esquive® WP on these diseases. Nearly 40,000 plants were observed and mapped each year in Trial Network

1. As previously reported elsewhere, it appears that the prevalence of symptoms of GTD is low in young vine plants (<8 years old). Overall prevalence rates of *Eutypa* dieback seemed to be extremely low in the network as a whole, while esca/BDA was found at average rates of 8.7% per site, on untreated plots, whatever their location or other characteristics.

The analysis of data gathered by the two trial networks demonstrated that Esquive® WP allows growers to significantly reduce the development of GTD, by an average of 50% a year, whatever the form of esca/BDA, compared with untreated control plots. The great potential of *T. atroviride* I-1237 as a biocontrol agent of a large number of GTD pathogens, already characterized *in vitro* by Agrauxine (Mounier and Dubournet, 2014a,b; Mounier *et al.*, 2014; Cordier and Romand, 2009), has now been confirmed in the vineyard. *T. atroviride* I-1237 effectively infiltrates pruning wounds and thereby limits later contamination or growth of pathogenic fungi in vine plants. According to observations made on trial sites since 2007, the effect of Esquive® WP on apoplectic forms of esca/BDA after 2 years of treatment allows on average 57 plants ha^{-1} to remain productive when compared with untreated controls. This efficacy of Esquive® WP on apoplectic forms of esca/BDA is cumulative each year as the treatment is continued. After 4 years of treatment, 103 plants ha^{-1} (on average) are still producing on treated plots compared with untreated control plots. This may allow growers to avoid having to plant new vines to replace apoplectic vines.

In association with preventive measures, Esquive® WP can therefore provide an opportunity to control GTD in vineyards, opening up new horizons in the management of GTD and in the development of sustainable control programmes.

Agrauxine has relied for many years on renowned GTD specialist laboratories for trials, including INRA Bordeaux, IFV, the University of California at Davis, the University of Lisbon and the University of Florence. In 2014 as well, trials have been conducted on Agrauxine's behalf by these laboratories as well as by new partners in Reims and in Spain. The purpose, as always, is to enhance the knowledge of how *T. atroviride* strain I-1237 works in vine plants. Within the framework of Agrauxine's national trial network, 40,000 plants have been closely monitored year to year. In 2012, a new network of 17 plots was added to that number, thanks to the involvement of Bayer Cropscience.

The use of *T. atroviride* strain I-1237 must be considered today as a necessary first line of defence in the fight against GTD, along with other preventive measures.

Conflict of Interest

The authors declare a conflict of interest as both Agrauxine (and Bayer) are companies.

References

Christen, D., Tharin, M., Perrin-Cherioux, S., Abou-Mansour, E., Tabacchi, R. and Défago, G. (2005) Transformation of Eutypa dieback and Esca disease pathogen toxins by antagonistic fungal strains reveals a second detoxification pathway not present in *Vitis vinifera*. *Journal of Agricultural and Food Chemistry* 53, 7043–7051.

Cordier, C. and Romand, C. (2009) Maladie du bois de la vigne, vers la lutte biologique? *Phytoma* 624/625, 34–38.

Deswarte, C., Eychenne, J., Davy de Virville, J., Roustan, J.P., Moreau, F. and Fallot, J. (1996) Protonophoric activity of eutypine, a toxin from *Eutypa lata*, in plant mitochondria. *Archives of Biochemistry and Biophysics* 334, 200–205.

Doublet, B. (2013) Maladie du bois. Etat des lieux issu des dispositifs d'observation au vignoble. In: Larignon, P. (ed.) *Les Maladies du Bois de la Vigne, 19–20 Mars 2013, Angers: Compte Rendu*. Institut Français de la Vigne et du Vin Angers, Beaucouzé, France, pp. 7–8.

Grosman, J. (2008) Observatoire national des maladies du bois: bilan de 4 années d'observation. In: *Euroviti: Colloque Viticole et Oenologique. Les Maladies de Bois: Observations au Vignoble at Recherches en Cours, Angers – France, 17 Janvier 2008*. ENTAV–ITV, Beaucouse, Angers Expo Congrès, Angers and Institut Français de la Vigne et du Vin, Le Grand du Roi, France. Available at: http://www.techniloire.com/documents/124963587/Colloque%20EUROVITIbis1.pdf (accessed 30 September 2015).

Larignon, P., Muller, M. and Santenac, G. (2013) Les traitements à l'eau chaude et les maladies du bois. In: Larignon, P. (ed.) *Les Maladies du Bois de la Vigne, 19–20 Mars 2013, Angers: Compte Rendu*. Institut Français de la Vigne et du Vin Angers, Beaucouzé, France, p. 26.

Lecomte, P., Darrieutort, G., Liminana, J.M., Louvet, G., Tandonnet, J.P., Guérin-Dubrana, L., Goutouly, J.P., Gaudillère, J.P. and Blancard, D. (2008) Eutypiose et Esca. 1 – Eléments de réflexion pour mieux appréhender ces phénomènes de dépérissement. *Phytoma* 615, 42–48.

Lecomte, P., Darrieutort, G., Laveau, C., Blancard, D., Louvet, G., Goutouly, J.P., Rey, P. and Guérin-Dubrana, L. (2011) Impact of biotic and abiotic factors on the development of Esca decline disease. *IOBC-WPRS Bulletin* 67, 171–180.

Mounier, E. and Dubournet, P. (2014a) Esquive® WP, pour limiter les maladies du bois. *Viticulture en Val de Loire* 07/2014, pp. 10–13.

Mounier, E. and Dubournet, P. (2014b) Maladies du bois de la vigne: résultats d'un outil polyvalent. *Phytoma* 675, 44–47.

Mounier, E., Blal, B. and Pajot, E. (2011) Protéger la vigne contre les maladies du bois: nouvelles précisions sur l'efficacité d'un moyen de lutte biologique, *Trichoderma atroviride*. *Phytoma* 647, 38–41.

Mounier, E., Cortes, F., Cadiou, M., Dubournet, P. and Pajot, E. (2014) Limiter le développement des maladies du bois de la vigne et l'érosion du potentiel productif des parcelles: les recherches les plus récentes autour du *Trichoderma atroviride* I-1237 et intérêt de l'Esquive WP. *Revue des Œnologues* 150, 9–12.

Philippe, I., Fallot, J., Petitprez, M. and Dargent, R. (1992) Effet de l'eutypiose sur les feuilles de *Vitis vinifera* cv. Sauvignon: étude cytologique. *Vitis* 31, 45–53.

Viguès, V., Serrano, E., Dumas, C., Coarer, M., Yobregat, O. and Larignon, P. (2007) Les maladies du bois peuvent commencer en pépinières. Contaminations des plants des vignes par les champignons associés aux maladies du bois, une enquête en Midi-Pyrénées. *Phytoma* 609, 20–23.

15

A Model for Biological Control Studies of Grapevine Trunk Diseases Under Laboratory Conditions

R.J.G. Pierron, J. Pouzoulet, A. Meziane, N. Mailhac and A. Jacques*

Equipe Agrophysiologie et Agromolécules, Département des Sciences Agronomiques et Agroalimentaires, Université de Toulouse, Toulouse, France

Introduction

The aetiology of grapevine trunk diseases (GTD) is poorly understood (Mugnai *et al.*, 1999; Surico, 2001; Larignon *et al.*, 2009; Lecomte *et al.*, 2012; Bertsch *et al.*, 2013). Several fungal species are associated with decay in or discoloration of the wood of grapevine trunks (Mugnai *et al.*, 1999; Lecomte *et al.*, 2012). Nevertheless, esca-associated fungi are also present in healthy wood, independently of the expression of foliar symptoms (Bruez *et al.*, 2014). To date, there is no resistant cultivar, although tolerance has been observed according to the expression of foliar symptoms under field conditions (Grosman and Doublet, 2012). This cultivar tolerance may be related to the dimensions of xylem vessels (Pouzoulet *et al.*, 2014), which seems to condition grapevine–microbe interactions. These interactions are poorly described (Bertsch *et al.*, 2013), especially their molecular aspects.

Phytopathogens such as *Phaeomoniella chlamydospora* (Gams, Crous, Wingfield and Mugnai) are often described as colonizers of the xylem vessels (Del Rio *et al.*, 2001; Surico *et al.*, 2008; Landi *et al.*, 2012). This colonization is associated with the presence of elements that are secreted by fungi, such as naphthalenone pentaketides (Abou-Mansour *et al.*, 2004; Bruno and Sparapano, 2006a; Andolfi *et al.*, 2011) or exopolysaccharides (Bruno and Sparapano, 2006a,c; Andolfi *et al.*, 2011), which are detected in xylem sap (Bruno and Sparapano, 2006b; Bruno *et al.*, 2007). The genome of *P. chlamydospora* is now available (Antonielli *et al.*, 2014) and will be helpful for studying the regulation of peptides involved in the virulence of this species. Infection may originate from the soil (Mori *et al.*, 2001; Gubler *et al.*, 2004; Retief *et al.*, 2006), may occur via aerial dispersal (Larignon *et al.*, 2009) or insect vectors (Rooney-Latham and Eskalen, 2005), or by contamination during nursery processes (Halleen *et al.*, 2003; Retief *et al.*, 2006; Waite and Morton, 2007; Gramaje and Armengol, 2011) and vineyard management (Mugnai *et al.*, 1999). However, if all cuttings have received

*Corresponding author: alban.jacques@purpan.fr

the same care, some will present foliar or apoplectic esca/black dead arm (BDA) symptoms, whereas other individuals may remain healthy for their entire lifespan (Hofstetter *et al.*, 2012).

Pathogenic fungi considered as pioneers of grapevine trunk colonization are general opportunists of wounding injuries (Mostert *et al.*, 2006). The contamination may occur in the above-ground or below-ground tissues of *Vitis vinifera* L., and visible symptoms of wilting can appear in the leaves after 30 years of incubation in the trunk (Mugnai *et al.*, 1999; Larignon *et al.*, 2009; Lecomte *et al.*, 2012). Consequently, the model selected for biological control studies is a complete plant, allowing several combinations of treatments according to the different plant compartments.

This chapter proposes a model to investigate interplays structuring the microbial ecology in the trunk of *V. vinifera*, with special emphasis on the biological control of *P. chlamydospora*. A model was elaborated with plants developed from cuttings. The aim of this 'nodal model' was to develop a quick and easy method to test the effect of any compound to enhance defence and resistance of the grapevine against fungal agents associated with GTD. First, plant defence responses were studied following treatments with different elicitors by measuring the induction of several genes in trunk tissues by quantitative reverse transcriptase PCR (RT-qPCR). Second, the development of *P. chlamydospora* was quantified 15 days post-inoculation by macroscopic observations and PCR quantification of *P. chlamydospora*. By using the model developed in this study, molecular tools also allowed observation of the impact of elicitors on fungal performance, even if wood discoloration or necrosis had not yet appeared.

Materials and Methods

Fungal isolates

The reference strain of *P. chlamydospora* (CBS239.74), deposited in the Centraalbureau voor Schimmelcultures (CBS, Utrecht, The Netherlands) fungal bank, was grown on a potato dextrose agar (PDA) medium (Merck, Germany), at 26°C in the dark.

Grapevine cuttings

V. vinifera cv. Cabernet Sauvignon (clone 15) canes from mother vines located near Gaillac (France, 43° 939246′ N, 1° 968121′ E) were cut so as to obtain cuttings with three dormant buds. Plant material with bark injuries and wood discoloration was rejected. Cane pieces of similar lengths and diameters were surface sterilized in a bath of tap water containing 10 ml 2.5% bleach (1 min), and the plants were rinsed three times with normal tap water. The cuttings were next immersed in a bath of Cryptonol® (0.05%, Chauvin, France) overnight at 4°C. The following day, they were removed from the bath and again washed three times with tap water. The cuttings were then placed in autoclaved moist glass wool (121°C, 15 min) in plastic cases, covered with film and placed under a 16 h day/8 h night light cycle at 28°C for 2 weeks to induce bud break and rooting. After 2 weeks, the most similar plants were selected and planted in pots containing a soil mixture of peat/river sand/perlite (1:1:1) that had been autoclaved twice at 120°C for 15 min. The plants were watered every 2 days with sterile tap water.

Grapevine treatments

In the nodal model technique, branch I of the young plant was pruned (Fig. 15.1A, B). The area surrounding branch I was first surface sterilized with 80% ethanol on a tissue, and the blades of a clipper were surface sterilized with 80% ethanol before cutting the branch near the trunk (Fig. 15.1C). The pruning wound then received the elicitor and/or the phytopathogenic agent. This was done by making a small 1-mm depth hole in the wounding damage with a 3-mm driller surface sterilized with ethanol 80% (Fig. 15.1D). This operation allowed any solution applied to the injury to stay in place longer. The elicitor solution (10 µl chitosan, 2% in sterile water; laminarin, 1 g l^{-1} in sterile water; or chitin, 2% in sterile water) was deposited in the drill hole. The control plants were similarly damaged but only treated with 10 µl of sterile water.

The next (optional) treatment was the deposit of 10 µl of a spore solution of *P. chlamydospora* (10^6 spores ml^{-1}) in phosphate-buffered saline (PBS, 10 mM at pH 7.2) (Fig. 15.1E) or, for the control, 10 µl PBS alone. The wound was protected by parafilm

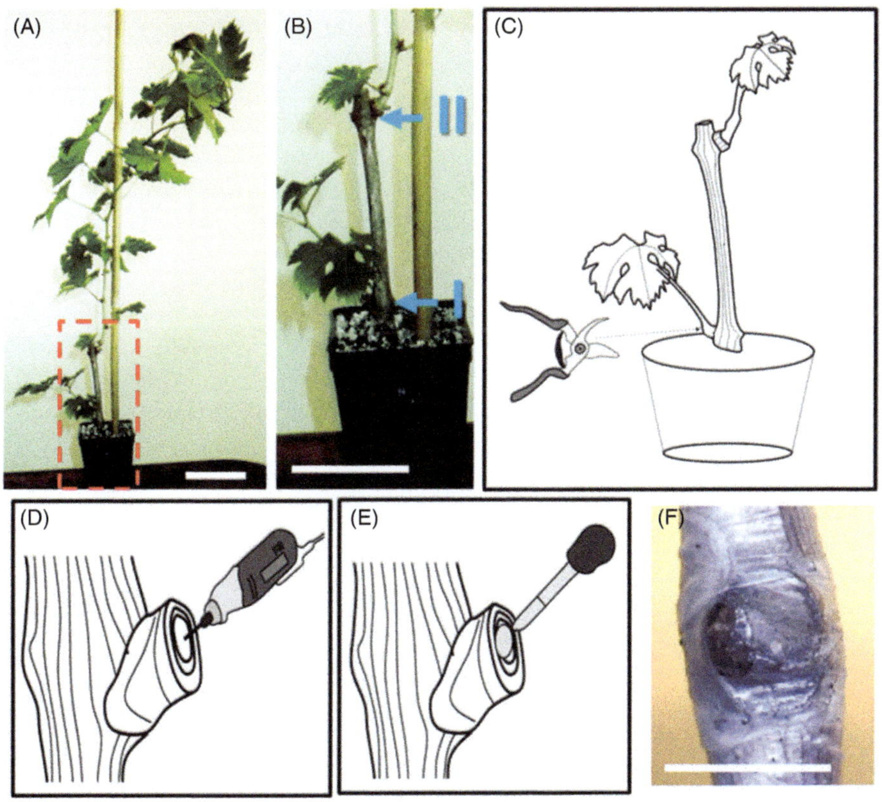

Fig. 15.1. The 'nodal model'. (A) A 3-week-old vine plant in soil showing two developed branches; (B) a closer view of the plant showing branches I and II; (C) branch I is pruned using a surface sterilized commercial clipper; (D) a hole of 3 mm wide and 1 mm deep is made using a drill; (E) the test elicitor compound and/or fungal spores are deposited in the hole; (F) after treatment, the wound is protected with parafilm until the day of sampling. Scale bars: 5 cm (A and B), 1 cm (F). Figure adapted from Pierron *et al.* (2015).

until harvest (Fig. 15.1F). Wood tissues were harvested as follows. The piece of wood corresponding to 5 mm above the wound and 5 mm down from the wound was cut from the branch using blades that had been surface sterilized with 80% ethanol. The samples were frozen in liquid nitrogen and stored at –80°C.

RNA extraction of treated trunk section

Trunk sections containing the nodal wound were ground in liquid nitrogen using a Retsch MM300 mortar grinder (60 s, 25 oscillations s^{-1}, twice) in a 35-ml stainless steel grinding jar with 20 mm stainless steel balls (all three from Retsch, Germany). The resulting powder (1 g) was incubated at 65°C in an RNA extraction buffer (cetyltrimethyl ammonium bromide (CTAB) 2%, polyvinylpolypyrrolidone (PVPP) 2%, Tris 300 mM, ethylenediamine tetraacetic acid (EDTA) 25 mM, NaCl 2 M, 2% β-mercaptoethanol, pH 8; Qiagen, California, USA). A half volume of a phenol:chloroform:isoamyl alcohol solution (25:24:1) was added and the mixture was incubated on ice. The mixture was centrifuged and the supernatant mixed with a 0.6 volume of isopropanol and a 0.1 volume of Na acetate (pH 5.2) in a new tube. Samples were stored at –80°C overnight, and then centrifuged until an RNA pellet formed at the bottom of the tube. The pellet was dissolved in 1.5 ml of RLT extraction buffer (a guanidine thiocyanate containing lysis buffer) supplied by the manufacturer (RNeasy plant mini Kit; Qiagen) and 0.5 ml ethanol. The subsequent steps were performed using buffers, materials and the protocol supplied with the RNeasy plant mini kit. The final elution volume was 50 µl. Samples were stored at –80°C. Total RNA was quantified by optical density measurement using a Biophotometer plus® (Eppendorf AG, Germany). The absence of contaminating genomic DNA was ensured by a DNase reaction (1 u l^{-1}, 30 min at 37°C, DNase I, RNase free kit, Fermentas, Canada) and checked by a PCR test with the total RNA extract as a template and primers of the reference gene *EF1α* (Table 15.1). The PCR products were loaded in a TAE/agarose gel (1.5% agarose, 40 mM Tris + 20 mM acetic acid +1 mM EDTA (TAE), pH 8.0). DNA was visualized under ultraviolet (UV) light after staining with ethidium bromide.

Table 15.1. Primer sequences of *Vitis vinifera* defence-related genes used in this study.

Gene	Primer sequence	Reference
EF1-α	F 5′-GAACTGGGTGCTTGATAGGC-3′ R 5′-AACCAAAATATCCGGAGTAAAAGA-3′	Terrier *et al.*, 2005
Glu1	F 5′-TACCTCTTTGCCATGTTTGATG-3′ R 5′-AGTACTTCGGCTGTTTGTTTGG-3′	Belhadj *et al.*, 2006
PIN	F 5′-GCAGAAACCATTAAGAGGGAGA-3′ R 5′-TCTATCCGATGGTAGGGACACT-3′	
Chit3	F 5′-AGATGGCATAGACTTCGACA-3′ R 5′-TGATGCTGTGGTCAAAGTAC-3′	Busam *et al.*, 1997
LOX	F 5′-CTGGGTGGCTTCTGCTCTC-3′ R 5′-GATAAGCCGCAGATTCATGC-3′	Aziz *et al.*, 2003
PR10-1	F 5′-CGTTAAGGGCGGCAAAGAG-3′ R 5′-GCATCAGGGTGTGCCAAGA-3′	Robert *et al.*, 2001

RT-qPCR analysis of RNA from treated trunk section

RT-qPCR amplifications were performed using the Maxima® First strand cDNA synthesis kit (Fermentas, Canada), starting from 1 µg total RNA. cDNA synthesis was performed at 50°C for 60 min and then at 70°C for 15 min. PCR experiments were conducted with an ABI 7500 Real-Time PCR cycler (Applied Biosystems, Massachusetts, USA), using ABI SDS software v.1.4 at the default setting. The cycling program was: (i) an initial denaturation step at 95°C for 5 min; (ii) 40 cycles of 5 s at 95°C for denaturation, followed by 35 s at 65°C for both annealing and extension; and (iii) an additional melting analysis of 40 min performed from 60 to 95°C. The primer sequences used are indicated in Table 15.1.

DNA extraction of treated trunk section

Samples were lyophilized for 24 h in a Heto C110 Lyophilisator (Kalisto, France) and ground at room temperature using a Retsch MM300 (60 s, 25 oscillations s^{-1}, twice) in a 35-ml stainless steel grinding jar with 20 mm stainless steel balls. The powder was incubated at 65°C in a DNA extraction buffer (CTAB 2%, PVPP 2%, Tris 100 mM, EDTA 20 mM, NaCl 1.4 M, 5 µl β-mercaptoethanol, 5 µl RNAse A, pH 8; Qiagen). A half volume of a chloroform/isoamyl alcohol solution (24:1) was added and the mixture was incubated on ice. The mixture was then centrifuged and the supernatant was mixed with AP2 buffer from the DNeasy plant mini kit (Qiagen). The subsequent steps were performed using buffers, materials and the protocol supplied with the DNeasy plant mini kit. The final elution volume was 50 µl. Samples were stored at –20°C (as described in Pouzoulet *et al.*, 2013).

PCR assays for quantification of *P. chlamydospora* in treated plants

Reactions were carried out in a final volume of 25 µl, and reaction mixes contained 12.5 µl of 2X Plexor™ Master Mix (Promega, Wisconsin, USA). Experiments were conducted with an ABI 7500 Real-Time PCR cycler (Applied Biosystems), using ABI SDS software v.1.4 with default settings. The cycling program consisted of: (i) an initial denaturation step at 95°C for 5 min; (ii) 40 cycles of 5 s at 95°C for denaturation, followed by 35 s at 65°C for both annealing and extension; and (iii) an additional melting analysis of 40 min performed from 60 to 95°C. Plexor™ Analysis Software version 1.5.6.2 (Promega) was used for data analysis (as described in Pouzoulet *et al.*, 2013).

Results

The 'nodal model'

Principles: Summarizing briefly (see Materials and Methods for full details), young grapevine plants were grown from cane cuttings with three eyes (dormant buds) placed in autoclaved rock wool under controlled conditions until budding occurred and the roots were sufficiently formed. At this stage, the plants were placed in a pot containing sand, perlite and peat and watered every 2 days. Despite the apical dominance of branch II, the lower branch (I) developed sufficiently and could be removed, thus mimicking a pruning site (Fig. 15.1C). The nodal wound so generated could be used as a model to evaluate the effects of different elicitors on plant defence mechanisms and induced resistance. For this, the wound was treated first with the test molecule(s), and then contaminated with a known amount of fungal spores of a pathogen(s) known to be involved in GTD (Fig. 15.1D, E and F). The effects on plant defence mechanisms and induced resistance were analysed over several weeks using molecular tools.

Effect of elicitors on development of *V. vinifera* cv. Cabernet Sauvignon and *P. chlamydospora*

Several molecules may induce plant defence mechanisms and thereby may protect the plant either by pre-inducing or priming that defence. To illustrate the use of our nodal model plant, we conducted a study of three molecules with elicitor or plant growth-promoting properties: chitosan (Trotel-Aziz *et al.*, 2006), laminarin (Aziz *et al.*, 2003) and chitin (Zhang *et al.*, 2002). The phenotypic changes induced by those compounds and their impact on the early development of *P. chlamydospora* were investigated.

Expression study of defence genes

Figure 15.2 shows the effects of treatment with three elicitors on the induction factors of five genes associated with the defence pathway in grapevine wood: *Glu1* (glucanase; Belhadj *et al.*, 2006); *PR10-1* (pathogenesis-related protein; Robert *et al.*, 2001); *Chit3* (chitinase; Busam *et al.*, 1997); *LOX* (lipoxygenase; Aziz *et al.*, 2003); and *PIN* (protease inhibitor; Belhadj *et al.*, 2006). The model nodal wound was treated with chitosan, laminarin or chitin. In order to identify only the induction due to the elicitor treatment, the induction in a damaged plant receiving the elicitor was compared with that in a damaged plant treated only with water. In each case, the induction of the gene of interest was measured relative to the induction of a reference gene. The reference used was *EF1*β, a well-known housekeeping gene recently shown to be one of the most relevant reference genes for expression studies in treated wood with *P. chlamydospora* (Terrier *et al.*, 2005; Borges *et al.*, 2014).

The analyses of gene induction analyses were done at 2 and 5 days post-inoculation, timings that were chosen as likely to correspond with the time lapse between pruning and the vine dresser's second passage to treat plants in the field.

Figure 15.2A shows the induction factors in model plants treated with chitosan as elicitor. The *LOX* gene is strongly induced at 5 days post-inoculation (dpi). In plants treated with laminarin (Fig. 15.2B) *Chit3* is quite strongly induced and *LOX* strongly induced 5 dpi. Finally, Fig. 15.2C shows a strong induction of both *LOX* and *Chit3* by chitin 2 and 5 dpi. All three elicitors strongly induce the *LOX* gene 5 dpi, and this is a marker of jasmonic acid pathway. The *PR10-1*, *Glu1* and *PIN* genes are not clearly induced by any of the three treatments (Fig. 15.2).

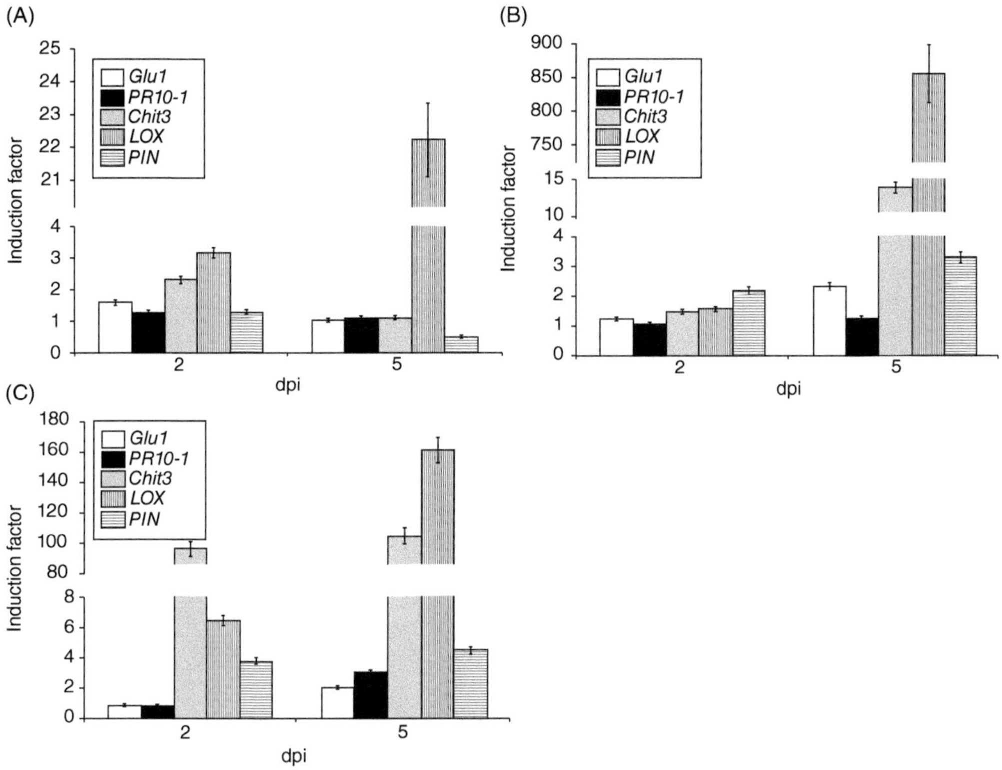

Fig. 15.2. Defence genes induced by elicitor treatments in nodal models of vine plants. Induction factors for each gene were calculated by the $2^{-\Delta\Delta Ct}$ (or comparative C_T) method with the housekeeping gene *EF1α* as a reference gene (this analyses the comparative expression of the two genes in cDNA). Water was the reference treatment for the time point of induction. Data are shown at 2 (the first block of results) and 5 (the second block of results) days post-inoculation (dpi) with the elicitor. *Glu1* is the gene for glucanase 1, *PR10-1* that for pathogenesis-related protein 1, *Chit3* that for chitinase 3, *LOX* that for lipoxygenase and *PIN* that for the host protease inhibitor. The elicitor treatments are (A) chitosan (2%), (B) laminarin (1 g l^{-1}) and (C) chitin (2%).

Quantification of fungal development

A molecular detection method was developed by our research group that allowed the simultaneous detection and quantification of *Phaeoacremonium aleophilum* and *Phaeomoniella chlamydospora*, thereby enabling the separate quantification of *P. chlamydospora*. This molecular technique is based on the Plexor™ technique of multiplex quantitative PCR (Pouzoulet *et al.*, 2013).

At 15 dpi, fungal development in the nodal model plant was almost completely prevented by chitosan treatment compared with the water control (Fig. 15.3A). It is noticeable that the induction of defence-related genes (as shown in Fig. 15.2) does not correlate automatically with resistance. First, this means that the expression of other genes that were not monitored may participate in resistance to *P. chlamydospora* to a larger extent; second, it indicates that pathogen performance is essential to the demonstration of

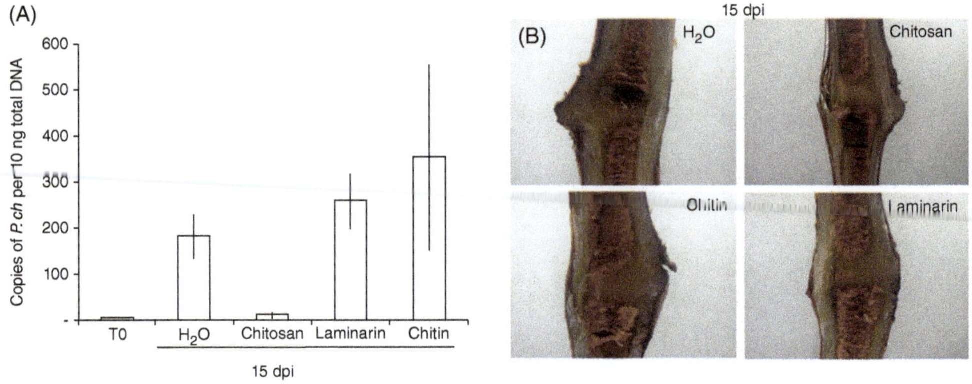

Fig. 15.3. Effect of elicitor pretreatment on the development of the pathogen *Phaeomaniella chlamydospora* (*P. ch*) in nodal models of vine plants. (A) Number of copies of *P. chlamydospora* per 10 ng of total extracted DNA from pruned tissues 15 days post-treatment (dpi) with 10,000 spores of *P. chlamydospora*; T0 indicates time zero. (B) Macroscopic view of injured node wounds at 15 dpi in control and elicitor-treated model plants.

resistance. A strong point of this model is its sensitivity. As shown in Fig. 15.3A, it enables the observation of the impact of a compound on fungal performance at only 15 dpi, whereas at this stage, no differences are distinguishable by the macroscopic observation of plant sections subjected to control (water) and elicitor treatments (Fig. 15.3B).

Discussion

The treatment of GTD represents a major problem for vine dressers, and these diseases are widely sprayed. However, the microbial ecology of grapevine trunk is poorly understood (Bertsch *et al.*, 2013). To further this knowledge, we conducted a grapevine–microbe interaction study on young plants, which has been reported in this chapter. The model that we have developed allows the measurement of gene expression at the site of treatment with an elicitor. To do this, cDNA is synthesized from mRNA and quantified by RT-qPCR. Gene induction is one of the earliest observations of plant responses that follows a challenge. It results in phenotypic changes that may affect phytopathogens or other members of the plant-associated microbial community (Utsumi *et al.*, 2010). Because wood is a difficult tissue to handle in molecular biology, the means by which trunk defence is regulated have previously been unclear, but the use of this model has enabled the observation of responses in wood tissue 2 and 5 dpi. Thus, the wood appears to be active, which raises further questions, notably on the kinetics of trunk response, its specificity and its impact on esca/BDA-associated fungi. Responses differ according to the elicitors employed, but the new model has given promising results and is easy to set up and maintain under laboratory conditions. An *in vitro* study by Bénard-Gellon *et al.* (2015) proposes another model to characterize effectors secreted by *Botryosphaeriaceae* species, and this could perhaps be tested *in planta* using the techniques presented in this chapter.

In addition, our model system allows measurement of the resistance resulting from defence induction, i.e. the impact of plant phenotypic changes due to the elicitor on pathogenic fungi. In this case, the development of *P. chlamydospora* was monitored by qPCR (Pouzoulet *et al*., 2013). The pathogen showed less development at 15 dpi in plants with pruning damage and induced by chitosan than in control pruned plants that just received water. More interestingly, it was possible to assess the impact of elicitation by three well-known compounds even if no wood discoloration or necrosis had appeared. This sensitivity implies an important economy of time in screening processes. In addition, the model is easily adaptable to other esca/BDA associated fungi (*Neofusicoccum parvum*, *Diplodia seriata*, *Phaeoacremonium aleophilum*, *Eutypa lata*), elicitors or BCAs (biocontrol agents).

Our plants are initially surface sterilized but far from sterile, so it is possible that other microorganisms may interfere in the relationship between elicitors and the pathogen. This could explain why several biological agents that are very efficient in the laboratory are not effective in field experiments, as has been observed in the control of soil-borne plant pathogenic fungi (Hubbard *et al*., 1983). The entire native microflora that inhabits the trunk is present in our model, and corresponds to the flora of a 1-year-old cane at the time of pruning, i.e. during dormancy. Consequently, the (laboratory) model is especially of use for investigating nursery concerns: clarifying infection processes and understanding defence mechanisms in the woody cane in order to optimize them.

To conclude, the nodal model that we have developed and described here has enabled study of the impact of elicitors and/or BCAs on grapevine defence regulation in wood. Furthermore, the model allows us to follow the consequences of the fungal pathogen development that is associated with the early steps of GTD. By combining molecular biology techniques (RT-qPCR and qPCR), a comprehensive screening of elicitors/BCAs can be realized and could therefore inform on the efficiency of pathogen control and the mechanism underlying this control.

Acknowledgements

Romain Pierron is supported by INP-EI PURPAN and Région Midi-Pyrénées. Jérôme Pouzoulet was supported by Loire Vini Viti Distribution (LVVD, Z.A. du Landreau, 49610 Mozé sur Louet, France). We are grateful to Christina Toma for assistance with English.

References

Abou-Mansour, A., Couché, E. and Tabacchi, R. (2004) Do fungal naphthalenones have a role in the development of Esca symptoms? *Phytopathologia Mediterranea* 43, 75–82.

Andolfi, A., Mugnai, L., Luque, J., Surico, G., Cimmino, A. and Evidente, A. (2011) Phytotoxins produced by fungi associated with grapevine trunk diseases. *Toxins* 3, 1569–1605.

Antonielli, L., Compant, S., Strauss, J., Sessitsch, A. and Berger, H. (2014) Draft genome sequence of *Phaeomoniella chlamydospora* strain RR-HG1, a grapevine trunk disease (Esca)-related member of the Ascomycota. *Genome Announcements* 2(2): e00098-14.

Aziz, A., Poinssot, B., Daire, X., Adrian, M., Bézier, A., Lambert, B., Joubert, J.-M. and Pugin, A. (2003) Laminarin elicits defense responses in grapevine and induces protection against *Botrytis cinerea* and *Plasmopara viticola*. *Molecular Plant-Microbe Interactions* 16, 1118–1128.

Belhadj, A., Saigne, C., Telef, N., Cluzet, S., Bouscaut, J., Corio-Costet, M.-F. and Merillon, J.-M. (2006) Methyl jasmonate induces defense responses in grapevine and triggers protection against *Erysiphe necator*. *Journal of Agricultural and Food Chemistry* 54, 9119–9125.

Bénard-Gellon, M., Farine, S., Goddard, M.L., Schmitt, M., Stempien, E., Pensec, F., Laloue, H., Mazet-Kieffer, F., Fontaine, F., Larignon, P. *et al.* (2015) Toxicity of extracellular proteins from *Diplodia seriata* and *Neofusicoccum parvum* involved in grapevine *Botryosphaeria* dieback. *Protoplasma* 252, 679–687.

Bertsch, C., Ramírez-Suero, M., Magnin-Robert, M., Larignon, P., Chong, J., Abou-Mansour, E., Spagnolo, A., Clément, C. and Fontaine, F. (2013) Grapevine trunk diseases: complex and still poorly understood. *Plant Pathology* 62, 243–265.

Borges, A.F., Fonseca, C., Ferreira, R.B., Lourenço, A.M. and Monteiro, S. (2014) Reference gene validation for quantitative RT-PCR during biotic and abiotic stresses in *Vitis vinifera*. *PLoS ONE* 9(10): e111399.

Bruez, E., Vallance, J., Gerbore, J., Lecomte, P., Da Costa, J.P., Guerin-Dubrana, L. and Rey, P. (2014) Analyses of the temporal dynamics of fungal communities colonizing the healthy wood tissues of Esca leaf-symptomatic and asymptomatic vines. *PLoS ONE* 9(5), e95928.

Bruno, G. and Sparapano, L. (2006a) Effects of three Esca-associated fungi on *Vitis vinifera* L.: I. Characterization of secondary metabolites in culture media and host responses to the pathogens in calli. *Physiological and Molecular Plant Pathology* 69, 209–223.

Bruno, G. and Sparapano, L. (2006b) Effects of three Esca-associated fungi on *Vitis vinifera* L.: II. Characterization of biomolecules in xylem sap and leaves of healthy and diseased vines. *Physiological and Molecular Plant Pathology* 69, 195–208.

Bruno, G. and Sparapano, L. (2006c) Effects of three Esca-associated fungi on *Vitis vinifera* L.: III. Enzymes produced by the pathogens and their role in fungus-to-plant or in fungus-to-fungus interactions. *Physiological and Molecular Plant Pathology* 69, 209–223.

Bruno, G., Sparapano, L. and Graniti, A. (2007) Effects of three Esca-associated fungi on *Vitis vinifera* L.: V. Changes in the chemical and biological profile of xylem sap from diseased cv. Sangiovese vines. *Physiological and Molecular Plant Pathology* 71, 210–229.

Busam, G., Kassemeyer, H.H. and Matern, U. (1997) Differential expression of chitinases in *Vitis vinifera* L. responding to systemic acquired resistance activators or fungal challenge. *Plant Physiology* 115, 1029–1038.

Del Rio, J., Gonzalez, A., Fuster, M.D., Botia, J.M., Gomez, P., Frias, V. and Ortuño, A. (2001) Tylose formation and changes in phenolic compounds of grape roots infected with *Phaeomoniella chlamydospora* and *Phaeoacremonium* species. *Phytopathologia Mediterranea* 40, 394–399.

Gramaje, D. and Armengol, J. (2011) Fungal trunk pathogens in the grapevine propagation process: potential inoculum sources, detection, identification, and management strategies. *Plant Disease* 95, 1040–1055.

Grosman, J. and Doublet, B. (2012) Maladies du bois de la vigne: synthèse des dispositifs d'observation au vignoble, de l'observatoire 2003–2008 au réseau d'épidémiosurveillance actuel. *Phytoma* 651, 31–35.

Gubler, W.D., Baumgartner, K., Browne, G.T., Eskalen, A., Latham, S.R., Petit, E. and Bayramian, L.A. (2004) Root diseases of grapevines in California and their control. *Australasian Plant Pathology* 33, 157–165.

Halleen, F., Crous, P.W. and Petrini, O. (2003) Fungi associated with healthy grapevine cuttings in nurseries, with special reference to pathogens involved in the decline of young vines. *Australasian Plant Pathology* 32, 47–52.

Hofstetter, V., Buyck, B. and Croll, D. (2012) What if Esca disease of grapevine were not a fungal disease? *Fungal Diversity* 54, 51–67.

Hubbard, J.P., Harman, G.E. and Hadar, Y. (1983) Effect of soilborne *Pseudomonas* spp. on the biological control agent, *Trichoderma hamatum,* on pea seeds. *Phytopathology* 73, 655–659.

Landi, L., Murolo, S. and Romanazzi, G. (2012) Colonization of *Vitis* spp. wood by sGFP-transformed *Phaeomoniella chlamydospora*, a tracheomycotic fungus involved in Esca disease. *Phytopathology* 102, 290–297.

Larignon, P., Fontaine, F., Farine, S., Clément, C. and Bertsch, C. (2009) Esca and black dead arm: two major actors of grapevine trunk diseases. *Comptes Rendus Biologies* 332, 765–783.
Lecomte, P., Darrieutort, G., Liminana, J.-M., Comont, G., Muruamendiaraz, A., Legorburu, F.-J., Choueiri, E., Jreijiri, F., El Amil, R. and Fermaud, M. (2012) New insights into Esca of grapevine: the development of foliar symptoms and their association with xylem discoloration. *Plant Disease* 96, 924–934.
Mori, B., Surico, G., Mugnai, L. and Troccoli, L. (2001) *Phaeomoniella chlamydospora*–grapevine interaction: histochemical reactions to fungal infection. *Phytopathologia* 40, 400–406.
Mostert, L., Groenewald, J.Z., Summerbell, R.C., Gams, W. and Crous, P.W. (2006) Taxonomy and pathology of *Togninia* (Diaporthales) and its *Phaeoacremonium* anamorphs. *Studies in Mycology* 54, 1–113.
Mugnai, L., Graniti, A. and Surico, G. (1999) Esca (black measles) and brown wood-streaking: two old and elusive diseases of grapevines. *Plant Disease* 83, 404–418.
Pierron, R., Gorfer, M., Berger, H., Jacques, A., Sessitsch, A., Strauss, J. and Compant, S. (2015) Deciphering the niches of colonisation of *Vitis vinifera* L. by the Esca-associated fungus *Phaeoacremonium aleophilum* using a *gfp* marked strain and cutting systems. *Plos ONE* 10(6): e0126851.
Pouzoulet, J., Mailhac, N., Couderc, N., Besson, X., Daydé, J., Lummerzheim, M. and Jacques, A. (2013) A method to detect and quantify *Phaeomoniella chlamydospora* and *Phaeoacremonium aleophilum* DNA in grapevine-wood samples. *Applied Microbiology and Biotechnology* 97, 10163–10175.
Pouzoulet, J., Pivovaroff, A.L., Santiago, L.S. and Rolshausen, P.E. (2014) Can vessel dimension explain tolerance toward fungal vascular wilt diseases in woody plants? Lessons from Dutch elm disease and Esca disease in grapevine. *Frontiers in Plant Science* 12, 253.
Retief, E., McLeod, A. and Fourie, P.H. (2006) Potential inoculum sources of *Phaeomoniella chlamydospora* in South African grapevine nurseries. *European Journal of Plant Pathology* 115, 331–339.
Robert, N., Ferran, J., Breda, C., Coutos-Thévenot, P., Boulay, M., Buffard, D. and Esnault, R. (2001) Molecular characterization of the incompatible interaction of *Vitis vinifera* leaves with *Pseudomonas syringae* pv. *pisi*: expression of genes coding for stilbene synthase and class 10 PR protein. *European Journal of Plant Pathology* 107, 249–261.
Rooney-Latham, S. and Eskalen, A. (2005) Occurrence of *Togninia minima* perithecia in Esca-affected vineyards in California. *Plant Disease* 89, 867–871.
Surico, G. (2001) Towards commonly agreed answers to some basic questions on Esca. *Phytopathologia Mediterranea* 40, 487–490.
Surico, G., Mugnai, L. and Marchi, G. (2008) The Esca disease complex. In: Ciancio, A. and Mukerji, K.G. (eds) *Integrated Management of Diseases Caused by Fungi, Phytoplasma and Bacteria*. Springer, Dordrecht, The Netherlands, pp. 119–136.
Terrier, N., Glissant, D., Grimplet, J., Barrieu, F., Abbal, P., Couture, C., Ageorges, A., Atanassova, R., Léon, C., Renaudin, J.-P. *et al.* (2005) Isogene specific oligo arrays reveal multifaceted changes in gene expression during grape berry (*Vitis vinifera* L.) development. *Planta* 222, 832–847.
Trotel-Aziz, P., Couderchet, M., Vernet, G. and Aziz, A. (2006) Chitosan stimulates defense reactions in grapevine leaves and inhibits development of *Botrytis cinerea*. *European Journal of Plant Pathology* 114, 405–413.
Utsumi, S., Ando, Y. and Miki, T. (2010) Linkages among trait-mediated indirect effects: a new framework for the indirect interaction web. *Population Ecology* 52, 485–497.
Waite, H.E. and Morton, L.U. (2007) Hot water treatment, trunk diseases and other critical factors in the production of high-quality grapevine planting material. *Phytopathologia Mediterranea* 46, 5–17.
Zhang, B., Ramonell, K., Somerville, S. and Stacey, G. (2002) Characterization of early, chitin-induced gene expression in *Arabidopsis*. *Molecular Plant-Microbe Interactions* 15, 963–970.

16 Detection of Bacterial Endophytes in *Vitis vinifera* L. and Antibiotic Activity Against Grapevine Fungal Pathogens

S. Lo Piccolo, A. Alfonzo, S. Burruano and G. Moschetti*

Department Scienze Agrarie e Forestali, University of Palermo, Palermo, Italy

Introduction

Endophytic bacteria are ubiquitous in most plant species, where they reside latently or actively colonize plant tissues either locally or systematically. Several endophytic bacteria have been reported to have beneficial effects on their host plants or they may have no observable effects (Hallmann *et al.*, 1997). In fact, bacterial endophytes may act as plant growth promoters (Tan *et al.*, 2014) and confer resistance to stress (Thomas, 2004). Moreover, they can also protect their hosts from bacterial and fungal pathogens (Van Buren *et al.*, 1993; Hinton and Bacon, 1995; Ho *et al.*, 2014), as well as from phytophagous insects and nematodes (Hallmann *et al.*, 1995; Shi *et al.*, 2013). Endophytic bacteria are also potential sources of novel natural molecules to be employed in medicine (Christina *et al.*, 2013), agriculture and industry (Strobel *et al.*, 2004).

Endophytic bacteria are known to reside inside the roots, stems and leaves of plants, and 82 genera have been found in a broad range of plants (Lodewyckx *et al.*, 2002). It has been shown that these microorganisms live within cells, in the intercellular spaces or in the vascular system (Rosenblueth and Martínez-Romero, 2006; Ryan *et al.*, 2007).

The occurrence of bacteria belonging to the genera *Pseudomonas*, *Enterobacter* and *Bacillus* as endophytes in grapevine tissues has been reported (Bell *et al.*, 1995; Thomas, 2004; Salomon *et al.*, 2014). Preliminary investigations on the endophytic microorganisms in different cultivars of healthy Sicilian grapevines have also allowed the isolation of bacteria belonging to the genus *Bacillus* from different organs – buds, leaves, stems and shoots (unpublished data).

In this study, we used fluorescence *in situ* hybridization (FISH) in combination with confocal laser scanning microscopy (CLSM) to detect and localize endophytic bacteria in symptomless and symptomatic leaf tissues of *Vitis vinifera*. In addition, we also tested the antagonistic activity of *Bacillus amyloliquefaciens* AG1, previously isolated from grape wood affected by the esca syndrome, against grapevine fungal pathogens.

*Corresponding author: giancarlo.moschetti@unipa.it

Materials and Methods

FISH experiment

Leaves of asymptomatic *V. vinifera* and of vines with esca syndrome were collected in Sicilian vineyards and used for FISH experiments.

Decolorization of leaf tissues (leaf discs) was carried out to minimize autofluorescence during the subsequent FISH and CLSM investigations by using the modified clearing procedure described by Burruano *et al.* (2008). The leaf discs (0.7 cm diameter) were soaked for 2 weeks in 80% ethanol in water, boiled in an aqueous solution of KOH (10%) for 20 min, transferred to a 30% H_2O_2 solution at room temperature until complete clarification (20–40 min) was achieved, and then acidified by rapid immersion in 10% HCl in water, and rinsed in sterile distilled water.

The universal oligonucleotide probe EUB338 (Amann *et al.*, 1990), which is complementary to a highly conserved region of the 16S rRNA of almost all members of the domain *Bacteria*, was used. The probe was synthesized by Invitrogen (Milan, Italy) and labelled with fluorescein at the 5' end. FISH experiments were performed according to Amann *et al.* (1990).

Microscopic observations were carried out with an epifluorescence microscope (Axiophot, Carl Zeiss, Oberkochen, Germany) coupled to an AxioCam MRc5 (Zeiss) camera. Images were captured using the software AxioVs40 V 4.6.1.0. A laser scanning confocal microscope (CLSM 510; Carl Zeiss) was also employed to determine the spatial localization of bacteria detected by FISH inside the foliar tissues.

Analysis of antagonistic activity

The antifungal activity of *B. amyloliquefaciens* AG1 (Alfonzo *et al.*, 2009) was tested against the following grapevine fungal pathogens: *Alternaria alternata*, obtained from grapevine leaves showing necrotic concentric spots; *Aspergillus carbonarius*, *A. ochraceus* and *Penicillium verrucosum*, isolated from grapes affected by secondary rot; *Botrytis cinerea*, isolated from grapes with grey mould; *Cladosporium viticola*, agent of grapevine leaf spots; *Fomitiporia mediterranea* (FOMED), obtained from grapevines with white rot symptoms; *Fusarium oxysporum*, isolated from the grapevine rhizosphere; *Lasiodiplodia theobromae*, isolated from vine wood with cortical cankers; *Phaeoacremonium aleophilum* (PAL) and *Phaeomoniella chlamydospora* (PCH), both isolated from grapevine showing esca symptoms; *Phoma glomerata* from grape leaves showing necrotic spots; and *Verticillium dahliae*, obtained from decaying vine.

The antagonistic activity of *B. amyloliquefaciens* AG1 was assayed using both direct and indirect methods. The direct method consisted of a dual culture (fungus–bacterium) in Petri dishes containing malt extract agar (MEA) (Kim and Chung, 2004). In the indirect method, the antifungal activity of the crude protein extract (CPE) of *B. amyloliquefaciens* AG1, obtained from the supernatant fluid of a liquid culture of the bacteria by precipitation with ammonium sulfate (Leelasuphakul *et al.*, 2008), was evaluated according to Villani *et al.* (1993), with some modifications. The fungal colonies were grown on potato dextrose agar (PDA), until full invasion of the plate had occurred. Five millilitres of peptone solution (1 g l^{-1}) were added to sporulating and non-sporulating FOMED colonies to facilitate the dispersion of propagules. The fungal suspensions were collected and those of sporulating fungi were sterile filtered to separate the conidia from the mycelial mass. The FOMED suspension contained only mycelium. Aliquots of each fungal suspension were separately placed in Petri dishes, covered and mixed with 14 ml of MEA (43°C). After solidification, ten wells (6 mm diameter) were cut and filled as follows: nine with 50 μl of CPE serially diluted twofold; and one used as negative control and filled with Tris-HCl buffer (1 M, pH 7.5). CPE activity was expressed in arbitrary units (AU ml^{-1}) according to the following formula: AU ml^{-1} = radius of inhibition (mm)/(well capacity (ml) × CPE concentration (mg/ml^{-1})) (Villani *et al.*, 1994).

The molecular mass of the antibiotic substances (peptides/proteins) in the CPE from *B. amyloliquefaciens* AG1 was estimated by Tricine-sodium dodecyl sulfate-polyacrylamide gel electrophoresis (Tricine-SDS-PAGE) according to Schägger and Von Jagow (1987). After electrophoresis, the gel was cut into two vertical parts. One half, containing the sample and molecular weight markers, was stained with Coomassie Brilliant Blue R-250. The other half of the gel was placed in a sterile Petri dish (20 cm diameter) and overlaid with 10 ml MEB soft agar (MEB plus 0.8% agar) containing about 10^6 colony forming units (cfu) ml^{-1} of *A. carbonarius* used as an indicator organism. The plate was incubated at 25°C for 7 days and examined for inhibition of fungal growth (Bhunia *et al.*, 1987).

To identify the substances involved in antibiosis, mass spectrometry (MS) analysis of the Coomassie Brilliant Blue-stained spots, excised from SDS gel, was performed with an ESI Q-ToF Ultima (Micromass, Manchester, UK).

Results

In situ detection of bacteria in grapevine leaves

The suitability of the clearing procedure of foliar discs for FISH experiments performed in grapevine leaf tissues is shown in Fig. 16.1, which reports a comparison of microscopic observations under transmitted light (Fig. 16.1A), fluorescent light (Fig. 16.1B) and superimposed transmitted and fluorescent lights (Fig. 16.1C).

Transmitted light allowed the visualization of the foliar tissue, while the probe-hybridized bacterial cells were detected by fluorescent light. In particular, in

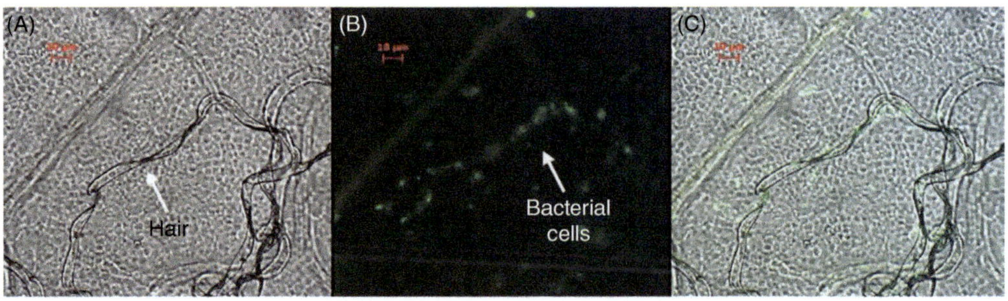

Fig. 16.1. Fluorescence *in situ* hybridization (FISH) on cleared foliar disc: hair under transmitted light (A), bacterial cells under fluorescent light (B), few instances of fluorescence in correspondence of foliar veins and hairs with overlapped lights (C).

asymptomatic leaves, many bacterial colonies were observed in the veins (Fig. 16.2A), cells and hairs (Fig. 16.2B) and intercellular spaces (Fig. 16.2C), and along the cut edges (Fig. 16.2D) of leaf fragments. Instead, in leaves of plants with esca syndrome, few epiphytic bacterial cells were observed (Fig. 16.2E). The observations made using CLSM confirmed that these bacteria lived within the leaf tissues examined (data not shown).

Antifungal activity of *B. amyloliquefaciens* AG1

Direct assays showed considerable biocontrol activity by *B. amyloliquefaciens* AG1 against all of the tested fungal species related to esca trunk disease (Fig. 16.3).

Antifungal metabolites produced by *B. amyloliquefaciens* AG1 were successfully isolated by preparing a crude extract from a liquid culture of the bacteria containing approximately 64 mg protein l^{-1}. Serial dilutions of this CPE tested in indirect assays (see Materials and Methods) revealed the inhibitory effect of this CPE on all of the fungi tested. However, the highest antifungal activity was against *V. dahliae* (12,500 AU ml^{-1}), followed by *P. verrucosum* (10000 AU ml^{-1}) (Fig. 16.4; Table 16.1).

The inhibitory activity of the substance isolated from the *B. amyloliquefaciens* AG1 by Tricine-SDS-PAGE against *A. carbonarius* was demonstrated by the zone of inhibition of the growth of *A. carbonarius* (Fig. 16.5A). This zone corresponded with

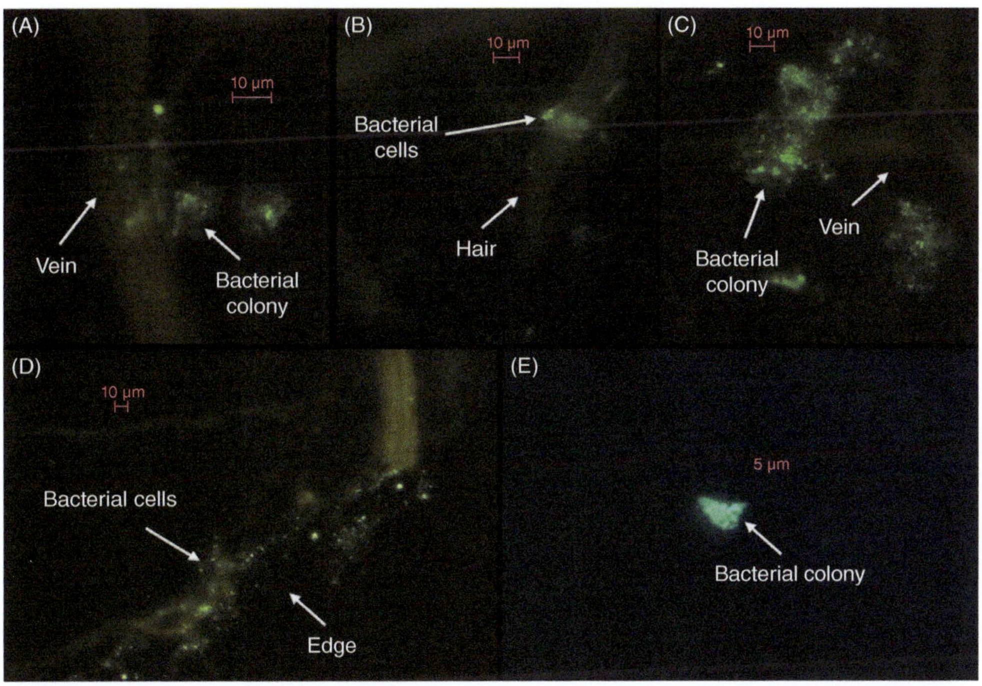

Fig. 16.2. Bacterial colonies/cells detected by fluorescence *in situ* hybridization (FISH) in vein (A), hair (B), intercellular space (C) and a cut edge (D) of an asymptomatic grapevine leaf; epiphytic bacterial colony detected in leaves from vines with esca syndrome (E).

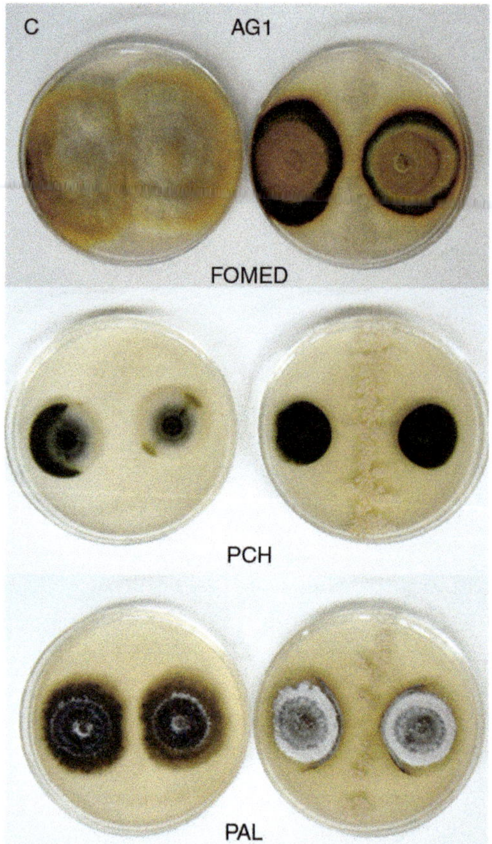

Fig. 16.3. The left side of the figure (controls, C) shows the growth of three esca-related pathogens – FOMED (*Fomitiporia mediterranea*), PCH (*Phaeomoniella chlamydospora*) and PAL (*Phaeoacremonium aleophilum*) inoculated onto both sides of a Petri dish. The right-hand side (AG1) shows the inhibition of the growth of these three pathogens by *Bacillus amyloliquefaciens* AG1 when this was streaked between the two halves of the plates (i.e. in dual culture).

the only band obtained in the gel stained with Coomassie Brilliant Blue (Fig. 16.5B, lane 1) with an approximate molecular weight in the range of 2510–3480 Da (Fig. 16.5B, lane 2).

MS analysis of this band allowed the identification of the substances involved in antibiosis as two tryptic peptides that correspond to the *N*-terminal sequence of subtilisin BPNγ, an extracellular alkaline serine protease.

Discussion

In this study, the FISH/CLSM approach was successfully applied to visualize endophytic bacteria in grapevine leaves, including also those that it is not possible to cultivate in sterilized medium.

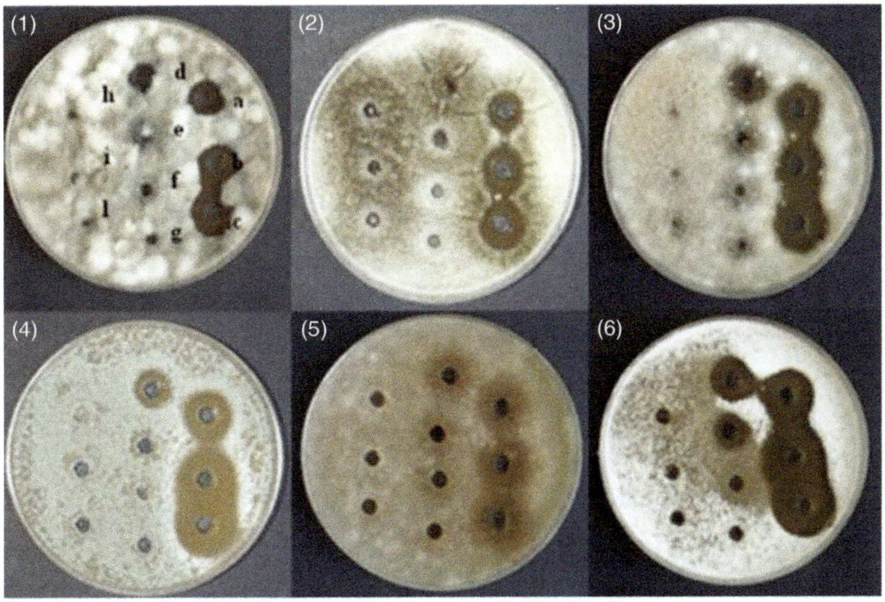

Fig. 16.4. Antifungal effect of crude protein extract (CPE) from a culture of *Bacillus amyloliquefaciens* AG1 against: (1) *Alternaria alternata*; (2) *Aspergillus carbonarius*; (3) *Aspergillus ochraceus*; (4) *Penicillium verrucosum*; (5) *Phoma glomerata*; and (6) *Verticillium dahliae*. The wells (b)–(l) (see (1)) contained 50 µl of CPE twofold serially diluted; well (a) contained a control buffer solution.

Table 16.1. Sensitivity of different fungal species and pathogens to the crude protein extract (CPE) from a culture *Bacillus amyloliquefaciens* AG1, as determined by the critical dilution assay.

Fungal species	AU ml^{-1}
Alternaria alternata	7500
Aspergillus carbonarius	5000
Aspergillus ochraceus	7500
Botrytis cinerea	7500
Cladosporium viticola	1250
Fomitiporia mediterranea	2000
Fusarium oxysporum	3750
Lasiodiplodia theobromae	5000
Phaeoacremonium aleophilum	5000
Phaeomoniella chlamydospora	5000
Phoma glomerata	2500
Penicillium verrucosum	10,000
Verticillium dahliae	12,500

The high presence of microbial colonies in the leaf tissues of asymptomatic grapevines suggests a potential use of these endophytic bacteria as plant growth-promoting agents. In contrast, the lower presence of microcolonies in the leaves of plants with esca-related fungi in the trunk suggests an impact of these fungi on the leaf microflora.

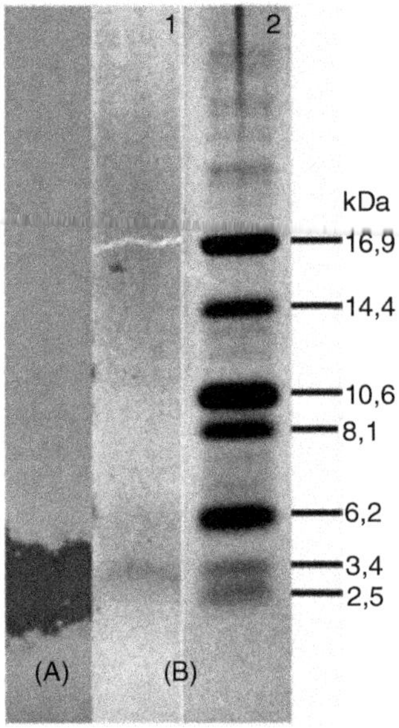

Fig. 16.5. Tricine-SDS-PAGE of the inhibitory (antibiotic) substance isolated from *Bacillus amyloliquefaciens* AG1 and the associated antifungal bioassay using *Aspergillus carbonarius* as indicator organism: (A) zone of growth inhibition in the antifungal assay; (B) corresponding active substances from *B. amyloliquefaciens* AG1 (lane 1), and protein molecular weight markers (lane 2).

Moreover, the lower presence of microcolonies in Esca fungi-infected leaves suggests a possible competition between the bacterial endophytes and pathogenic fungi. Further studies, including use of species-specific FISH probes, are needed to understand the role of bacterial endophytes in plant–microbe interactions in *V. vinifera*.

Our study also reports the antifungal activity of *B. amyloliquefaciens* AG1, an endophyte in symptomatic grapevine wood, against several fungal pathogens of *V. vinifera* in direct plate assays. The antifungal metabolites produced by this bacterium were characterized using crude protein extracts prepared from liquid cultures. The results of indirect assays of these extracts demonstrated that *B. amyloliquefaciens* AG1 has a wide spectrum of antifungal activity as it inhibited the mycelial growth *in vitro* of all of the fungi assayed, although they did exhibit differential sensitivity.

MS analysis of a specific fraction of protein associated with antifungal activity allowed the identification of the chemical compounds involved in antibiosis as two tryptic peptides that are part of the *N*-terminal sequence of subtlisin BPN′. The subtilisins are a prototypical group of bacterial serine proteases produced by various species of *Bacillus* (Rao *et al.*, 1998), and *B. amyloliquefaciens* in particular is reported as producer of subtilisin BPN′ (Wells *et al.*, 1983).

Our results, in confirming endophytes as sources of new molecules to be employed in various fields (industry, medicine, agriculture etc.), suggest a potential role of *B. amyloliquefaciens* AG1 as a biocontrol agent against trunk fungal pathogens as well as other fungal phytopathogens.

References

Alfonzo, A., Conigliaro, G., Ventorino, V., Mondello, V., Burruano, S. and Moschetti, G. (2009) Antimycotic activity of *Bacillus amyloliquefaciens* against fungi of vineyards soil origin. In: *Proceedings of the 2nd International Symposium 'Wine Microbiology and Safety: from the Vineyard to the Bottle', Martina Franca (TA), Italy, 19–20 November 2009. Book of Abstracts*, p. 58. Available at: http://www.unipa.it/persone/docenti/m/giancarlo.moschetti/?pagina=pubblicazione&idPubblicazione=60178 (accessed 7 January 2016).

Amann, R.I., Binder, B.J., Olson, E.J., Chisholm, S.W., Devereux, R. and Stahl, D.A. (1990) Combination of 16S rRNA-target oligonucleotide probes with flow cytometry for analyzing mixed microbial populations. *Applied and Environmental Microbiology* 56,1919–1925.

Bell, C.R., Dickie, G.A., Harvey, W.L.G. and Chan, J.W.Y.F. (1995) Endophytic bacteria in grapevine. *Canadian Journal of Microbiology* 41, 46–53.

Bhunia, A.K., Johnson, M.C. and Ray, B. (1987) Direct detection of an antimicrobial peptide of *Pediococcus acidilactici* in sodium dodecyl sulfate polyacrylamide gel electrophoresis. *Journal of Industrial Microbiology* 2, 319–322.

Burruano, S., Alfonzo, A., Lo Piccolo, S., Conigliaro, G., Mondello, V., Torta, L., Moretti, M. and Assante, G. (2008) Interaction between *Acremonium byssoides* and *Plasmopara viticola* in *Vitis vinifera*. *Phytopathologia Mediterranea* 47, 122–131.

Christina, A., Christapher, V. and Bhore, S. (2013) Endophytic bacteria as a source of novel antibiotics: an overview. *Pharmacognosy Reviews* 7, 11–16.

Hallmann, J., Kloepper, J.W., Rodriguez-Kabana, R. and Sikora, R.A. (1995) Endophytic rhizobacteria as antagonist of *Meloidogyne incognita* on cucumber. *Phytopathology* 85, 1136.

Hallmann, J., Quadt-Hallmann, A., Mahaffee, W.F. and Kloepper, J.W. (1997) Bacterial endophytes in agricultural crops. *Canadian Journal of Microbiology* 43, 895–914.

Hinton, D.M. and Bacon, C.W. (1995) *Enterobacter cloacae* is an endophytic symbiont of corn. *Mycopathologia* 129, 117–125.

Ho, Y.-N., Chiang, H.-M., Chao, C.-P., Su, C.-C., Hsu, H.-F., Guo, C.-T., Hsieh, J.-L. and Huang, C.-C. (2014) *In planta* biocontrol of soilborne Fusarium wilt of banana through a plant endophytic bacterium, *Burkholderia cenocepacia* 869T2. *Plant and Soil* 387, 295–306.

Kim, P.I. and Chung, K. (2004) Production of an antifungal protein for control of *Colletotrichum lagenarium* by *Bacillus amyloliquefaciens* MET0908. *FEMS Microbiology Letters* 234, 177–183.

Leelasuphakul, W., Hemmanee, P. and Chuenchitt, S. (2008) Growth inhibitory properties of *Bacillus subtilis* strains and their metabolites against the green mold pathogen (*Penicillium digitatum* Sacc.) of citrus fruit. *Postharvest Biology and Technology* 48, 113–121.

Lodewyckx, C., Vangronsveld, J., Porteous Moore, F., Moore, E.R.B., Taghavi, S. and Van der Lelie, D. (2002) Endophytic bacteria and their potential applications. *Critical Review in Plant Sciences* 21, 583–606.

Rao, M.B., Tanksale, A.M., Ghatge, M.S. and Deshpande, V.V. (1998) Molecular and biotechnological aspects of microbial proteases. *Microbiology and Molecular Biology Reviews* 62, 597–635.

Rosenblueth, M. and Martłnez-Romero, E. (2006) Bacterial endophytes and their interactions with hosts. *Molecular Plant-Microbe Interactions* 19, 827–837.

Ryan, R.P., Germaine, K., Franks, A., Ryan, D.J. and Dowling, D.N. (2007) Bacterial endophytes: recent developments and applications. *FEMS Microbiology Letters* 278, 1–9.

Salomon, M.V., Bottini, R., De Souza Filho, G.A., Cohen, A.C., Moreno, D., Gil, M. and Piccoli, P. (2014) Bacteria isolated from roots and rhizosphere of *Vitis vinifera* retard water losses, induce abscisic acid accumulation and synthesis of defense-related terpenes in *in vitro* cultured grapevine. *Physiologia Plantarum* 151, 359–374.

Schägger, H. and Von Jagow, G. (1987) Tricine-sodium dodecyl sulfate-polyacrylamide gel electrophoresis for the separation of proteins in the range from 1 to 100 kDa. *Analytical Biochemistry* 166, 368–379.

Shi, Y.W., Zhang, X. and Louc, K. (2013) Isolation, characterization, and insecticidal activity of an endophyte of drunken horse grass, *Achnatherum inebrians*. *Journal of Insect Science* 13, 151.

Strobel, G., Daisy, B., Castillo, U. and Harper, J. (2004) Natural products from endophytic microorganisms. *Journal of Natural Products* 67, 257–268.

Tan, K.Z., Radziah, O., Halimi, M.S., Khairuddin, A.R., Habib, S.H. and Shamsuddin, Z.H. (2014) Isolation and characterization of rhizobia and plant growth-promoting rhizobacteria and their effects on growth of rice seedlings. *American Journal of Agricultural and Biological Sciences* 9, 342–360.

Thomas, P. (2004) Isolation of *Bacillus pumilus* from *in vitro* grapes as a long-term alcohol-surviving and rhizogenesis inducing covert endophyte. *Journal of Applied Microbiology* 97, 114–123.

Van Buren, A., Andre, C. and Ishmaru, C.A. (1993) Biological control of the bacterial ring rot pathogen by endophytic bacteria isolated from potato. *Phytopathology* 83, 1406.

Villani, F., Salzano, G., Sorrentino, E., Pepe, O., Marino, P. and Coppola, S. (1993) Enterocin 226 NWC, a bacteriocin produced by *Enterococcus faecalis* 226, active against *Listeria monocytogenes*. *Journal of Applied Bacteriology* 74, 380–387.

Villani, F., Pepe, O., Mauriello, G., Salzano, G., Moschetti, G. and Coppola, S. (1994) Antimicrobial activity of *Staphylococcus xylosus* from Italian sausages against *Listeria monocytogenes*. *Letters in Applied Microbiology* 18, 159–161.

Wells, J.A., Ferrari, E., Henner, D.J., Estell, D.A. and Chen, E.Y. (1983) Cloning, sequencing, and secretion of *Bacillus amyloliquefaciens* subtilisin in *Bacillus subtilis*. *Nucleic Acids Research* 11, 7911–7925.

17 Impact of Grapevine Preharvest Treatments with Elicitor on the Occurrence and Toxigenesis of Ochratoxigenic Fungi

C. Dachoupakan, C. Strub, V. Martinez,
J.-C. Baccou and S. Schorr-Galindo*

Joint Research Unit on Integrated Approach to Food Quality – Food Safety Team, Université de Montpellier, Montpellier, France

Introduction

Ochratoxin A (OTA), International Union of Pure and Applied Chemistry (IUPAC) name: L-phenylalanine-*N*-[(5-chloro-3,4-dihydro-8-hydroxy-3-methyl-1-oxo-1H-2-benzopyran-7-yl) carbonyl]-(*R*)-isocoumarin (Ringot *et al.*, 2006), is a mycotoxin, a product of the secondary metabolism of moulds, and is one of the most common naturally occurring mycotoxins that contaminates a wide range of different plant products including cereals, coffee beans, cocoa, nuts, spices, dried fruits, beer and wine (Miraglia *et al.*, 2002). OTA is a compound with recognized nephrotoxic activity, which is possibly involved in Balkans endemic nephropathy (BEN) (Vrabcheva *et al.*, 2000; Mantle, 2002), and leads to the inhibition of immune responses transmitted by B- and T-lymphocytes (Petzinger and Weidenbach, 2002). In 1993, the International Agency for Research on Cancer (IARC) classified OTA in group 2B as a possible human carcinogen (IARC, 1993).

In Europe, wine, and especially red wine, has been determined to represent the second major source of human exposure to OTA, following cereals and preceding coffee and beer (Walker, 1999). The presence of OTA in wine and grape juice was reported for the first time by Zimmerli and Dick (1996), who had previously determined its amounts (at ppt level) in human blood, serum, milk and various foodstuffs (Zimmerli and Dick, 1995). Since then, the occurrence of OTA in wine has been reported in various other studies, predominantly with European wines (Visconti *et al.*, 1999; Cabañes *et al.*, 2002; Delage *et al.*, 2003; Blesa *et al.*, 2004; Ratola *et al.*, 2004). In 2004, the European Commission (EC) set the maximum limits for OTA in wine and on grapes at 2.0 μg kg^{-1} (Commission Regulation (EC) No. 123/2005).

Aspergillus section Nigri (black aspergilli), mainly *A. carbonarius* and the species included in the *A. niger* aggregate, were considered to be the principal sources of OTA contamination in grapes, dried grapes (raisins) and wine (Abarca *et al.*, 2001; Esteban *et al.*, 2004; Samson *et al.*, 2004b; Bejaoui *et al.*, 2006; Perrone *et al.*, 2006; Perrone,

*Corresponding author: sabine.galindo@umontpellier.fr

2007). In addition, black aspergilli are the causal agents of several plant diseases, and are considered to be opportunistic grape pathogens. These species usually attack damaged berries that have been infested by bunch pests (insects, mites and nematodes) or have been subjected to physical or physiological damage (Delage *et al.*, 2003; Rousseau, 2004; Varga and Kozakiewicz, 2006).

The control of the vine diseases still largely depends on the use of chemical fungicides applied before harvest. After cereals and rice, the grapevine is the world's third most fungicide-consuming crop, with these chemicals mainly being used against mildew, powdery mildew and grey rot (Lepoivre, 2003). These products have various levels of toxicity towards the user, the consumer and the environment. The presence of residues of chemical fungicide on the grape and in wine involves health risks. Moreover, the emergence of pathogen resistance has been correlated with use of such chemical products (Wightwick *et al.*, 2008). However, in plants that are treated by signalling molecules, such as an elicitor, their natural defence mechanisms can be stimulated. These molecules have been developed as one of the solutions to limiting the excessive use of chemical pesticide treatments in the vineyard. For example, chitosan has the potency to induce responses in grapevine leaves which, in turn, might improve resistance to grey mould disease (Trotel-Aziz *et al.*, 2006). The exogenous application of methyl jasmonate also acts as an elicitor against fungal pathogens of grapevine such as powdery mildew (Belhadj *et al.*, 2006). β-Aminobutyric acid (Slaughter *et al.*, 2008) and sulfated laminarin (Trouvelot *et al.*, 2008) induced the activation of plant defences against the oomycete responsible of downy mildew, and treatment with FEN 560, a product approved in the European Union (EU) for the treatment of grapevines (AMM 2050030), which was derived from fenugreek, is also efficient against this disease (Martinez *et al.*, 2002). The biocontrol/biodegradation of OTA (and its production by *Aspergillus*) using microorganisms on grapes and in wines has been investigated by various researchers (Abrunhosa *et al.*, 2014; Fiori *et al.*, 2014; Petruzzi *et al.*, 2014; Kogkaki *et al.*, 2015), and the control of OTA production in grapes has been reviewed by Ponsone *et al.* (2012).

This chapter aims to compare the effect of two preharvest grape treatments (chemical and biological) on the OTA content of grape juices and the occurrence of potentially ochratoxigenic strains of *Aspergillus* section Nigri group, isolated from grapes. It also presents results on the measurement of the effect of the biological treatment on grapes, especially its ability to induce plant defence responses. To this end, the grape berry is characterized in terms of its morphology (diameter and skin thickness), colour and content of the main soluble chemical compounds of grape juice (sugars, acids and polyphenols). The study was conducted over 4 years (2004–2007) using a sampling pattern of different grape varieties recovered from different French vineyards in the southern Mediterranean region of Languedoc–Roussillon, where most of the grapes are contaminated by ochratoxigenic fungi and have the highest concentrations of OTA in France (Bejaoui *et al.*, 2006).

Materials and Methods

Vineyard, strain origins and study area

The characteristics of the grape samples chosen for the study of the black aspergilli diversity – region and French Département of origin, grape variety, site geographic coordinates and harvest years – are listed in Table 17.1. During the study years (2004–2007),

Table 17.1. Details of samples analysed from different production zones and grape varieties.

Production zone	French département	Grape variety	Latitude N	Longitude E	Harvest years
Béziers	Hérault	Syrah	43°3′	3°12′	2004, 2005
		Cabernet Franc	43°28′	3°13′	2005, 2006
		Grenache	43°3′	3°42′	2005, 2006, 2007
Vauvert	Gard	Carignan	43°7′	4°66′	2004, 2006
		Grenache	43°7′	4°66′	2006
		Syrah	43°7′	4°66′	2006

two different vineyards, located in the zones of Vauvert and Béziers in the Languedoc–Roussillon region of the Mediterranean area of France, were used. Different grape varieties were chosen from the different zones each year of the study to maximize variation in the biodiversity and toxigenesis of the fungal flora (Dachoupakan *et al.*, 2009). For each vineyard, 1 ha was treated with the biological treatment (FEN 560) (SOFT, Port La Nouvelle, France) (4.5 kg ha^{-1}), with three applications at 2-week intervals before harvest, and a nearby hectare followed a common chemical treatment: a mixture of the anti-powdery mildew fungicides Sofral® (10 kg ha^{-1}) and Copral® (3 kg ha^{-1}), with seven treatments at 2-week intervals from April to July.

Sampling

Grapes were collected each September from 2004 to 2007 during the harvest period. At each sampling time and in each vineyard, 3 kg of bunches were randomly collected per grape variety. The samples were taken to the laboratory in closed plastic bags transported in cool boxes. Mycological analyses were done immediately and the remaining samples were frozen at –20°C.

Three bunches of each sample were randomly selected to produce the grape juice. Batches of 100–150 g of grapes were crushed with a mortar and pressed in a laboratory-scale press for 5 min to obtain 50–60 ml of juice. Juice samples were filtered on Whatman filter paper type 591 before analysis.

Mycological analysis of grapes

To estimate the fungal population, 50 g of grapes were randomly selected in triplicate, mixed with 450 ml of sterile water and homogenized by a blender (Lab-Blender 400, Seward Medical UAC House, London) for 2 min. The suspension (1 ml) was diluted in 9 ml of sterile physiological water (NaCl, 8.5 g l^{-1}) with 0.01% Tween 80 (Merck Schuchardt, Darmstadt, Germany). Fungal populations were quantified by plating a sixfold dilution series of each suspension, by spreading 0.1 ml of each dilution on potato dextrose agar (PDA) plates (Biokar Diagnostics, Beauvais, France). The plates were incubated at 25°C for 5 days. The results were expressed in colony forming units (cfu) g^{-1} (ISO 7218:2007) (ISO, 2007).

From each sample, five berries were randomly selected and aseptically plated on PDA medium. The plates were incubated at 25°C for 7 days. The black spore producing filamentous fungi detected were isolated and identified to genus and species level based on morphological criteria; colour, density and colony appearance (layer colour, wrinkled, umbilical, thick or flat) and microscopical observations (conidial head, conidiophore and conidial characters) in accordance with the appropriate keys (Samson *et al.*, 1997, 2004a; Pitt and Hocking, 1999). The main black aspergilli found during the process of enumeration were also isolated.

OTA analysis

The method of Visconti *et al.* (1999) was used for the determination of OTA in grape juices. Juice (10 ml) was diluted with 10 ml of a mixture of 1% polyethylene glycol/5% $NaHCO_3$ in deionized water. An immunoaffinity column (IAC) (OchraTest™ WB, Vicam, Massachusetts, USA) was placed on a vacuum manifold (Visiprep Supelco, Sigma-Aldrich, France) and washed with 10 ml of washing solution (NaCl 2.5 %/0.5% $NaHCO_3$ in deionized water). The diluted sample (10 ml) was then applied to the IAC at a flow rate of 1 drop s^{-1}, followed by 5 ml of the washing solution and 5 ml of deionized water, and the column was then dried by air. The OTA was slowly eluted from the IAC by passing 2 ml of methanol at a flow rate of 1 drop s^{-1}. The eluted solution was evaporated under a nitrogen stream. Finally, the residue was dissolved in 250 ml of the mobile phase and injected into a high-performance liquid chromatography (HPLC) column.

OTA production by the black aspergilli isolates was determined on PDA medium (Suárez-Quiroz *et al.*, 2004b). Inoculates were prepared by growing the strains on culture medium at 25°C for 5 days. Conidial suspensions of each isolate were prepared in physiological water of 1% (v/v) Tween 80 and adjusted to 10^6 conidia ml^{-1}, and enumerated using a Thoma haemocytometer (0.0025 mm^2 × 0.1 mm). The culture medium plates were spot inoculated at the centre of the medium with 5 µl of the adjusted suspension and incubated at 25°C. After 3, 5, 10, 15 and 20 days of incubation, the diameter of the colony was measured and three agar plugs of 5 mm diameter were removed from the middle area of the colony. The plugs were weighed, introduced into 3-ml vials and extracted with 2.5 ml of solvent (methanol/formic acid, 25:1) with sonication for 15 min. The solvent was then evaporated under a nitrogen stream at 40°C (Suárez-Quiroz *et al.*, 2004a). The extracts were solubilized in the mobile phase (49.5% deionized water, 49.5% acetonitrile and 1% acetic acid), filtered on filter syringes Minisart SRP 4, 0.45 µmin diameter (Sartorius, Germany) and injected into an HPLC. The results are expressed as micrograms of OTA per gram of culture.

OTA was detected and quantified by HPLC (pump, Shimadzu LC/9A, Japan) with fluorescence detection (λ_{exc} 333 nm; λ_{em} 460 nm, Shimadzu RF-10AXL, Japan), using a C18 column (25 × 4.6 mm, 5 µm, Brownlee, PerkinElmer, Massachusetts, USA). The mobile phase was pumped at 1 ml min^{-1}. The injection volume was 20 µl and the retention time was around 8.6 min. The calibration curve was prepared with OTA solutions (Sigma-Aldrich) from 5 ng ml^{-1} to 55 ng ml^{-1} for extracted culture samples and from 0.05 ng ml^{-1} to 5 ng ml^{-1} for extracted wine samples in which there was a lower OTA content. The limit of detection (LOD) and limit of quantification (LOQ) of this method applied to grape juice and wine samples were set to 0.023 µg l^{-1} and 0.076 µg l^{-1}, respectively.

Grape berry and juice analysis

To determine the diameter of a berry, 100 randomly chosen berries from each sample were analysed using a camera (SPT-M328CE, Sony®, Japan) linked to a PC with PCI-IMAQ-1408 acquisition card (National Instruments, Texas, USA)-based NI 4.4.1 measurement and automation software (National Instruments). The images collected were then analysed using the image treatment software, NI Vision Assistant 8.5.1 (National Instruments); this treatment, based on the threshold brightness of the photos, was used to consider only the berries and to determine their size by using the software (Fig. 17.1). The values obtained are given in pixels. A prior calibration was carried out using a photo of a double decimetre.

To study the thickness of the skin of a berry, thin sections of the skin (about 0.5 × 1.0 cm) were taken with a scalpel, dried and measured under a scanning electron microscope (Quanta FEG, Type 200, D8399, FEI Company™, Oregon, USA). Ten measurements were made at various locations of the skin for each of two berries per bunch.

The pH of each juice sample was measured using a Series ph720 (WTW® InoLab, Germany) pH meter. All analyses were performed in duplicate.

The sugar (glucose and fructose) content of each juice sample was determined as follows. Each sample was diluted 1000 times to match the calibration curve. Sugars were determined by HPLC after filtration with a filter syringe (0.45 µm, Mini-Sart, Sartorius®, Germany). The HPLC system was composed of a pump (Shimadzu LC-10A, Japan) coupled with a refractometer (Shimadzu, RID-6A, Japan) and an oven for the column (Shimadzu, CTO-10AC, Japan), with a loop injection of 20 µl and a C 610H column (9 µm, 300 × 7.8 mm,) (Supelco®, Pennsylvania, USA). The column was kept at a constant temperature of 40°C. The mobile phase consisted of ultra-pure water containing 1% sulfuric acid with a flow rate of 0.6 ml min^{-1}. For acid analysis, the same HPLC system with the same column was used, but the mobile phase contained 0.5% sulfuric acid. The temperature of the column was kept constant at 50°C. Detection was performed with a UV detector set at 210 nm.

Evaluation of the colour of the juice was performed using an $L^*a^*\ b^*$ colorimetric method. This method determines three parameters: *L* (lightness), with values ranging from 0 (black) to 100 (white); point *a* on the green–red axis; and point *b* on the blue–yellow axis. The juice (2 ml) was centrifuged at 8000 *g* and filtered (0.45 µm,

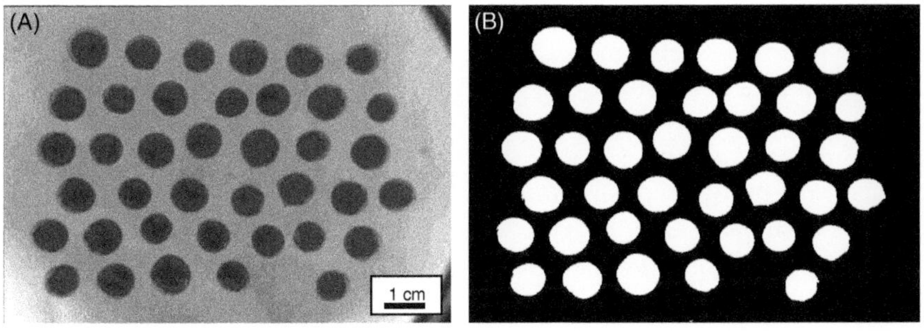

Fig. 17.1. (A) Acquisition photo of a grape sample obtained using NI (National Instruments) 4.1.1 measurement and automation software; (B) image treated by NI Vision Assistant 8.5.1.

Mini-Sart, Sartorius®). The analyses were made with a Tri-Stimulus colorimeter (Konica-Minolta, Japan) and the three colour parameters (L^*, a^*, b^*) were recorded. The values for each sample were averaged from three separate measurements.

The total concentration of polyphenols in grape juice was estimated using Folin–Ciocalteau reagent (Singleton *et al.*, 1999). A sodium carbonate solution (0.25 g ml^{-1}) was filtered on nylon membranes (45 µm, Sartorius®) and added to 200 ml of water. For the analysis, 20 µl of each juice sample was used tenfold prediluted.

Results

Occurrence of black aspergilli

Results from the enumeration of the total fungal population and the percentage occurrence of black fungi (aspergilli) detected in the grape samples from the two geographical study areas (Béziers and Vauvert) are presented in Table 17.2. Over 4 years of study, 60% of the samples showed a total number of mould flora that was higher for the biologically treated grapes than for the chemically treated grapes, but the occurrence of

Table 17.2. Enumeration of the total moulds (cfu g^{-1})[a], percentage occurrence of black aspergilli potentially able to produce ochratoxin A (OTA) and OTA content in the chemically treated grapes (C) and biologically treated grapes (B) from the two production zones from 2004 to 2007.

Samples						
Year	Zone	Variety	Treatment	Total moulds (cfu g^{-1})[b]	% Occurrence (black moulds/ total moulds)	OTA content of juice (µg l^{-1})
2004	Béziers	Syrah	C	$8.1 \pm 0.2 \times 10^3$	5.68	0.40 ± 0.01
			B	$1.5 \pm 0.1 \times 10^4$	5.27	0.21 ± 0.00
	Vauvert	Carignan	C	$8.8 \pm 0.4 \times 10^3$	67.04	0.08 ± 0.00
			B	$3.4 \pm 0.2 \times 10^4$	24.71	<LOQ[c]
2005	Béziers	Cabernet Franc	C	$4.1 \pm 1.2 \times 10^4$	73.17	1.53 ± 0.03
			B	$2.9 \pm 1.2 \times 10^5$	0.10	0.27 ± 0.03
		Grenache	C	$8.9 \pm 2.8 \times 10^3$	0	0.10 ± 0.00
			B	$9.7 \pm 3.2 \times 10^4$	0	0.26 ± 0.04
2006	Béziers	Cabernet	C	$3.2 \pm 0.9 \times 10^3$	0	<LOD[d]
			B	<150 cfu g^{-1}	0	<LOD
		Grenache Franc	C	<150 cfu g^{-1}	0	<LOD
			B	<150 cfu g^{-1}	0	<LOD
	Vauvert	Carignan	C	$3.5 \pm 0.2 \times 10^3$	0	<LOD
			B	$4.3 \pm 0.7 \times 10^3$	0	<LOD
		Grenache	C	$2.7 \pm 0.4 \times 10^3$	0	0.16 ± 0.02
			B	$1.9 \pm 0.1 \times 10^3$	0	<LOD
		Syrah	C	<150 cfu g^{-1}	0	<LOD
			B	$3.1 \pm 1.7 \times 10^3$	53.49	0.12 ± 0.05
2007	Béziers	Grenache	C	<150 cfu g^{-1}	0	<LOD
			B	$2.0 \pm 0.8 \times 10^3$	0	<LOD

[a]Counted by the spread plate technique; [b]average colony forming units from three samplings; [c]limit of quantification, LOQ = 0.076 µg l^{-1}; [d]limit of detection, LOD = 0.023 µg l^{-1}.

black fungi was higher for the chemically treated grapes than for the biologically treated grapes. Moreover, in the two production zones, the levels of contamination by mould flora and black fungi were higher in 2004 and 2005 than in 2006 and 2007, particularly so in 2005. The samples of Cabernet Franc in 2005 had the maximum total fungal flora with 2.9 ± 1.2 × 10^5 cfu g^{-1} from biologically treated grape samples and the highest percentage occurrence of black mould (73%) from the chemically treated grapes.

Grape berry and juice analyses

Berry size

The size of the berries was treated as a first indicator of the morphological differences between biologically and chemically treated grapes. Results from the measurement of the diameter of grape berries from the chemically and biologically treated grapes are presented in Fig. 17.2.

There are disparities between grape varieties, with, for example, an average size of 1.2 cm for Cabernet Franc berries from the Béziers zone harvested in 2006 and of about 1.6 cm for Carignan berries from the Vauvert zone in 2006. However, according to the standard deviations of the measurements, there was no significant difference in berry size detectable between the samples from the two treatments.

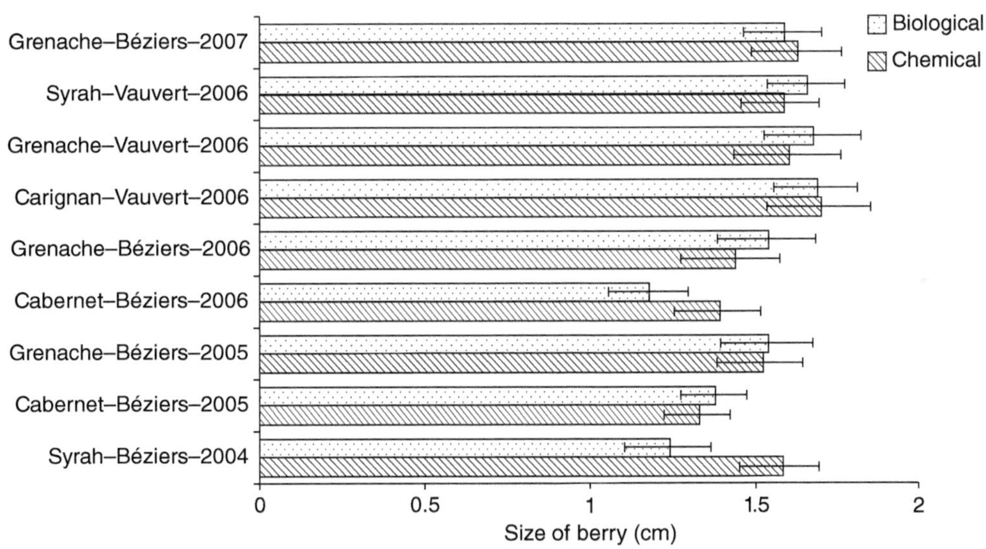

Fig. 17.2. Size of grape berries according to vine treatment (biological or chemical), grape variety, production zone and year.

Skin thickness

Results from the analyses of the skin thickness of grape berries as measured by scanning electron microscopy are shown in Figs 17.3 and 17.4. In all of the samples, the

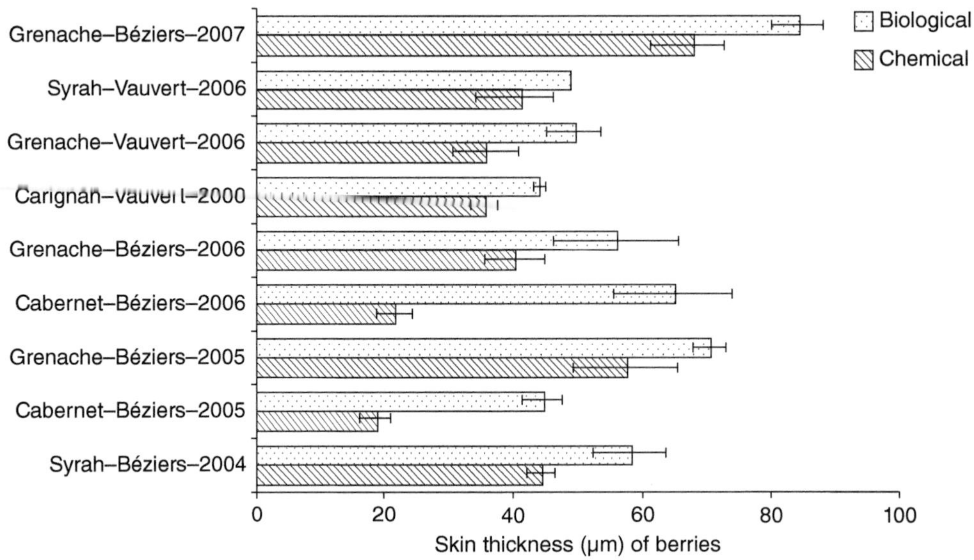

Fig. 17.3. Skin thickness of grape berries according to vine treatment (biological or chemical), grape variety, production zone and year.

grapes that had been biologically treated had thicker berry skin than those that had been chemically treated. The skin thickness of the chemically treated grapes ranged from 18 to 68 μm (average 40.3 μm), whereas that of the grapes that were biologically treated had a skin thickness in the range of 44–85 μm (average 57.9 μm). This was especially noteworthy for the Cabernet Franc grape samples from the Béziers area harvested in 2005 and 2006. Skin thicknesses were 2 to 3 times higher for grapes that had been biologically treated than those that had been chemically treated as shown in Fig. 17.4.

OTA

The OTA contents of juices from the biologically and chemically treated grapes from the two zones between 2004 and 2007 are presented in Table 17.2. Over the 4 years of the study, the OTA content of every juice sample was very low, particularly in 2006 and 2007. In 60% of the samples, the juices from the biologically treated grapes had an OTA content significantly lower than that from the chemically treated grapes. The maximum reductions in OTA content of juice from biologically vs chemically treated grapes was in Syrah grapes growing in the Béziers zone in 2004, Cabernet Franc grapes in the Béziers zone in 2005 and Grenache grapes in the Vauvert zone in 2006, where the reductions were 50, 80 and >80%, respectively.

pH

All samples of the grape juices displayed pH values ranging between 3 and 4 (Fig. 17.5). In 80% of the samples, the juices from chemically treated grapes were more acidic than those from the biologically treated grapes. However, in 2005, two juice samples

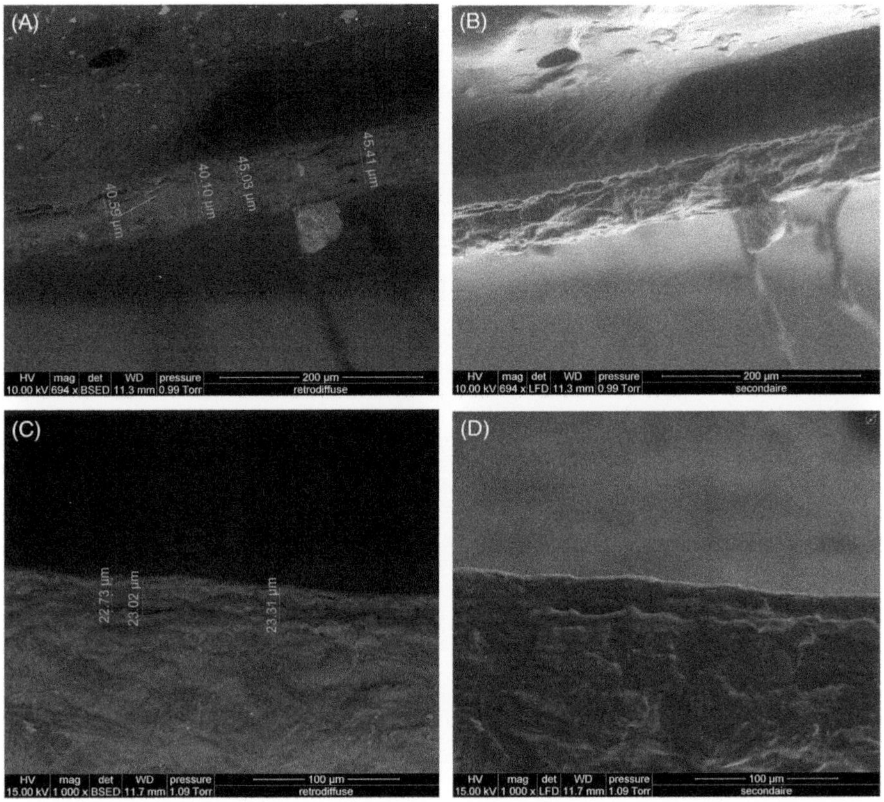

Fig. 17.4. Scanning electron micrographs of skin thickness of Cabernet Franc grape samples from the Béziers zone in 2005. (A, B): biologically treated grapes; (C, D): chemically treated grapes. (A) and (C) are back-scattered ('retrodiffuse') electron images and (B) and (D) are secondary ('secondaire') electron images.

from Béziers (from Grenache and Cabernet Franc grapes) had a lower pH from biologically treated grapes, particularly the juice from Cabernet Franc grapes (pH 3.25 for juice from biologically treated grapes juice and pH 3.71 for juice from chemically treated grapes).

Acids

The contents of the two main organic acids present in grape juice (malic acid and tartaric acid) were determined by HPLC, and the results are shown in Fig. 17.6 and Fig. 17.7.

In all of the juice samples, malic acid was the most abundant acid, with concentrations ranging from 3.08 to 9.93 g^{-1}, in contrast to tartaric acid concentrations of 0.54 to 7.36 g^{-1}. The highest concentration of the two acids was found in the juice samples from biologically treated Cabernet Franc grapes in Béziers in 2005. Few differences were visible in the malic acid content of the juices from biologically and chemically treated grapes, except in the samples from Cabernet Franc grapes in

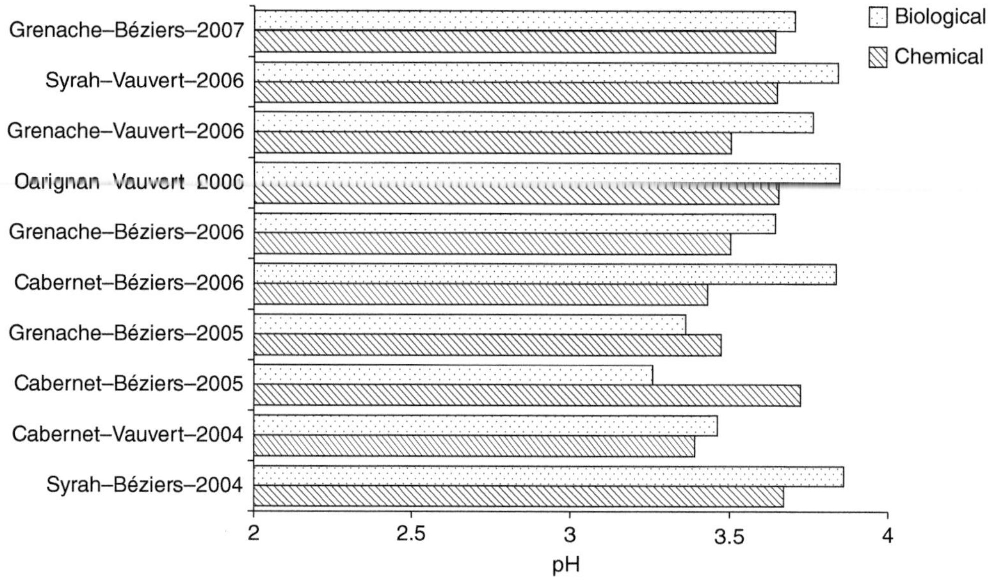

Fig. 17.5. pH of grape juice samples according to vine treatment (biological or chemical), grape variety, production zone and year.

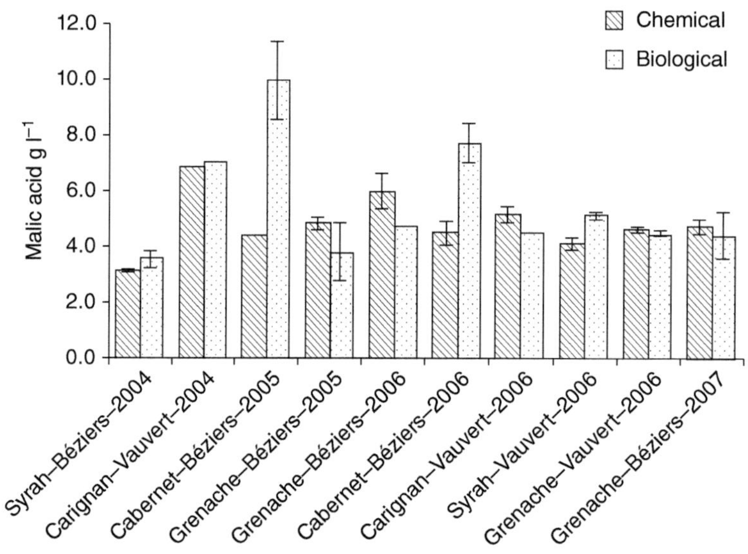

Fig. 17.6. Concentrations of malic acid in grape juice samples according to vine treatment (biological or chemical), grape variety, production zone and year.

Béziers in 2005 and 2006. For these samples, juices from the biologically treated grapes had a higher concentration of malic acid than those from the chemically treated grapes, particularly in 2005 (9.93 g^{-1} for the juice from biologically treated grapes vs 4.35 g^{-1} for the juice from chemically treated grapes). All of the juice samples from biologically treated grapes had tartaric acid contents higher than those

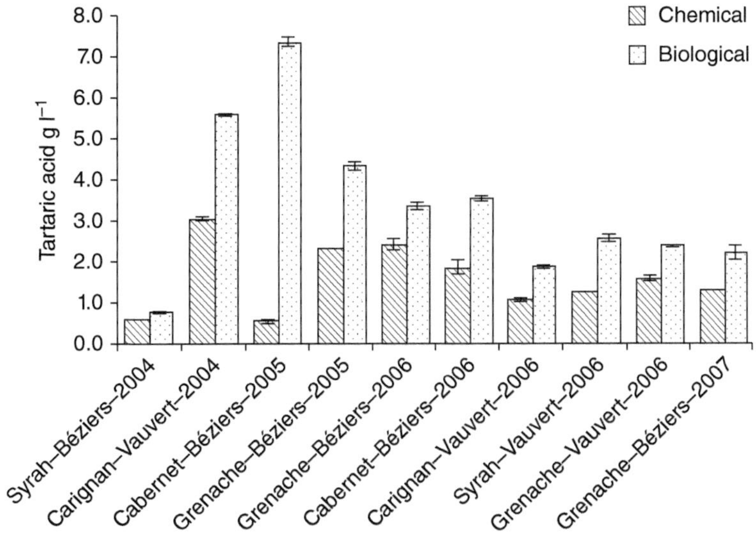

Fig. 17.7. Concentrations of tartaric acid in grape juice samples according to vine treatment (biological or chemical), grape variety, production zone and year.

from chemically treated grapes (Fig. 17.7), especially the juice from Cabernet Franc grapes in Béziers in 2005, which had tartaric acid concentrations of 0.54 and 7.36 g l^{-1}, respectively, in the juice from the chemically treated and biologically treated grapes.

Sugars

The concentrations of glucose and fructose in the juice samples are given in Figs 17.8 and 17.9. The results show similar differences in the glucose and fructose contents of the juice according to grape variety and type of grape treatment. However, there were significant differences between samples from different grape varieties and in different years, which ranged from 60.18 (Cabernet Franc, 2006) to 151.82 g l^{-1} (Grenache, 2006) for glucose, and from 62.45 (Cabernet Franc, 2006) to 164.60 g l^{-1} (Grenache, 2006) for fructose. Two juice samples had particularly significant differences depending on treatment: Cabernet Franc from Béziers in 2006 and Grenache from Vauvert in 2006; for the former, the juice from chemically treated grapes had higher glucose and fructose contents than that from the biologically treated grapes; whereas for the latter, the juice from the biologically treated grapes had a higher sugar content than that from the chemically treated grapes.

Juice colour

Figure 17.10 is a 3D scatterplot chart produced using STATISTICA 7 software that shows the characteristics of the juice from each grape sample according to the $L^*a^*b^*$ colour system. The colours of each juice are very different. The L^* term is a measure

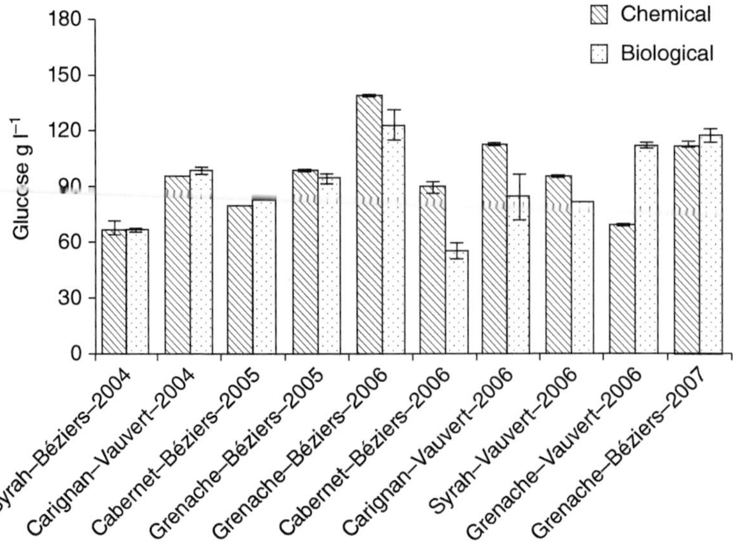

Fig. 17.8. Concentrations of glucose in grape juice samples according to vine treatment (biological or chemical), grape variety, production zone and year.

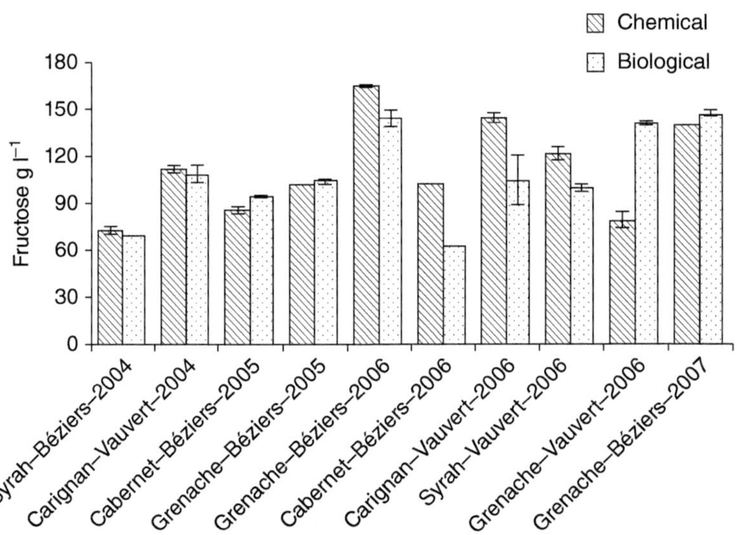

Fig. 17.9. Concentrations of fructose in grape juice samples according to vine treatment (biological or chemical), grape variety, production zone and year.

of brightness (or lightness) from black to white for more than 50% of the juice samples, the colour of the juice from biologically treated grapes is darker than that from chemically treated grapes, i.e. the L^* parameter had a low value (towards black) in juice samples from biologically treated grapes. In contrast, the a^* and b^* terms, which indicate green–red and blue–yellow, respectively, showed similar results between the grape juice samples from biologically and chemically treated grapes.

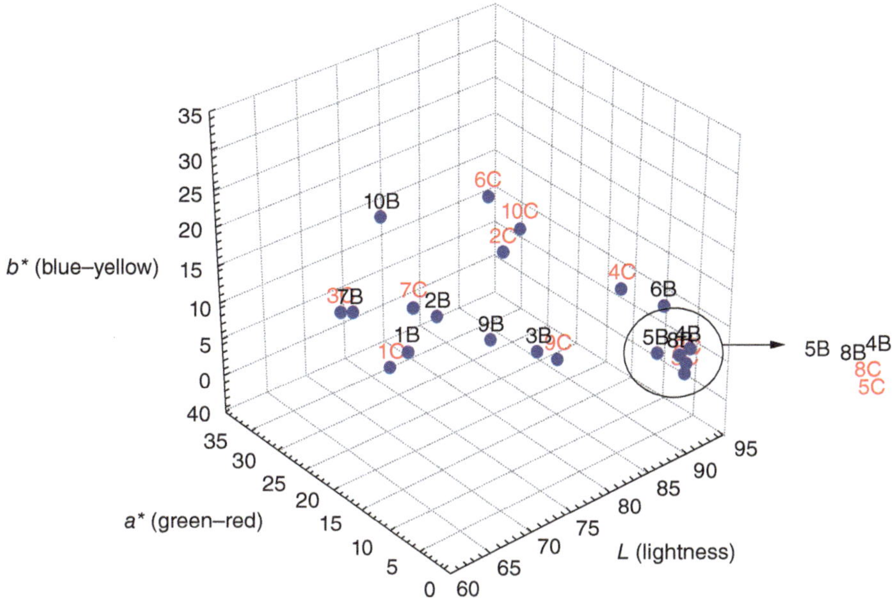

Fig. 17.10. 3D Scatterplot (using STATISTICA 7 software) of colours of juice samples according to vine treatment (biological or chemical), grape variety, production zone and year. Key: 1, Syrah–Béziers–2004; 2, Carignan–Vauvert–2004; 3, Cabernet Franc–Béziers–2005; 4, Grenache–Béziers–2005; 5, Cabernet Franc–Béziers–2006; 6, Grenache–Béziers–2006; 7, Carignan–Vauvert–2006; 8, Grenache–Vauvert–2006; 9, Syrah–Vauvert–2006; 10, Grenache–Béziers–2007; B, biologically treated grapes (in black); C, chemically treated grapes (in red).

Total polyphenols

Five out of ten juice samples had higher total polyphenol content (as gallic acid g l^{-1}) in juice samples from biologically treated grapes than in juice samples from chemically treated grapes (Fig. 17.11), particularly the samples from Cabernet Franc grapes from the Béziers area in 2005 and 2006.

Discussion

From these results, we can establish that grapes given the biological treatment are more heavily contaminated by fungal flora than those treated with traditional chemical products, but that the occurrence of black fungi is lower, and thus the juice from the biologically treated grapes has a reduced OTA content compared with that from the chemically treated grapes. Indeed, OTA reduction rates of 50 to 80% were obtained with juice from biologically treated grapes. This is particularly true for the Cabernet Franc variety grape samples in 2005; this sample had the highest OTA content in the juice from chemically treated grapes and a maximal reduction of 80% in the juice from the biologically treated grapes. The occurrence of black moulds (73%

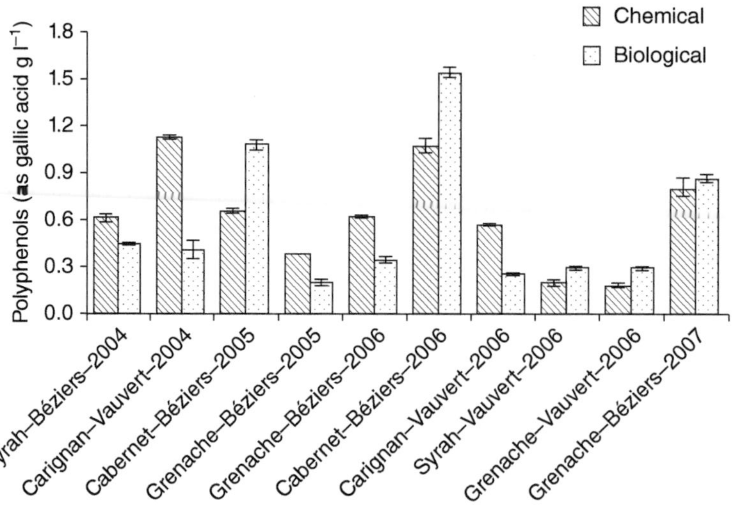

Fig. 17.11. Total polyphenol content in grape juice for the biologically and chemically treated vines.

in the chemically treated grapes and 0.10% in the biologically treated grapes) played the main role in determining the OTA content of these juice samples. The morphological, physico-chemical and biochemical parameters of the grape berries provided some insights into the mechanism of action of this treatment, which could result in the stimulation of plant defences.

The study of the morphology of the grape berries gave interesting results in showing that the skin of all of biologically treated grapes was clearly thicker than that of those of the chemically treated grapes, especially for the Cabernet Franc grapes in in 2005. This could explain the action of FEN 560 on the developing berries; the treatment results in an increase in the thickness of the skin of berries, which could favour the reduction of OTA through better resistance to lesions and insect perforations, which are known to be responsible for pulp contamination by moulds (Delage *et al.*, 2003; Rousseau, 2004); the increased grape skin thickness could also protect the grapes from physical and climatic injuries. Moreover, a thicker skin, particularly in the Cabernet Franc grape sample in 2005, could also explain the darker colour and the increase of total polyphenol content in the juice of biologically treated grapes. The precursors of polyphenols are located in the cell walls of grape skin cells, and vary according to grape variety and vine treatment (Pinelo *et al.*, 2005, 2006), and polyphenols in red wine have been reported to benefit human health, mainly via protection against coronary arterial and cerebrovascular diseases (Tomera, 1999).

The biological treatment seems to have allowed a stronger conservation of acids in the grapes, particularly of tartaric acid, which is important for the stabilization of grape juice and wine (Blouin and Cruège, 2003). The concentrations in the juices from chemically treated grapes remained relatively low (average 3.39 and 1.59 g l^{-1} for juice from biologically and chemically treated grapes, respectively). Cabernet Franc is the only grape variety that showed a significant modification of acid content (tartaric and malic acids) in biologically vs chemically treated grapes, especially in grapes harvested in 2005, which had concentrations approximately two and six times higher for malic and tartaric acids, respectively, in biologically treated grapes. This increased acid content is also related to the higher pH of the juice from biologically vs chemically treated grapes. These changes in grape acid content and pH may explain the reduction of OTA content in the juice from biologically treated grapes as well, acting via effects on the development and/or OTA production of ochratoxinogenic fungi. Previous work has demonstrated some influence of the type of grape treatment on the biodiversity of the black aspergilli (Dachoupakan *et al.*, 2009), with the occurrence of these fungi reduced in the case of biological compared with chemical treatment. Other results on the growth rate of strains of black aspergilli on potato dextrose agar (PDA) medium at pH 5.6 ± 0.1 (Samson *et al.*, 2004a), and on a synthetic grape medium (SGM) medium with a composition similar to that of grapes and with a pH of 4.2 ± 0.1 (as used in previous work by Bellí *et al.*, 2004, 2005; Leong *et al.*, 2006), showed that after 5 days of culture at 25°C, an acidic pH seemed to have an inhibiting effect on the growth and OTA production of some strains of black aspergilli isolated from grapes (data not shown). The acidity of the growth medium has previously been reported to influence the growth and OTA production of fungi; for example, a high concentration of chlorogenic acid in PDA medium inhibits the growth and toxigenesis of *A. ochraceus* (Suárez-Quiroz *et al.*, 2004a). In addition, O'Callaghan *et al.* (2006) reported that OTA production by *A. ochraceus* is strongly influenced by the pH of the culture medium, with large amounts of OTA produced at pH <7.0 but reduced amounts produced at higher pH values.

The results from the sugar analyses show no obvious effect of the biological vs chemical treatment of grapes on the sugar concentration in the grape juice. There are only two samples (Grenache grapes from the Vauvert zone and Cabernet Franc from the Béziers zone, both in in 2006) that show significant differences between the two treatments, but with opposite effects. Furthermore, there is no difference in the OTA content of the grape juice between the two treatments of Cabernet Franc grapes in Béziers in 2006, while the juice from the biologically treated Grenache grapes in Vauvert in 2006 showed a reduction of up to 80% in OTA content (to less than the LOD) compared with that from chemically treated grapes.

These first encouraging results will be confirmed by an ongoing study performed in an experimental field. This will be carried out over 3 successive years using the same sensitive grape varieties, with artificial contamination of grapes and monitoring of the resulting grape composition done in parallel. Negative controls with no artificial contamination or treatment could be included. This experiment will increase our knowledge of the use of the stimulation of plant defences to control grape diseases and reduce the presence of OTA in the end products. Moreover, the induction of plant defences could also result in changes in the production of metabolites in leaves. Preliminary results from headspace solid phase microextraction for gas chromatography mass spectrometry (HS-SPME-GC-MS) analyses of the volatile compounds of vine leaves show higher aldehyde contents after the stimulation of plant defences. Some of these compounds could have an antifungal action, resulting in a decrease in the growth and toxigenesis of ochratoxigenic fungi (to be published).

Further studies towards a better understanding of the mechanisms of and the interactions between product, plant and mould will increase our knowledge of the use of natural (biological) treatments as an elicitor to control grape diseases and to reduce the presence of OTA in manufactured products.

Acknowledgements

We wish to thank all of the wine producers in the two geographic areas for allowing us to conduct the treatment tests in their vineyards with FEN 560 and to take grape, grape juice and wine samples, and the Société Occitane de Fabrications et de Technologies (SOFT society, Port La Nouvelle, France) for the samples of FEN 560.

References

Abarca, M.L., Accensi, F., Bragulat, M.R. and Cabañes, F.J. (2001) Current importance of ochratoxin A-producing *Aspergillus* spp. *Journal of Food Protection* 64, 903–906.

Abrunhosa, L., Inês, A., Rodrigues, A.I., Guimarães, A., Pereira, V.L., Parpot, P., Mendes-Faia, A. and Venâncio, A. (2014) Biodegradation of ochratoxin A by *Pediococcus parvulus* isolated from Douro wines. *International Journal of Food Microbiology* 188, 45–52.

Bejaoui, H., Mathieu, F., Taillandier, P. and Lebrihi, A. (2006) Black aspergilli and ochratoxin A production in French vineyards. *International Journal of Food Microbiology* 111, S46–S52.

Belhadj, A., Saigne, C., Telef, N., Cluzet, S., Bouscaut, J., Corio-Costet, M.F. and Merillon, J.M. (2006) Methyl jasmonate induces defense responses in grapevine and triggers protection against *Erysiphe necator*. *Journal of Agricultural and Food Chemistry* 54, 9119–9125.

Bellí, N., Ramos, A.J., Sanchis, V. and Marín, S. (2004) Incubation time and water activity effect on ochratoxin A production by *Aspergillus* section Nigri strains isolated from grapes. *Letters in Applied Microbiology* 38, 72–77.

Bellí, N., Ramos, A.J., Sanchis, V. and Marín, S. (2005) *Aspergillus carbonarius* growth and ochratoxin A production on a synthetic grape medium in relation to environmental factors. *Journal of Applied Microbiology* 98, 839–844.

Blesa, J., Soriano, J.M., Molto, J.C. and Manes, J. (2004) Concentration of ochratoxin A in wines from supermarkets and stores of Valencian Community (Spain). *Journal of Chromatography A* 1054, 397–401.

Blouin, J. and Cruège, J. (2003) *Analyse et Composition des Vins*. Dunod-La Vigne, Paris.

Cabañes, F.J., Accensi, F., Bragulat, M.R., Abarca, M.L., Castellá, G., Minguez, S. and Pons, A. (2002) What is the source of ochratoxin A in wine? *International Journal of Food Microbiology* 79, 213–215.

Dachoupakan, C., Ratomahenina, R., Martinez, V., Guiraud, J.P., Baccou, J.C. and Schorr-Galindo, S. (2009) Study of the phenotypic and genotypic biodiversity of potentially ochratoxigenic black aspergilli isolated from grapes. *International Journal of Food Microbiology* 132, 14–23.

Delage, N., D'Harlingue, A., Colonna Ceccaldi, B. and Bompeix, G. (2003) Occurrence of mycotoxin in fruit juices and wine. *Food Control* 14, 225–227.

Esteban, A., Abarca, M.L., Bragulat, M.R. and Cabañes, F.J. (2004) Effects of temperature and incubation time on production of ochratoxin A by black aspergilli. *Research in Microbiology* 155, 861–866.

Fiori, S., Urgeghe, P.P., Hammami, W., Razzu, S., Jaoua, S. and Migheli, Q. (2014) Biocontrol activity of four non- and low-fermenting yeast strains against *Aspergillus carbonarius* and their ability to remove ochratoxin A from grape juice. *International Journal of Food Microbiology* 189, 45–50.

IARC (1993) Ochratoxin A. Summary of data reported and evolution. *IARC Summaries and Evaluations* 56, 489.

ISO (2007) *ISO 7218:2007(en): Microbiology of Food and Animal Feeding Stuffs – General Requirements and Guidance for Microbiological Examinations.* International Organization

for Standardization, Geneva, Switzerland. Available at: https://www.iso.org/obp/ui/#iso:std:36534:en (accessed 9 October 2013).
Kogkaki, E.A., Natskoulis, P.I., Magan, N. and Panagou, E.Z. (2015) Effect of interaction between *Aspergillus carbonarius* and non-ochratoxigenic grape-associated fungal isolates on growth and ochratoxin A production at different water activities and temperatures. *Food Microbiology* 46, 521–527.
Leong, S.L., Hocking, A.D. and Scott, E.S. (2006) Effect of temperature and water activity on growth and ochratoxin A production by Australian *Aspergillus carbonarius* and *A. niger* isolates on a simulated grape juice medium. *International Journal of Food Microbiology* 110, 209–216.
Lepoivre, P. (2003) *Phytopathologie*. De Boeck and Larcier Publishers, Brussels.
Mantle, P.G. (2002) Risk assessment and the importance of ochratoxins. *International Biodeterioration and Biodegradation* 50, 143–146.
Martinez, C., Baccou, J.C., Delpech, L., Delrieu, M. and Cassan, H. (2002) Procédé de lutte contre des agents pathogènes de plantes agronomiquement utiles à l'aide d'un nouvel éliciteur. Patent no. PCT/FR02/02056. [No further details available.]
Miraglia, M., Brera, C., Pazzaglini, B. and Grossi, S. (eds) (2002) *Reports on Tasks for Scientific Cooperation: Report of Experts Participating in Task 3.2.7, January 2002. Assessment of Dietary Intake by the Population in EU Member States*. Compiled for the Directorate-General Health and Consumer Protection of the European Commission, Istituto Superiore di Sanità, Rome. Available at: http://ec.europa.eu/food/food/chemicalsafety/contaminants/task_3-2-7_en.pdf (accessed 2 October 2015).
O'Callaghan, J., Patricia, C., Stapleton, P.C. and Dobson, A.D.W. (2006) Ochratoxin A biosynthetic genes in *Aspergillus ochraceus* are differentially regulated by pH and nutritional stimuli. *Fungal Genetics and Biology* 43, 213–221.
Perrone, G. (2007) Biodiversity of *Aspergillus* Sect. Nigri from grapes in Europe. In: Samson, R.E. and Varga, J. (eds) Aspergillus *Systematics in the Genomic Era [An International Workshop Held at the CBS Fungal Biodiversity Centre, Utrecht, The Netherlands on 12–14 April]. Studies in Mycology* 59, 53–66. Available at: http://www.cbs.knaw.nl/publications/sim59.pdf (accessed 2 October 2015).
Perrone, G., Mulè, G., Susca, A., Battilani, P., Pietri, A. and Logrieco, A. (2006) Ochratoxin A production and amplified fragment length polymorphism analysis of *Aspergillus carbonarius*, *Aspergillus tubingensis*, and *Aspergillus niger* strains isolated from grapes in Italy. *Applied and Environmental Microbiology* 72, 680–685.
Petruzzi, L., Corbo, M.R., Sinigaglia, M. and Bevilacqua, A. (2014) Yeast cells as adsorbing tools to remove ochratoxin A in a model wine. *International Journal of Food Science and Technology* 49, 936–940.
Petzinger, E. and Weidenbach, A. (2002) Mycotoxin in the food chain: the role of ochratoxins. *Livestock Production Science* 76, 245–250.
Pinelo, M., Rubilar, M., Jerez, M., Sineiro, J. and Núñez, M.J. (2005) Effect of solvent, temperature, and solvent-to-solid ratio on the total phenolic content and antiradical activity of extracts from different compounds of grape pomace. *Journal of Agricultural and Food Chemistry* 53, 2111–2117.
Pinelo, M., Arnous, A. and Meyer, A.S. (2006) Upgrading of grape skins: significance of plant cell-wall structural components and extraction techniques for phenol release. *Trends in Food Science and Technology* 17, 579–590.
Pitt, J.I. and Hocking, A.D. (1999) *Fungi and Food Spoilage*, 2nd edn. Aspen Publishers, Gaithersburg, Maryland.
Ponsone, M.L., Chiotta, M.L., Palazzini, J.M., Combina, M. and Chulze, S. (2012) Control of ochratoxin A production in grapes. *Toxins* 4, 364–372.
Ratola, N., Martins, L. and Alves, A. (2004) Ochratoxin A in wines – assessing global uncertainty associated with the results. *Analytica Chimica Acta* 513, 319–324.

Ringot, S., Chango, A., Schneider, Y.-J. and Larondelle, Y. (2006) Toxicokinetics and toxicodynamics of ochratoxin A, an update. *Chemico-Biological Interactions* 159, 18–46.
Rousseau, J. (2004) Ochratoxin A in wines: current knowledge. First part: factors favoring its emergence in vineyards and wines. *Vinidea.net. Wine Internet Technical Journal* No. 5. [No current details available.]
Rule (CE) No. 123/2005 (2005) Commission Regulation (EC) No 123/2005 of 26 January 2005 amending Regulation (EC) No 466/2001 as regards ochratoxin A (1) *Official Journal of the European Union* L25 (48), 3–5.
Samson, R.A. (1992) Current taxonomic schemes of the genus *Aspergillus* and its teleomorphs. In: Bennett, J.W. and Klich, M.A. (eds) *Aspergillus: Biology and Industrial Applications*. Butterworth–Heinemann, Boston, Massachusetts, pp. 355–390.
Samson, R.A., Hoekstra, E.S., Frisvad, J.C. and Filtenborg, O. (2004a) *Introduction to Food Borne Fungi*, 7th edn. Centraalbureau voor Schimmelcultures, Utrecht, The Netherlands.
Samson, R.A., Houbraken, J.A.M.P., Kuijpers, A.F.A., Frank, J.M. and Frisvad, J.C. (2004b) New ochratoxin A or sclerotium producing species in *Aspergillus* section Nigri. *Studies in Mycology* 50, 45–61.
Singleton, V.L., Orthofer, R. and Lamuela-Raventos, R.M. (1999) Analysis of total phenols and other oxidation substrates and antioxidants by means of Folin–Ciocalteu reagent. *Methods in Enzymology* 299, 152–178.
Slaughter, A.R., Hamiduzzaman, M.M., Gindro, K., Neuhaus, J.M. and Mauch-Mani, B. (2008) Beta-aminobutyric acid-induced resistance in grapevine against downy mildew: involvement of pterostilbene. *European Journal of Plant Pathology* 122, 185–195.
Suárez-Quiroz, M.L., González-Rios, O., Barel, M., Guyot, B., Schorr-Galindo, S. and Guiraud, J.P. (2004a) Effect of chemical and environmental factors on *Aspergillus ochraceus* growth and toxigenesis in green coffee. *Food Microbiology* 21, 629–634.
Suárez-Quiroz, M., González-Rios, O., Barel, M., Guyot, B., Schorr-Galindo, S. and Guiraud, J.-P. (2004b) Study of ochratoxin A-producing strains in coffee processing. *International Journal of Food Science and Technology* 39, 501–507.
Tomera, J.F. (1999) Current knowledge of the health benefits and disadvantages of wine consumption. *Trends in Food Science and Technology* 10, 129–138.
Trotel-Aziz, P., Couderchet, M., Vernet, G. and Aziz, A. (2006) Chitosan stimulates defense reactions in grapevine leaves and inhibits development of *Botrytis cinerea*. *European Journal of Plant Pathology* 114, 405–413.
Trouvelot, S., Varnier, A.L., Allègre, M., Mercier, L., Baillieul, F., Arnould, C., Gianinazzi-Pearson, V., Klarzynski, O., Joubert, J.M., Pugin, A. and Daire, X. (2008) A beta-1,3 glucan sulfate induces resistance in grapevine against *Plasmopara viticola* through priming of defense responses, including HR-like cell death. *Molecular Plant-Microbe Interactions* 21, 232–243.
Varga, J. and Kozakiewicz, Z. (2006) Ochratoxin, A in grapes and grape derived products. *Trends in Food Science and Technology* 17, 72–81.
Visconti, A., Pascale, M. and Centonze, G. (1999) Determination of ochratoxin A in wine by means of immunoaffinity column clean-up and high performance liquid chromatography. *Journal of Chromatography A* 864, 89–101.
Vrabcheva, T., Usleber, E., Dietrich, R. and Märtlbauer, E. (2000) Co-occurrence of ochratoxin A and citrinin in cereals from Bulgarian villages with a history of Balkan endemic nephropathy. *Journal of Agricultural and Food Chemistry* 48, 2483–2488.
Walker, R. (1999) *Mycotoxins of Growing Interest, Ochratoxins*. Paper presented at the Third Joint FAQ/WHO/UNEP International Conference on Mycotoxins, Tunis, Tunisia, 3–6 March 1999. Document MYC-CONF/99/5b, Food and Agriculture Organization of the United Nations, Rome. Available at: ftp://ftp.fao.org/es/esn/food/myco5b.pdf (accessed 2 October 2015).

Wightwick, A.M., Mollah, M.R., Partington, D.L. and Allinson, G. (2008) Copper fungicide residues in Australian vineyard soils. *Journal of Agricultural and Food Chemistry* 56, 2457–2464.
Zimmerli, B. and Dick, R. (1995) Determination of ochratoxin A at the ppt level in human blood, serum, milk and some foodstuffs by HPLC with enhanced fluorescence detection and immunoaffinity column cleanup: methodology and Swiss data. *Journal of Chromatography B* 666, 58–99.
Zimmerli, B. and Dick, R. (1996) Ochratoxin A in table wine and grape-juice: occurrence and risk assessment. *Food Additives and Contaminants* 13, 655–668.

18 An Experimental Approach to Understanding Grapevine Yellows Associated with Phytoplasma Infections

G. Brader,[1*] S. Compant,[1,2] H. Gangl[3] and A. Sessitsch[1]

[1] *AIT Austrian Institute of Technology GmbH, Health and Environment Department, Tulln an der Donau, Austria;* [2]*LGC UMR 5503 (CNRS/INPT/UPS), Département Bioprocédés et Systèmes Microbiens, Université de Toulouse, Castanet-Tolosan, France;* [3]*Bundesamt für Weinbau, Eisenstadt, Austria*

Introduction

Phytoplasmas are cell wall-less bacteria of the class *Mollicutes* with low G + C content and small genome sizes. They colonize the phloem of plants and are transmitted by phloem-sucking Hemiptera. Phytoplasmas are associated with more than 100 plant diseases worldwide. The symptoms of phytoplasma infections are phyllody/virescence, yellowing, witches' broom formations, stunting and (lethal) declines and diebacks. They affect various domesticated plants, and can cause significant economic losses, particularly in high-value crops (Hogenhout *et al.*, 2008; Bertaccini and Duduk, 2009). In 2001, a single outbreak of apple proliferation phytoplasma resulted in losses of more than €100 million in Italy and €25 million in Germany (Strauss, 2009). The treatment of most phytoplasma diseases is difficult and concentrates on the control of vectors transmitting the disease. Direct plant treatments with antibiotics are possible, but not desirable for sustainable agriculture and also very costly and often ineffective (Bertaccini and Duduk, 2009). The study of phytoplasma diseases is complicated by the difficulty of cultivating these organisms *in vitro*. Recently, axenic culture of various phytoplasma strains has been reported (Contaldo *et al.*, 2012), but plant infection with axenic cultures has not yet been reported and the suppliers of phytoplasma media have been not able to confirm axenic phytoplasma growth using additional analytical approaches (Windsor and Windsor, personal communication, 2014).

In Europe, two phytoplasmas, *Candidatus* Phytoplasma solani (Stolbur), associated with 'bois noir' (BN), and *C.* Phytoplasma vitis, associated with 'flavescence dorée' (FD), are the cause of most Phytoplasma-associated diseases affecting grapevine (Constable, 2010). Symptoms of grapevine yellows (GY, the broader group to which BN and FD belong) include irregular yellowing and reddening, backward curling and necrosis of leaves, drying and decline of berries and dieback of affected shoots and plants.

*Corresponding author: Guenter.Brader@ait.ac.at

Measures to limit phytoplasma diseases include insecticide treatments to reduce vector populations and agricultural practices such as pruning. Grapevines affected by phytoplasma may also undergo remission and recovery. The rate of these spontaneous recoveries varies between years and the reasons for these variations are unclear. Moreover, there are indications that certain abiotic stresses and inducing agents of systemic acquired resistance (SAR) can affect the severity of phytoplasma infections and may also affect recovery rates (Romanazzi *et al.*, 2009; Musetti, 2010).

All of these factors indicate that alterations in plant stress and systemic resistance might play an important role in the plant defence against phytoplasmas. Little is known of the impact of beneficial bacteria and plant defence signalling on remission and recovery from phytoplasma disease. Here, we will briefly discuss the experimental challenges of evaluating the impact of beneficial microorganisms and plant effectors on phytoplama infections and describe the use of tomatoes and tomato signalling mutants for studying plant–phytoplasma interactions.

Materials and Methods

DNA from leaves of *Convolvulus arvensis* (bindweed) and *Catharanthus roseus* (periwinkle), segments of *Cuscuta pentagona* (dodder) and grapevine veins were isolated using the PhytoPure DNA extraction kit (GE Healthcare, Little Chalfont, UK) according to the instructions of the manufacturer.

C. pentagona seeds were treated for 30 min with concentrated sulfuric acid and washed with water before being sown close to 1-week-old tomato seedlings of cultivars Castlemart and MoneyMaker.

Tomato leaf graftings were performed by cutting tomato leaves with a sterile, sharp razor blade and immediately clipping two leaves together with 2-mm plastic clips from HILD samen (Marbach, Germany), and keeping the grafts under high humidity for a week.

Nested PCR was performed with FirePol DNA Polymerase (Solis BioDyne, Tartu, Estonia) and TUF1 and TUF AY primers (Schneider *et al.*, 1997) for *C. pentagona*, *C. arvensis* and grapevine samples, and direct PCR was performed with TUF AY primers (annealing temperature 53°C) for the *C. roseus* samples (Langer and Maixner, 2004).

The 16S rDNA was amplified according to Lee *et al.* (1998).

Results and Discussion

BN occurs frequently in the vineyards of Seewinkel east of Lake Neusiedl in Burgenland, eastern Austria. In grapevines of cultivar 'Zweigelt', the symptoms are sporadically visible and C. Phytoplasma solani infections have been confirmed by 16S rDNA and Tu (elongation factor) *tuf* gene sequence analysis. Sequencing the amplified *tuf* genes of affected grapevines and of the bindweed growing nearby with TUF AY reverse primers (Schneider *et al.*, 1997) revealed that both plants are infected with phytoplasmas of the tuf b (genetic) type (Langer and Maixner, 2004). Grapevine *tuf* sequences also showed that apart from the tuf a and b discriminating HpaII site at 507 (T/C), an additional polymorphism exists at position 568 (A/G), which is 'G' in eight out of nine tuf a sequence database entries as well as in five additional tuf b type sequence database entries from Croatian grapevines. The remaining 31 database entries were 'pure' tuf b type. Indeed, sequencing of the additional markers, *secY*, *vmp1* and *Stamp*, has confirmed a closer relationship of this strain to nettle-associated tuf a strains, and the strain has been named CPsM4_At1 (Aryan *et al.*, 2014).

To evaluate differences between phytoplasma strains and to investigate the role of plant defence signalling and endophytic colonization on plant resistance to phytoplasmas, an experimental system is required, but this is difficult to achieve owing to the difficulties of artificially infecting model plants. Tomatoes are hosts for Stolbur phytoplasmas, which cause flower abnormalities in this host, and have the advantage that several mutants allow investigations of plant defence mechanisms. The first challenge is to infect tomatoes with different Stolbur strains. Previously, dodder plants have been used to transfer phytoplasmas over the species barrier (Bertaccini and Duduk, 2009). We used Stolbur strains infecting bindweed at two locations in eastern Austria, as well as a 16Sr XII-A (Stolbur)-infected periwinkle (infected with CA-1, Phytoplasma collection #92) to attempt the transfer to tomato with the aid of dodder. Dodders (which are parasitic plants) were germinated next to tomato seedlings (germination rate >90%), and 16% of the young dodders became successfully attached to tomato plants. The dodder was then allowed to attach to periwinkle cuttings and to bindweed growing in soil. In all cases, the dodders attached to the bindweed and the periwinkle, and were PCR positive for phytoplasma 8 days after attaching to the host plants. Tomato cultivars with phytoplasma infection can then be used for the graft transmission and propagation of phytoplasmas.

To allow higher throughput of experimental infections we also evaluated the possibility of leaf grafting with grafting clips by grafting the front half of tomato leaves on one cultivar (cvs Castlemart, MoneyMaker and jai1 – a jasmonate-insensitive mutant) as leaf scion to the rear half of another (Castlemart; Table 18.1; Fig. 18.1) as leaf stock. Leaf grafts survived for more than 6 weeks and allowed transfer of phytoplasmas (evaluated by PCR), indicating successful phloem vessel attachment using this approach.

Conclusion

Phytoplasma effectors have been shown to reprogramme plant development and defence with effects on plant hormone signalling (Kazan and Lyons, 2014; MacLean *et al.*, 2014). Study of the manipulation of hormone signalling by phytoplasmas will aid understanding of Stolbur disease in tomatoes and in the grapevine. Grapevines especially contain rich endophytic communities (Compant *et al.*, 2011), which may play a role in the phenomenon of recovery and remission of BN disease. Tomatoes can be systematically infected with phytoplasmas, which will allow study of the role of plant defence pathways and evaluation of the effect of beneficial microorganisms on phytoplasma infections.

Table 18.1. Leaf graftings with Castlemart as leaf stock.

Leaf scion	Successful/total grafts
MoneyMaker	12/20
Castlemart	12/25
jai1	14/30

Fig. 18.1. A 6-week-old leaf graft of jai1 on Castlemart.

Acknowledgements

We thank W.T. Lanini for kindly providing us with *Cuscuta pentagona* seeds, the Tomato Genetics Resource Center and Gregg Howe for tomato seeds, and A. Bertaccini for phytoplasma-infected *Catharanthus roseus* and DNA.

References

Aryan, A., Brader, G., Mörtel, J., Pastar, M. and Riedle-Bauer, M. (2014) An abundant *Candidatus* Phytoplasma solani tuf b strain is associated with grapevine, stinging nettle and *Hyalesthes obsoletus*. *European Journal of Plant Pathology* 140, 213–227.

Bertaccini, A. and Duduk, B. (2009) Phytoplasma and phytoplasma diseases: a review of recent research. *Phytopathologia Mediterranea* 48, 355–378.

Compant, S., Mitter, B., Colli-Mull, J.G., Gangl, H. and Sessitsch, A. (2011) Endophytes of grapevine flowers, berries, and seeds: identification of cultivable bacteria, comparison with other plant parts, and visualization of niches of colonization. *Microbial Ecology* 62, 188–197.

Constable, F.E. (2010) Phytoplasma epidemiology: grapevines as a model. In: Weintraub, P.G. and Jones, P. (eds) *Phytoplasmas: Genomes, Plant Hosts and Vectors*. CAB International, Wallingford, UK, pp. 188–212.

Contaldo, N., Bertaccini, A., Paltrinieri, S., Windsor, H.M. and Windsor, G.D. (2012) Axenic culture of plant pathogenic phytoplasmas. *Phytopathologia Mediterranea* 51, 607–617.

Hogenhout, S.A., Oshima, K., Ammar, E.D., Kakizawa, S., Kingdom, H.N. and Namba, S. (2008) Phytoplasmas: bacteria that manipulate plants and insects. *Molecular Plant Pathology* 9, 403–423.

Kazan, K. and Lyons, R. (2014) Intervention of phytohormone pathways by pathogen effectors. *The Plant Cell* 26, 2285–2309.

Langer, M. and Maixner, M. (2004) Molecular characterisation of grapevine yellows associated phytoplasmas of the Stolbur-group based on RFLP-analysis of non-ribosomal DNA. *Vitis* 43, 191–199.

Lee, I.M., Gundersen-Rindal, D.E., Davis, R.E. and Bartoszyk, I.M. (1998) Revised classification scheme of phytoplasmas based on RFLP analyses of 16S rRNA and ribosomal protein gene sequences. *International Journal of Systematic Bacteriology* 48, 1153–1169.

MacLean, A.M., Orlovskis, Z., Kowitwanich, K., Zdziarska, A.M., Angenent, G.C., Immink, R.G.H. and Hogenhout, S.A. (2014) Phytoplasma effector SAP54 hijacks plant reproduction by degrading MADS-box proteins and promotes insect colonization in a RAD23-dependent manner. *PLoS Biology* 12(4), e1001835.

Musetti, R. (2010) Biochemical changes in plants infected by phytoplasmas. In: Weintraub, P.G. and Jones, P. (eds) *Phytoplasmas: Genomes, Plant Hosts and Vectors*. CAB International, Wallingford, UK, pp. 132–146.

Romanazzi, G., D'Ascenzo, D. and Murolo, S. (2009) Field treatment with resistance inducers for the control of grapevine bois noir. *Journal of Plant Pathology* 91, 677–682.

Schneider, B., Gibb, K.S. and Seemüller, E. (1997) Sequence and RFLP analysis of the elongation factor Tu gene used in differentiation and classification of phytoplasmas. *Microbiology* 143, 3381–3389.

Strauss, E. (2009) Phytoplasma research begins to bloom. *Science* 325, 388–390.

19 Endophytic Bacteria Associated with Grapevine Plants: Putative Candidates for Phytoplasma Containment

D. Bulgari, P. Casati, F. Quaglino and P.A. Bianco*

Dipartimento di Scienze Agrarie ed Abientali, Università degli Studi di Milano, Milan, Italy

Introduction

Bacteria residing in plant tissues without inducing symptoms of disease are defined as endophytes (Wilson, 1995). They are assumed to enter plant tissues from the adjacent rhizosphere either by passive diffusion or by active selection. Bacterial endophytes have been associated with different parts of plants, such as the roots, tubers, stems and leaves (Compant *et al.*, 2010), where they mainly inhabit the vascular system, the intercellular spaces and/or the cell cytoplasm. Even if bacterial endophytes are commonly present in plants, the implication of their association with the host tissues is basically unknown. Although most of them exhibit no detectable impact on their hosts (Kado, 1992), it has been demonstrated that bacterial endophytes can produce beneficial effects on their host plants, such as growth promotion, or reduction of the deleterious effects of plant pathogens (Compant *et al.*, 2010; Bulgarelli *et al.*, 2013). This ability to suppress plant pathogens can be used to develop new methods for pathogen containment. Among these pathogens, phytoplasmas are emerging worldwide as causal agents of important diseases, and endophytes could be used to contain them.

Phytoplasmas are pleomorphic, cell wall-less bacteria that exist as obligate trans-kingdom parasites inhabiting phloem sieve cells in plants and diverse cell types in insect vectors. They are classified in the class *Mollicutes*, which includes bacteria with a single membrane that have diverged from a Gram-positive ancestor, such as *Clostridium* or *Lactobacillus* spp., through genome reduction (Weisburg *et al.*, 1989). Based on their molecular and biological features, phytoplasma strains have been grouped within the genus '*Candidatus* Phytoplasma' and more than 35 species have been distinguished (IRPCM Phytoplasma/Spiroplasma Working Team, 2004; Duduk and Bertaccini, 2011). Plants infected by phytoplasmas are characterized by a broad range of symptoms that include virescence, phyllody, flower sterility, witches' broom (proliferation of the axillary buds), abnormal internode elongation and stunting (Bertaccini and Duduk, 2009). Phytoplasmas cause economically relevant yield losses in different annual and perennial crops worldwide, including fruit species and woody trees (Razin *et al.*, 1998; Lee *et al.*, 2000).

*Corresponding author: piero.bianco@unimi.it

Flavescence dorée (FD) and bois noir (BN), the two main diseases of the grapevine yellows (GY) complex associated with genetically distinct phytoplasmas, have been demonstrated as seriously damaging wine production in European countries and countries outside Europe (Batlle *et al.*, 1997; Gajardo *et al.*, 2009; Belli *et al.*, 2010). Until now, no vine (*Vitis vinifera* L.) cultivars have been found as either hosts or non-hosts that are resistant to phytoplasma infection (Laimer *et al.*, 2009). Neither are there any effective chemicals available for controlling phytoplasmas directly. Insecticide treatments against *Scaphoideus titanus*, the FD phytoplasma insect vector, are so far the only effective measure for controlling FD epidemics. Unfortunately, this strategy is less efficient for BN containment owing to the biological complexity of the BN phytoplasma life cycle, whose vector, *Hyalesthes obsoletus*, can also feed on other plant species (Langer and Maixner, 2004). However, the use of insecticides entails undesirable effects such as residual toxicity, environmental pollution and extra costs for the producers. Moreover, the guidelines of the European Community Common Agricultural Policy (CAP) (Directive 128/2009) require the use of sustainable agricultural practices as a means of environmental safeguarding. Thus, other sustainable approaches to control FD and BN are desirable (Bianco *et al.*, 2011).

Intriguingly, both FD and BN diseased plants may spontaneously recover, and remission of symptoms in the diseased plants may or may not involve the elimination of the pathogen from the host (Schmid, 1965). Recovery has been reported in apple, grape and apricot plants affected by phytoplasmas (Osler *et al.*, 1993; Musetti *et al.*, 2004). Physiological mechanisms and possible biological factors involved in this phenomenon are not yet understood. Recent studies highlighted a significant increase in the reactive oxygen species (ROS), pathogen related (PR) proteins and H_2O_2 accumulation through a selective down-regulation of enzymatic H_2O_2 scavengers (Musetti *et al.*, 2007; Albertazzi *et al.*, 2009). Recently, the idea that the recovery might be related to the activity of grapevine microbiota has arisen.

To date, analyses of endophytic bacterial species have been mainly performed through cultivation-dependent approaches in a large range of monocotyledonous and dicotyledonous plant species (see among others, Brooks *et al.*, 1994; Gutierrez-Zamora and Martínez-Romero, 2001). The development of cultivation-independent molecular fingerprinting methods based on 16S rRNA gene automated analysis and shotgun metagenomic sequencing has allowed more specific, replicable and detailed descriptions to be obtained of the diversity in complex bacterial communities. Length heterogeneity (LH)-PCR can be successfully used to describe the bacterial communities (Brusetti *et al.*, 2006), and has been previously applied for the characterization of endophytic community of potatoes (Sessitsch *et al.*, 2002) and sugarbeet (Dent *et al.*, 2004). Recently, these techniques have been applied to investigating how the community composition of endophytes is influenced by different parameters, such as plant genotype, growth stage, physiological status, tissue, environmental conditions and agricultural practices. Furthermore, the effects of pathogens on endophytic bacterial communities associated with plants have been examined (Trivedi *et al.*, 2010; Bulgari *et al.*, 2012, 2014).

In the study reported here, the endophytic bacterial communities associated with healthy and phytoplasma-infected grapevine plants have been characterized by LH-PCR and sequencing of 16S rRNA gene libraries from the total leaf DNA, and by bacterial cultivation. In addition, we used LH-PCR to compare bacterial community composition in healthy, GY-diseased and recovered grapevine plants. On the basis of our findings, we discuss the possible role of endophytic bacteria in the recovery phenomenon.

Materials and Methods

Grapevine leaf collection and DNA extraction

In September 2006, leaf samples were collected in a vineyard in the Lombardy region of north-western Italy. In the following year and from the same vineyard, leaf samples with the same chronological age were collected from each of 60 plants belonging to the same clone of the Barbera variety. The influence of environmental factors (such as climatic conditions, vineyard management, grapevine variety) was reduced by collecting samples in the same vineyard and on the same day. The plants from which leaves were collected were chosen from grapevines that were asymptomatic (20) and symptomatic (20) for phytoplasma disease, and from vines that had recovered (20); they were sampled on the basis of previous GY surveys carried out on this vineyard since 2000. Recovered plants had not showed GY symptoms for at least for 2 years after the remission. Additionally, four healthy, four diseased and four recovered grapevines, selected from among the plants previously examined, were also collected in October 2007 and September 2008.

Grapevine tissues were surface sterilized by washing with sterile water, followed by 70% ethanol for 3 min, 2% sodium hypochlorite for 5 min, 70% ethanol for 30 s and DNaseZap (Applied Biosystems) according to the manufacturer's instructions; finally they were washed with sterile water. Aliquots of sterile water, used in the final rinse step, were set on tryptic soy agar medium (TSA; Sigma). TSA plates were examined for bacterial growth after 5 days of incubation at 30°C, and the grapevine leaves that were not contaminated were used for further analysis. Bacterial DNA from entire grapevine leaves (20 g) was extracted by the method described by Bulgari *et al.* (2009).

Detection of phytoplasmas

Molecular identification of phytoplasmas was carried out on leaf veins separated by a sterilized blade. Total plant DNA was extracted using the DNeasy plant mini kit (Qiagen) according to the manufacturer's instructions. Phytoplasma amplicons were obtained by nested PCR by means of the universal primer pair P1/P7 (Deng and Hiruki, 1991), followed by the universal internal primer pair R16F2n/R16R2 (Lee *et al.*, 1998) (see Table 19.1). PCR reactions were executed in an automated thermal cycler (Mastercycler gradient, Eppendorf) with *Taq* polymerase (Invitrogen) as previously described (Deng and Hiruki, 1991; Gundersen and Lee, 1996). Nested PCR products (5 µl) were electrophoresed through 1% agarose gel and digested using the restriction enzyme *Mse*I (New England Biolabs) (Lee *et al.*, 1998). Restriction products were separated by electrophoresis through 5% polyacrylamide gel and visualized with a UV trans-illuminator.

Identification of endophytic bacteria by 16S rRNA gene library

The bacterial 16S rRNA genes of healthy plants were amplified through two distinct PCRs with the primer pairs 27F/1495R and 8F-I/907R-I (see Table 19.1). For the PCR with the 27F/1495R primer pair, the reaction mixture (50 µl) contained 1× PCR buffer,

Table 19.1. Primers used for PCR and sequencing.

Name	Primer sequence (5′–3′)	Gene target	Taxon target	Reference
P1	AAGAGTTTGATCCTGGCTCAGGATT	16S rRNA	Bacteria	Deng *et al.*, 1991
P7	CGTCCTTCATCGGCTCTT	23S rRNA	Bacteria	Schneider *et al.*, 1995
R16F2n (F2n)	ACGACTGCTGCTAAGACTGG	16S rRNA	Phytoplasmas	Lee *et al.*, 1998
R16R2 (R2)	TGACGGGCGGTGTGTACAAACCCCG	16S rRNA	Phytoplasmas	Lee *et al.*, 1998
16S-27F	AGAGTTTGATCCTGGCTCAG	16S rRNA	Bacteria	Lane, 1991
16S-1495R	CTACGGCTACCTTGTTACGA	16S rRNA	Bacteria	Lane, 1991
8F-I	GGATCCAGACTTTGATYMTGGCTAGI	16S rRNA	Bacteria	Ben-Dov *et al.*, 2006
907R-I	CCGTCAATTCMTTTGAGTTTI	16S rRNA	Bacteria	Ben-Dov *et al.*, 2006
338R	GCTGCCTCCCGTAGGAGT	16S rRNA	Bacteria	Ritchie *et al.*, 2000

2 mM $MgCl_2$, 0.2 mM dNTPs, 0.25 μM of each primer, and 1.25 U of AccuPrime™ *Taq* DNA polymerase, High Fidelity (Invitrogen). The initial denaturation of 2 min at 95°C was followed by 35 cycles of 1 min at 95°C, 2 min at 50°C and 3 min at 72°C, and a final extension of 7 min at 72°C. For the 8F-I/907R-I PCR, the reaction mixture (50 μl) contained 1 × PCR buffer, 2 mM $MgCl_2$, 1.5 mM dNTPs, 0.2 μM of each primer, 0.5 μg/ml BSA (bovine serum albumin) and 1.25 U of AccuPrime™ *Taq* DNA polymerase, High Fidelity. The initial denaturation of 4 min at 95°C was followed by 30 cycles of 30 s at 95°C, 2 min at 50°C and 3 min at 72°C, and a final extension of 7 min at 72°C.

The PCR products from both the amplifications were separated in 1% agarose gels. The bands were excised and purified with the QIAEX II kit (Qiagen), according to the manufacturer's instructions. Purified products were cloned in the plasmid vector pCR2.1-TOPO (Invitrogen) and propagated in *Escherichia coli*, as described by Shuman (1994). The plasmid DNA was extracted from *E. coli* colonies with the Qiagen Plasmid mini kit and sequenced in an ABI 3730 sequencer (Primm). Sequences were identified by comparison with the US National Institutes of Health (NIH) GenBank sequence database with the Basic Local Alignment Search Tool (BLAST) software at the US National Center for Biotechnology Information (NCBI) server (http://www.ncbi.nlm.nih.gov/BLAST). Rarefaction analysis was done using PAST 1.88 software (available at: http://folk.uio.no/ohammer/past/).

Development of an LH-PCR database

Cultivable bacteria associated with healthy, diseased and recovered grapevine plants were characterized by cultivation-dependent methods. Grapevine leaves (2 g) were sterilized as previously described. The sterilized leaves were ground in a pre-cooled mortar and mixed with a 'Ringer solution' (Oxoid). A partial volume (100 µl) of homogenized leaves, serially diluted in the Ringer solution, was incubated on TSA and LB medium at 30°C for 5 days. Bacterial colonies were selected on the basis of phenotypic traits and isolated. Each bacterium isolated was tested for catalase activity by laying a drop of 3% hydrogen peroxide directly in contact with the colonies.

The 16S rRNA gene PCR products, amplified with the 27F and 1495R primers, were cloned and sequenced with an ABI 3730 sequencer (Primm). Sequences were identified by comparison with the GenBank sequence database with the BLAST software. The cloned sequence for each bacterial strain was clustered in a neighbour-joining phylogenetic dendrogram bootstrapped 1000 times with the software MEGA4 (Tamura *et al.*, 2007). LH-PCR analyses were carried out on grapevine total DNA and on the 16S rRNA genes from both representative bacterial isolated strains and clones. Additional LH-PCR analyses were performed in order to obtain the characteristic peak(s) of the phytoplasma amplicons.

The LH-PCR reaction was done with the primers 27F/338R, as explained by Brusetti *et al.* (2006), with the primer 27F labelled at its 5′ end with the 6-FAM phosphoramidite dye. Quantified PCR products (100 ng) were added to 0.8 µl of 500 ROX-labelled internal size standard (Applied Biosystems) and 15 µl of deionized formamide. The samples were denatured at 95°C for 8 min, rapidly put into ice for 5 min, and loaded onto the ABI Prism 310 Genetic Analyzer (Applied Biosystems). The LH-PCR data were analysed with GeneScan 3.1.2 software (Applied Biosystems); a threshold of 50 fluorescent units was used. Peak sizing and peak matrix analyses were done with the GeneScan 3.1.2 software. For all of the bacterial strains and clones, the same PCR amplification was run three times, and three separate PCRs were also run to confirm the LH-PCR peak sizing through different PCR reactions.

The length of the LH-PCR fragments of the bacterial clones and isolates were used as a reference to tentatively attribute the single peaks in the LH-PCR profiles of the whole bacterial grapevine communities to the identified bacterial species.

LH-PCR profiles from healthy, diseased and recovered grapevine plants

To study the structure and diversity of the endophytic bacteria in healthy, diseased and recovered grapevines, total DNA extracted from 20 g of entire leaves was analysed by LH-PCR. The LH-PCR database developed as described above was used as reference to tentatively attribute the single peaks in the LH-PCR profiles of the whole bacterial grapevine communities. The LH-PCR reaction was done as described above.

Statistical analyses

The diversity and structure of the endophytic bacterial community associated with the leaves from 60 grapevines collected in September 2007 were analysed by univariate and multivariate indices. Bacterial richness was calculated as the presence of single peaks in the plants analysed.

To evaluate any possible relation between endophytic bacterial diversity and the health status of the grapevines, LH-PCR profiles from samples collected in October 2007 and September 2008 were processed by the same univariate analysis.

The repeatability of LH-PCR analyses was tested by contingency table analyses and the results were displayed as a mosaic plot.

The bacterial diversity associated with healthy, diseased and recovered grapevines was assessed using the univariate Jaccard similarity index. The multivariate correspondence analyses (CA) were performed to identify the LH-PCR peaks that determined a possible relation between bacterial diversity and the health status of the grapevine plants. The Jaccard analysis was performed using the NTSYSpc statistical software (Numerical Taxonomy System, version 2.01, Applied Biostatistics, 1986–1997) while CA and contingency table analyses were calculated using JMP software (JMP, version 7, SAS Institute, Cary, New Carolina, 1989–2007).

Results

Phytoplasma detection

On the basis of the *Mse*I enzymatic restriction patterns obtained from the R16F2n/R16R2 nested PCR products, FD (subgroups 16SrV-C/-D) and BN (subgroup 16SrXII-A) phytoplasmas were identified in single infection in the symptomatic grapevine plants (data not shown). No phytoplasmas were detected in asymptomatic (healthy and recovered) grapevine plants.

Characterization of the bacterial community associated with grapevine leaf tissues by cultivation-independent method

The diversity of the microbiota associated with healthy grapevine plants was investigated by sequence analysis of two 16S rRNA gene libraries, based on amplification with bacterial primer pairs 27F/1495R (library 271-A) and 8F-I/907R-I (library 271-B). A total of 162 clones (94 from library 271-A and 68 from library 271-B) were sequenced. The coverage of each single library was of 99 and 93%, respectively, for libraries 271-A and 271-B; the coverage of the pooled libraries was 98%.

Some 98% of the 159 clones yielded best matches with bacteria of the family *Enterobacteriaceae* within the α-*Proteobacteria*. Among these, 137 sequences (86%) were more than 97% identical to *Pantoea agglomerans*, which was the dominant bacterium in both of the libraries. Other bacterial species found in the libraries were *Ewingella americana*, retrieved in both libraries, and *Erwinia persicina* and *P. ananatis*, found in libraries 271-A and 271-B, respectively.

LH-PCR reference database

As previously described, to create the LH-PCR database for identifying bacterial species associated with LH-PCR peaks, endophytic bacteria were isolated from the leaf tissues of two healthy, two diseased and two recovered representative grapevine plants and were characterized by LH-PCR. Sixteen phenotypically diverse colonies were isolated from healthy grapevine plants, nine colonies from diseased plants and five colonies from recovered plants.

Sequences obtained by the sequencing of the 27F/1495R PCR products were closely related (sequence identity ≥97%) to: (i) the γ-*Proteobacteria* genera *Enterobacter*, *Pectobacterium*, *Sphingomonas* and *Stenotrophomona*; (ii) the α-*Proteobacteria* genera *Methylobacterium*, *Brevundimonas* and *Agrobacterium*; (iii) the β-*Proteobacteria* genus *Burkholderia*; (iv) the *Firmicutes* genera *Brevibacillus*, *Staphylococcus*, *Paenibacillus*, *Enterococcus*, *Oceanobacillus* and *Bacillus*; and (v) the *Actinobacteria* genera *Acaricomes*, *Curtobacterium* and *Streptomyces* (Fig. 19.1 and Table 19.2). The distribution of cultivable endophytic bacteria in the grapevines analysed is reported in Table 19.2. Summarizing, nine bacterial isolates were specifically associated with healthy plants, seven with diseased plants and four with recovered plants. Furthermore, *Acaricomes phytoseuli* was identified in both healthy and recovered plants, and *Bacillus subtilis* and *Staphylococcus epidermidis* were isolated from both healthy and diseased plants.

The bacteria were also tested for the possibility that they could detoxify hydrogen peroxide. All of the strains isolated from diseased and recovered grapevine plants had the catalase enzyme.

Diversity of the grapevine endophytic bacteria by LH-PCR

To investigate the bacterial diversity in grapevine leaves, LH-PCR peak profiles of bacteria from grapevine leaves collected in September 2007 were compared with those of the LH-PCR reference database. A total of 26 LH-PCR peaks were detected within 272–356 bp in the whole bacterial community associated with the analysed plants.

Twenty LH-PCR peaks were found in healthy grapevine plants (13 peaks that were specifically associated), eight in diseased grapevines (one peak that was specifically associated) and ten in recovered grapevines (three peaks that were specifically associated). Comparison between the LH-PCR profiles from September 2007 and the LH-PCR database allowed the attribution of ten of the 26 LH-PCR peaks in the range 304–317 bp and 338–367 bp to be assigned to 33 bacterial species (Table 19.2). Peaks at 317 bp (chloroplast and *Burkholderia fungorum*) and at 356 bp (*Bacillus* sp.) were detected in each of the examined grapevines. *Pectobacterium* sp. (342 bp) was associated with healthy, diseased and recovered plants, with a prevalence in recovered grapevines. Peaks attributed to *Bacillus pumilus* (304–350 bp) and to *Staphylococcus pasteuri* (350 bp) were identified only in healthy grapevines; a peak associated with *Brevundimonas* sp. (349 bp) was detected only in diseased grapevine plants; peak attributed to *Burkholderia* sp. (338 bp) was identified only in recovered grapevines. Furthermore, *P. agglomerans* (345–347 bp), *P. ananatis* (347 bp), *Brevibacillus brevis* (345 bp), *Enterobacter* sp. (345 bp) and *Curtobacterium* sp. (345 bp) were associated with both healthy and recovered plants.

Statistical analyses of LH-PCR data

Having analysed the diversity and structure of the endophytic bacterial community associated with healthy, diseased and recovered grapevine plants by LH-PCR, contingency

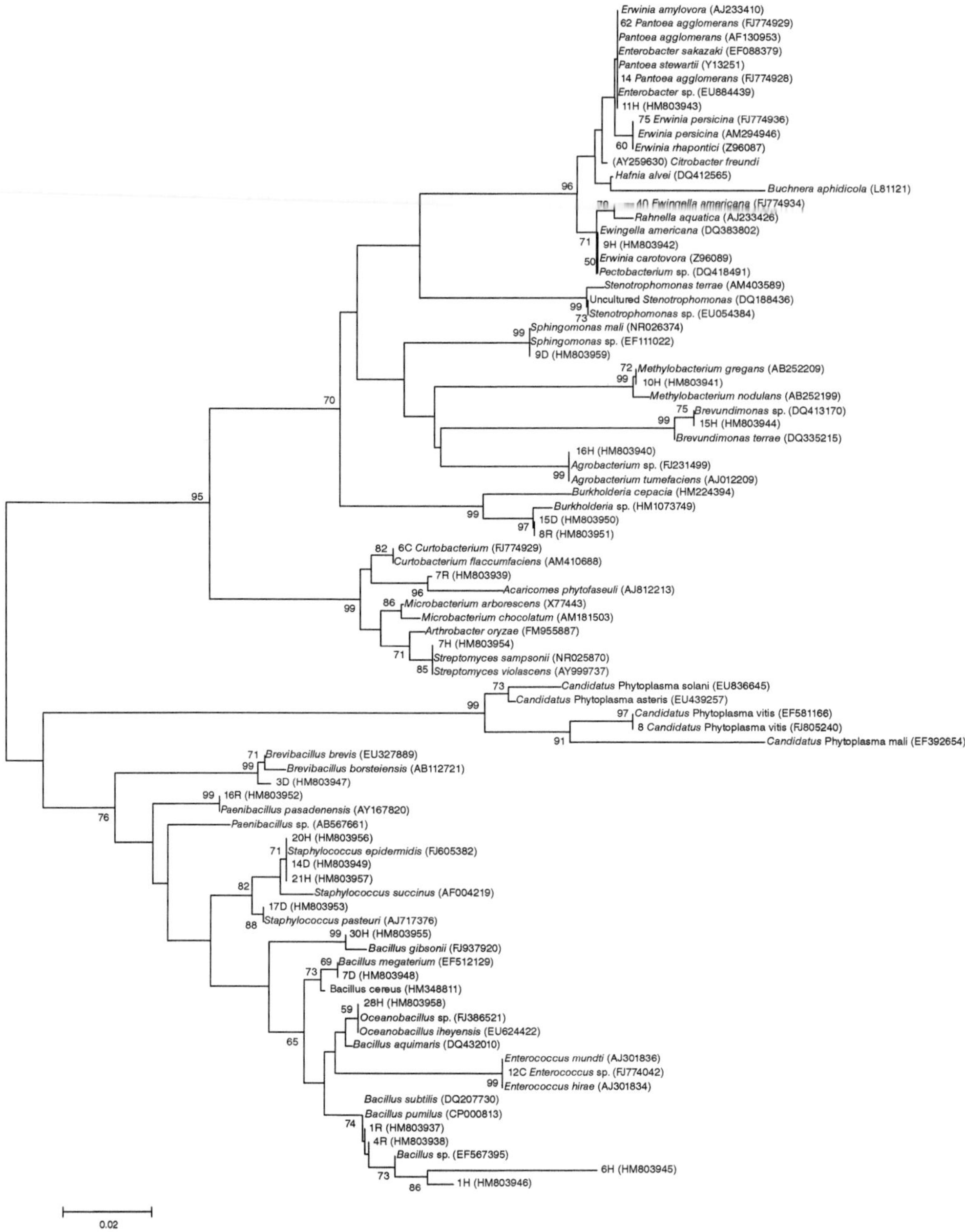

Fig. 19.1. Phylogenetic relationships based on partial 16S rDNA sequences obtained from the endophytic bacteria associated with grapevine tissues (in bold) and closely related sequences, retrieved from the US NIH (National Institutes of Health) genetic sequence database, GenBank. Bootstrap values are displayed at tree nodes.

table analyses, combined with chi squared (χ^2)-tests, were graphically portrayed in a mosaic plot. Two of the three independent PCRs shared the same mosaic plot profile, while the third was statistically different from the others with a significant value of $\chi^2_{\alpha\ 0.05} = 0.045$.

Table 19.2. Length heterogeneity (LH)-PCR fragment reference database representing bacteria isolated from leaves of healthy (H), diseased (D) and recovered (R) grapevines.

ID[a]	Acc. no.[b]	GenBank closest relative	GenBank Acc. no.	% match	Fragment length (bp)	Plant
1R	HM803937	*Bacillus pumilus*	FJ263042	100	304 (350)	R
14I	HM803949	*Sphingomonadaceae* bacterium	DQ520832	99	310 (314)	D
16S	HM803940	*Agrobacterium* sp.	AM269517	99	312	H
10S	HM803941	*Methylobacterium gregans*	AB252209	98	314	H
9I	HM803959	*Sphingomonas* sp.	EF111022	98	314	D
15I	HM803950	*Burkholderia fungorum*	HM113360	99	317	D
8R	HM803951	*Burkholderia* sp.	EU723189	99	338	R
16R	HM803952	*Paenibacillus pasadenensis*	AY167820	99	338	R
7S	HM803954	*Streptomyces violascens*	EU273550	95	340	H
30S	HM803955	*Bacillus gibsonii*	EU373538	99	341	H
9S	HM803942	*Pectobacterium* sp.	DQ418491	99	342	H
3I	HM803947	*Brevibacillus brevis*	FJ598016	98	345	D
11S	HM803943	*Enterobacter* sp.	EU884439	99	345	H
7R	HM803939	*Acaricomes phytoseuli*	AJ812213	99	348 (350)	H/R
15S	HM803944	*Brevundimonas* sp.	DQ413170	97	349 (351)	H
17I	HM803953	*Staphylococcus pasteuri*	GQ503327	99	350 (352)	D
1S	HM803946	*Bacillus subtilis*	FJ263034	99	351	H/D
4R	HM803938	*B. subtilis*	GQ150489	99	351	R
6S	HM803945	*Bacillus* sp. ZW2531	EF567395	97	352	H
20S	HM803956	*Staphylococcus epidermidis*	FJ613565	96	352	H/D
21S	HM803957	*Staphylococcus* sp.	GU003862	99	352	H
7I	HM803948	*Bacillus megaterium*	FJ460474	99	352 (355)	D
28S	HM803958	*Oceanobacillus* sp.	FJ386521	99	367	H

[a]Identification no.; [b]Accession no.

Jaccard analyses indicated significant differences ($r = 0.97$) among the endophytic bacterial communities associated with healthy, diseased and recovered grapevines collected in September 2007 (Fig. 19.2A). In detail, the bacterial community associated with healthy grapevines shared a similarity of >10% with those present in symptomatic and recovered plants, whereas the endophytes associated with symptomatic plants showed >50% similarity with the bacteria present in recovered grapevines. Moreover, four diseased plants shared 80% of similarity with recovered plants. In addition, the Jaccard analysis performed on plants collected in October 2007 and September 2008 displayed the same clustering.

Multivariate correspondence analysis (CA) was carried out to evaluate the microbial complexity in 20 healthy, 20 diseased and 20 recovered grapevines. The percentage of variation in species data was represented by the two ordination axes: C1 77% (77.25%) and C2 23% (22.75%) (Fig. 19.2B). Three different clusters of samples are clearly shown in the CA diagram. Healthy grapevine plants clustered in the upper part of diagram (C1 0.68, C 2 0), diseased plants, which had both coordinates negative, were placed in the lower left part of the plot, while recovered grapevines had C1 negative and C2 positive. The CA also highlighted which peaks are strictly associated with each kind of sample and which of them determine the clustering.

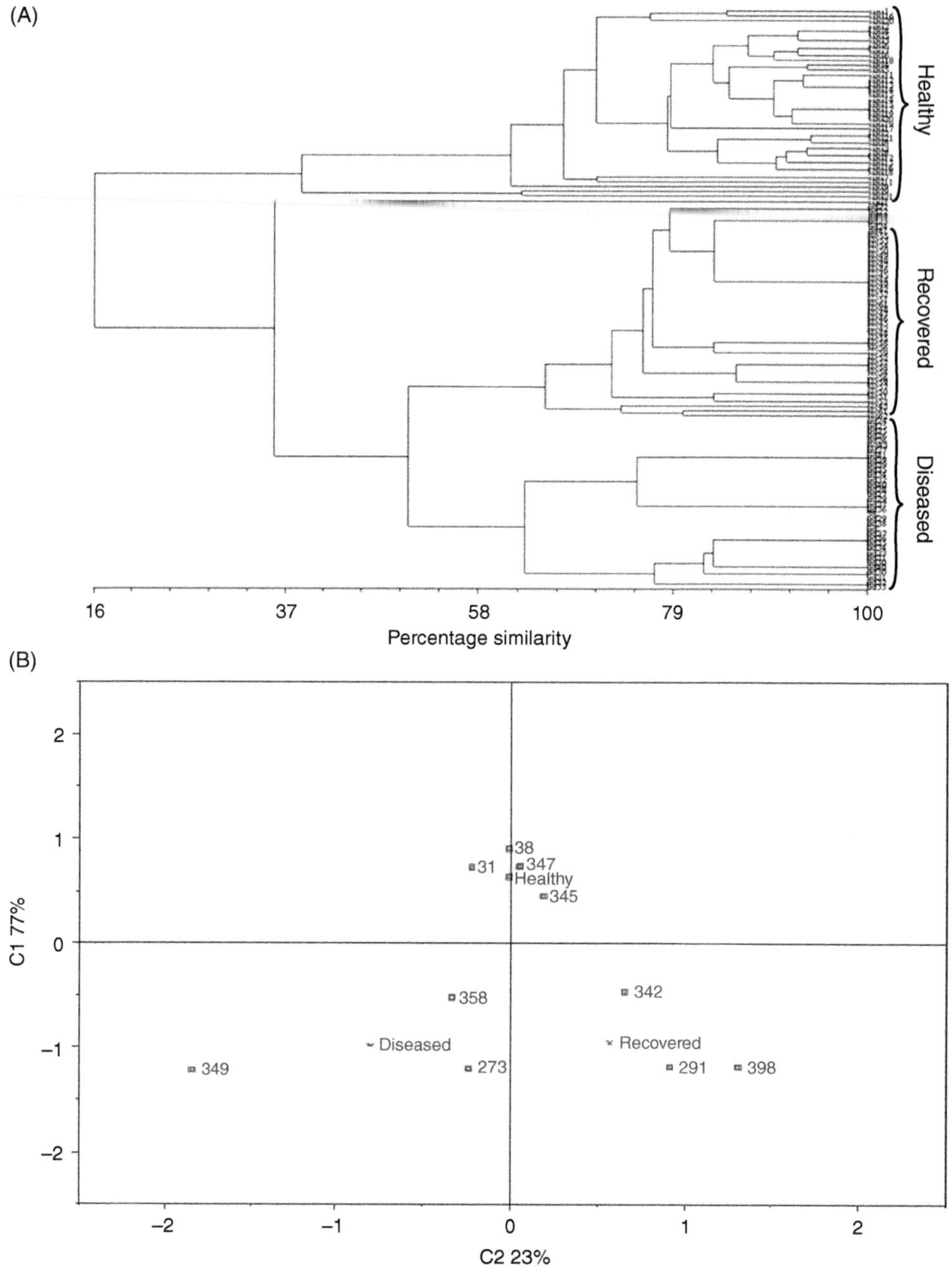

Fig. 19.2. Length heterogeneity (LH)-PCR profiles statistically analysed using: (A) the Jaccard index with $r = 0.97$; and (B) correspondence analyses. In (A), the numbers to the right represent the healthy (numbers prefixed sana), recovered (numbers prefixed rec) and diseased (numbers prefixed inf) profiles. In (B), the percentage variation in species data is represented by the two ordination axes C1 77% (77.25%) and C2 23% (22.75%). Three different clusters of samples are clearly shown in the CA diagram.

The LH-PCR peaks that determine the main separation on C1 are of 291, 294 and 338 bp (associated with diseased and recovered plants) and of 278, 280, 285, 287, 303, 306, 331, 334 and 350 bp (associated with healthy plants), while the separation between recovered and diseased plants on C2 was determined by the peaks at 349, 294 and 338 bp.

Discussion

The diversity of microbiota associated with grapevine leaves was investigated using both cultivation-dependent and cultivation-independent methods in order to increase the range of diversity explored in a sample.

Library analysis of a PCR-amplified 16S rDNA fragment showed best sequence matches with *Proteobacteria*, family *Enterobacteriaceae*, with a dominance of the genus *Pantoea*. Nucleotide sequences of 16S rDNA from bacteria isolated through cultivation showed the best sequence matches with *Bacillus*, *Curtobacterium*, *Stenotrophomonas*, *Methylobacterium*, *Pectobacterium*, *Enterobacter*, *Brevundimonas*, *Agrobacterium*, *Burkholderia*, *Brevibacillus*, *Staphylococcus*, *Sphingomonas*, *Acaricomes* and *Enterococcus*. Bacteria of these genera are commonly described as endophytes associated with different plant species (Perin *et al.*, 2006; Sun *et al.*, 2008; Ulrich *et al.*, 2008). Interestingly, some of them are reported as biocontrol agents against diverse plant pathogens (Chen *et al.*, 1995; Choudhary and Johri, 2009).

In this study, LH-PCR assays were used along with statistical analyses to investigate the association between endophytic bacteria and the recovery phenomenon from by comparing the endophytic bacterial diversity associated with healthy, diseased and recovered grapevine plants. Jaccard coefficient and correspondence analyses carried out on LH-PCR profiles highlighted that the composition of the endophytic bacterial community associated with healthy, diseased and recovered grapevines was significantly different. Interestingly, consistent data were observed in grapevines collected in the same vineyard but in different periods. This evidence suggests that the presence/absence of phytoplasmas can influence microbial community composition more than environmental factors. LH-PCR electropherogram analysis also allowed the determination of bacterial richness, which was higher (with a major number of LH-PCR peaks) in healthy plants than in phytoplasma-infected and recovered plants. However, bacterial richness in diseased and recovered plants was quite similar (with comparable numbers of LH-PCR peaks). This evidence was also confirmed by cultivation-dependent assays performed on grapevines collected in September 2007.

The diverse composition of the endophytic bacterial community in diseased plants could be the result of (i) a direct interaction between phytoplasmas and endophytic bacteria, or (ii) a phytoplasma-mediated plant response that restructured the endophytic bacterial community.

Phytoplasmas are obligate bacterial parasites characterized by a reductively evolved genome (Oshima *et al.*, 2004; Hogenhout *et al.*, 2008; Kube *et al.*, 2012). In the grapevine, phytoplasma infection causes various metabolic changes such as an inhibition of photosystem I activity, and an alteration of carbohydrate metabolism. Phytoplasmas also induce anatomical changes in the phloem tissues such as callose deposition on the sieve plates, leading to collapse of the sieve elements (Albertazzi *et al.*, 2009; Hren *et al.*, 2009). As endophytic bacteria mainly reside in the leaf intercellular spaces and vascular elements (Rosenblueth and Martínez-Romero, 2006), phytoplasmas could compete with them for carbon sources and for inhabitation of a favourable niche. Conversely, the higher endophyte richness present in some plants could prevent phytoplasma infection for the same competitive reasons.

The other hypothesis to explain the restructuring of endophytic bacterial communities living in diseased and recovered plants is that phytoplasmas induce plant defence responses that affect bacterial community composition. In particular, H_2O_2, in combination with other ROS, are essential components in the signal transduction cascade(s) leading to defence reactions, such as the hypersensitive response and the salicylic acid (SA)-mediated defence pathway (Foyer and Noctor, 2005). Recent research has shown that SA-dependent and SA-independent pathways reduce endophytic bacterial diversity and the ability of endophytic bacteria to colonize plants. In agreement with data reported by Trivedi *et al.* (2010) about '*Candidatus* Liberibacter asiaticus', phytoplasma infection could restructure the endophytic community, by selecting only few bacterial strains. Intriguingly, in this study, phytoplasmas apparently induced a rearrangement of the bacterial community, selecting endophytic strains capable of detoxifying H_2O_2, possibly because they could overcome ROS stress generated by the plant responses following infection. Moreover, some of these strains, i.e. *Burkholderia* sp., *B. pumilus*, *Paenibacillus pasadenensis* and uncultured *Bacillus*, are known as resistance inducers and we exclusively detected them in recovered plants. *Burkholderia* is a common genus in agricultural soil and some species are able to colonize a wide range of host plants (Compant *et al.*, 2008). *Burkholderia* sp. is also capable of controlling several plant pathogens owing to its ability to produce antifungal compounds (El-Banna and Winkelmann, 1998; Benítez and McSpadden Gardener, 2009; Suárez-Moreno *et al.*, 2012). A number of *Bacillus* strains express activities that suppress pathogens; e.g. *B. pumilus* caused hyphal inhibition activity by secreting two chitinases (Dehestani *et al.*, 2010; Shali *et al.*, 2010). This bacterium and other *Bacillus* strains, such as *B. subtilis* (isolated here), *B. cereus*, *B. pasteuri*, *B. sphaericus* and *Bacillus* spp. elicited significant reduction of disease severity on a diversity of hosts by causing induced systemic resistance (ISR) (Choudhary and Johri, 2009).

Once ISR is established, it protects plants from pathogens for a considerable part of their lifetime; this can happen because of a decrease of plant susceptibility and disease severity, or a reduction in the number of diseased plants (Van Loon, 2007). Mark *et al.* (2005) showed that disease suppression can be the result of alterations in microbial populations in the rhizosphere or the outcome of a change in the expression of resistance traits by these organisms. A similar result was presented in a study about recovery from dumping-off in turfgrass (Van Luijk, 1938). In the case of grapevine-associated phytoplasmas, it was observed that recovered plants maintain their health for at least 2 years (Morone *et al.*, 2007). What is more, recovered grapevines are not easily reinfected by phytoplasmas. The evidence that recovery in grapevines is stable and that ISR-inducing bacteria are present in recovered grapevines further indicate a possible involvement of endophytes in recovery from GY.

In conclusion, from the above data, it can be speculated that the alterations induced by phytoplasmas in the grapevine endophytic bacterial community – by selecting those bacterial strains more resistant to ROS and able to elicit plant defence responses, including ROS – may ultimately lead to recovery of the grapevines (Fig. 19.3). This view is supported by previously reported findings showing that recovered grapevine

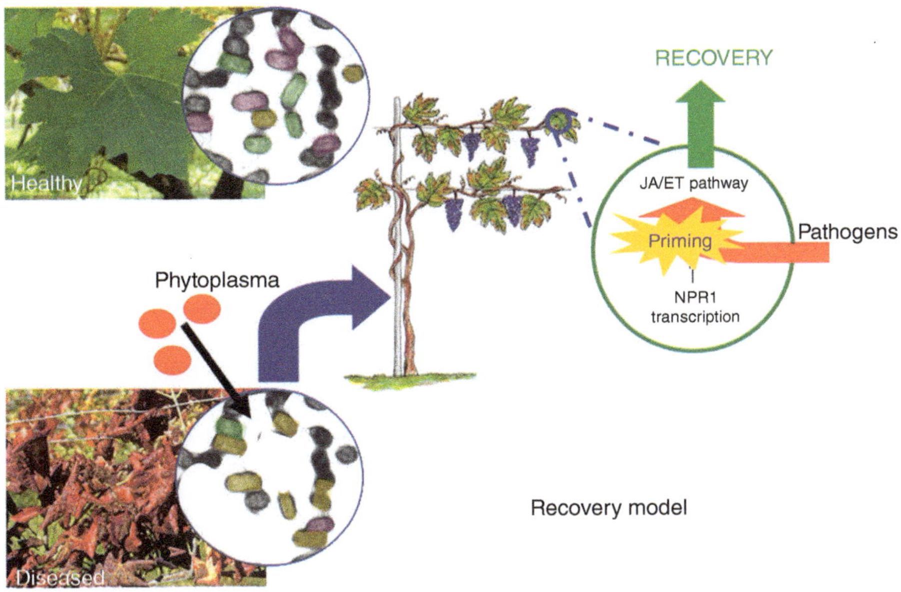

Fig. 19.3. Hypothetic recovery model from phytoplasma infection. The infection restructures the microbial community by selecting bacterial endophytes able to: (i) compete for carbon sources and/or to inhabit a favourable niche; and (ii) detoxify hydrogen peroxide and other reactive oxygen species (ROS) produced by infected plants. In diseased plants, specific bacterial endophytes elicit plant defence responses (priming status), ultimately leading to recovery. Key: JA/ET pathway: jasmonic acid/ethylene pathway; NPR1, non-expressor of pathogenesis-related (PR) gene 1.

plants have higher levels of ROS than diseased and healthy plants (Musetti *et al.*, 2007). In order to verify this hypothesis, future studies will focus on determining the relative abundance of putative recovery inducers within the microbial community living in grapevines.

The possibility that endophytic bacteria are involved in the recovery phenomenon opens new perspectives in the control of these detrimental diseases, and could be a starting point for developing environmentally friendly biocontrol strategies to be used in vineyards.

References

Albertazzi, G., Milc, J., Caffagni, A., Francia, E., Roncaglia, E., Ferrari, F., Tagliafico, E., Stefani, E. and Pecchioni, N. (2009) Gene expression in grapevine cultivars in response to bois noir phytoplasma infection. *Plant Science* 176, 792–804.

Batlle, A., Lavina, C., Kuszala, C., Clair, D., Larrue, J. and Boudon-Padieu, E. (1997) Detection of flavescence dorée phytoplasma in grapevine in northern Spain. *Vitis* 36, 211–212.

Belli, G., Bianco, P.A. and Conti, M. (2010) Grapevine yellows in Italy: past, present and future. *Journal of Plant Pathology* 92, 303–326.

Ben-Dov, E., Shapiro, O.H., Siboni, N. and Kushmaro, A. (2006) Advantage of using inosine at the 3′ terminal of 16S rRNA gene universal primers for the study of microbial diversity. *Applied and Environmental Microbiology* 72, 6902–6909.

Benítez, M.S. and McSpadden Gardener, B.B. (2009) Linking sequence to function in soil bacteria: sequence-directed isolation of novel bacteria contributing to soilborne plant disease suppression. *Applied and Environmental Microbiology* 75, 915–924.

Bertaccini, A. and Duduk, B. (2009) Phytoplasma and phytoplasma diseases: a review of recent research. *Phytopathologia Mediterranea* 48, 355–378.

Bianco, P.A., Bulgari, D., Casati, P. and Quaglino, F. (2011) Conventional and novel strategies for the phytoplasma diseases containment. *Phytopathogenic Mollicutes* 1, 77–82.

Brooks, D.S., Gonzalez, C.F., Appel, D.N. and Filer, T.H. (1994) Evaluation of endophytic bacteria as potential biological control agents for oak wilt. *Biological Control* 4, 373–381.

Brusetti, L., Borin, S., Mora, D., Rizzi, A., Raddadi, N., Sorlini, C. and Daffonchio, D. (2006) Usefulness of length heterogeneity-PCR for monitoring lactic acid bacteria succession during maize ensiling. *FEMS Microbiology Ecology* 56, 154–164.

Bulgarelli, D., Schlaeppi, K., Spaepen, S., Ver Loren van Themaat, E. and Schulze-Lefert, P. (2013) Structure and functions of the bacterial microbiota of plants. *Annual Review of Plant Biology* 64, 807–838.

Bulgari, D., Casati, P., Brusetti, L,. Quaglino, F., Daffonchio, D. and Bianco, P.A. (2009) Endophytic bacterial diversity in grapevine (*Vitis vinifera* L.) leaves described by 16S rRNA gene sequence analysis and length heterogeneity-PCR. *Journal of Microbiology* 47, 393–401.

Bulgari, D, Bozkurt, A.I., Casati, P., Caglayan, K., Quaglino, F. and Bianco, P.A. (2012) Endophytic bacterial community living in roots of healthy and '*Candidatus* Phytoplasma mali'-infected apple (*Malus domestica*, Borkh.) trees. *Antonie Van Leeuwenhoek* 102, 677–687.

Bulgari, D., Casati, P., Quaglino, F. and Bianco, P.A. (2014) Endophytic bacterial community of grapevine leaves influenced by sampling date and phytoplasma infection process. *BMC Microbiology* 14:198.

Chen, C., Bauske, E.M., Musson, G., Rodrìguez-Kàbana, R. and Kloepper, J.W. (1995) Biological control of *Fusarium* wilt on cotton by use of endophytic bacteria. *Biological Control* 5, 83–91.

Choudhary, D.K. and Johri, B.N. (2009) Interactions of *Bacillus* spp. and plants – with special reference to induced systemic resistance (ISR). *Microbiological Research* 164, 493–513.

Compant, S., Clément, C. Nowak, J., Coenye, T., Clément, C. and Ait Barka, E. (2008) Diversity and occurrence of *Burkholderia* spp. in the natural environment. *FEMS Microbiology Reviews* 32, 607–626.

Compant, S., Clément, C. and Sessitch, A. (2010) Plant growth-promoting bacteria in the rhizo- and endosphere of plants: their role, colonization, mechanisms involved and prospective for utilization. *Soil Biology and Biochemistry* 42, 669–678.

Deng, S. and Hiruki, C. (1991) Genetic relatedness between two non culturable mycoplasma like organisms revealed by nucleic acid hybridization and polymerase chain reaction. *Phytopathology* 81, 1475–1479.

Dent, K.C., Stephen, J.R. and Finch-Savage, W.E. (2004) Molecular profiling of microbial community associated with seeds of *Beta vulgaris* subsp. *vulgaris* (sugar beet). *Journal of Microbiological Methods* 56, 17–26.

Dehestani, A., Kazemitabar, K., Ahmadian, G., Jelodar, N.B., Salmanian, A.H., Seyedi, M., Rahimian, H. and Ghasemi, S. (2010) Chitinolytic and antifungal activity of a *Bacillus pumilus* chitinase expressed in *Arabidopsis*. *Biotechnology Letters* 32, 539–546.

Duduk, B. and Bertaccini, A. (2011) Phytoplasma classification: taxonomy based on 16S ribosomal gene, is it enough? *Phytopathogenic Mollicutes* 1, 3–13.

El-Banna, N. and Winkelmann, G. (1998) Pyrrolnitrin from *Burkholderia cepacia*: antibiotic activity against fungi and novel activities against *Streptomyces*. *Journal of Applied Microbiology* 85, 69–78.

Foyer, C. and Noctor, G. (2005) Oxidant and antioxidant signaling in plants: a re-evaluation of the concept of oxidative stress in a physiological context. *Plant, Cell and Environment* 28, 1056–1071.

Gajardo, A., Fiore, N., Prodan, S., Paltrinieri, S., Botti, S., Pino, A.M., Zamorano, A., Montealegre, J. and Bertaccini, A. (2009) Phytoplasmas associated with grapevine yellows disease in Chile. *Plant Disease* 93, 789–796.

Gundersen, D.E. and Lee, I.-M. (1996) Ultrasensitive detection of phytoplasmas by nested-PCR assays using two universal primer pairs. *Phytopathologia Mediterranea* 35, 144–155.

Gutierrez-Zamora, M.L. and Martínez-Romero, E. (2001) Natural endophytic association between *Rhizobium etli* and maize (*Zea mays* L.). *Journal of Biotechnology* 91, 117–126.

Hogenhout, S., Oshima, K., Ammar, E.D., Kakizawa, S., Kingdom, H. and Namba, S. (2008) Phytoplasmas: bacteria that manipulate plants and insects. *Molecular Plant Pathology* 9, 403–423.

Hren, M., Nikolic, P., Rotter, A., Blejec, A., Terrier, N., Ravnikar, M., Dermastia, M. and Gruden, K. (2009) 'Bois noir' phytoplasma induces significant reprogramming of the leaf transcriptome in the field grown grapevine. *BMC Genomics* 10, 460.

IRPCM Phytoplasma/Spiroplasma Working Team (2004) '*Candidatus* Phytoplasma', a taxon for the wall-less, non-helical prokaryotes that colonize plant phloem and insects. *International Journal of Systematic and Evolutionary Microbiology* 54, 1243–1255.

Kado, C.I. (1992) Plant pathogenic bacteria. In: Ballows, A., Truper, G.G., Dworkin, M., Harder, W. and Schleifer, K.-H. (eds) *The Prokaryotes*. Springer, New York, pp. 660–662.

Kube, M., Mitrovic, J., Duduk, B., Rabus, R. and Seemüller, E. (2012) Current view on phytoplasma genomes and encoded metabolism. *Scientific World Journal* 2012: 185942.

Laimer, M., Lemaire, O., Herrbach, E., Goldschmidt, V., Minafra, A., Bianco, P.A. and Wetzel, T. (2009) Resistance to viruses, phytoplasmas and their vectors in the grapevine in Europe: a review. *Journal of Plant Pathology* 91, 7–23.

Lane, D.J. (1991) 16S/23S rRNA sequencing. In: Stackebrandt, E. and Goodfellow, M. (eds) *Nucleic Acid Techniques in Bacterial Systematics. Modern Microbiological Methods*, Volume 4, Wiley, Chichester, UK, pp. 115–175.

Langer, M. and Maixner, M. (2004) Molecular characterization of grapevine yellows associated phytoplasmas of the Stolbur-group based on RFLP-analysis of non-ribosomal DNA. *Vitis* 43, 191–199.

Lee, I.-M., Gundersen-Rindal, D.E., Davis, R. and Bartoszyk, I.M (1998) Revised classification scheme of phytoplasmas based on RFLP analyses of 16S rRNA and ribosomal protein gene sequences. *International Journal of Systematic Bacteriology* 48, 1155–1169.

Lee, I.-M., Davis, R. and Gundersen-Rindal, D.E. (2000) Phytoplasma phytopathogenic mollicules. *Annual Review of Microbiology* 54, 221–255.

Mark, G.L., Dow, J.M., Kiely, P.D., Higgins, H., Haynes, J., Baysse, C., Abbas, A., Foley, T., Franks, A., Morrissey, J. and O'Gara, F. (2005) Transcriptome profiling of bacterial responses to root exudates identifies genes involved in microbe–plant interactions. *Proceedings of the National Academy of Sciences of the United States of America* 102, 17454–17459.

Morone, C., Boveri, M., Giosuè, S., Rossi, V., Scapin, I. and Marzachì, C. (2007) Epidemiology of flavescence dorée in vineyards in northwestern Italy. *Phytopathology* 97, 1422–1427.

Musetti, R., Sanità di Toppi, L., Emacora, P. and Favali, M.A. (2004) Recovery in apple trees infected with apple proliferation phytoplasma: an ultrastructure and biochemical study. *Phytopathology* 94, 203–208.

Musetti, R., Marabottini, R., Badiani, M., Martini, M., Sanità di Toppi, L., Borselli, S., Borgo, M. and Osler, R. (2007) On the role of H_2O_2 in the recovery of grapevine (*Vitis vinifera* cv Prosecco) from flavescence dorée disease. *Functional Plant Biology* 34, 754–758.

Osler, R., Carraro, R., Loi, N. and Refatti, E. (1993) Symptom expression and disease occurrence of a yellows disease of grapevine in north-eastern Italy. *Plant Disease* 77, 496–498.

Oshima, K., Kakizawa, S., Nishigawa, H., Jung, H.Y., Wei, W., Suzuki, S., Arashida, R., Nakata, D., Miyata, S., Ugaki, M. and Namba, S. (2004) Reductive evolution suggested from the complete genome sequence of a plant pathogenic phytoplasma. *Nature Genetics* 36, 27–29.

Perin, L., Martínez-Aguilar, L., Paredes, G., Baldani, J.I., Estrada-de los Santos, P., Reis, V.M. and Caballero-Mellando, J. (2006) *Burkholderia silvatlantica* sp. nov., a diazotrophic bacterium associated with sugar cane and maize. *International Journal of Systematic and Evolutionary Microbiology* 56, 1931–1937.

Razin, S., Yogev, D. and Naot, Y. (1998) Molecular biology and pathogenicity of mycoplasmas. *Microbiology and Molecular Biology Reviews* 62, 1094–1156.

Ritchie, N.J., Schutter, M.E., Dick, R.P. and Myrold, D.D. (2000) Use of length heterogeneity PCR and fatty acid methyl ester profiles to characterize microbial communities in soil. *Applied and Environmental Microbiology* 66, 1668–1675.

Rosenblueth, M. and Martínez-Romero, E. (2006) Bacterial endophytes and their interactions with hosts. *Molecular Plant-Microbe Interactions* 19, 827–837.

Shali, A., Ghasemi, S., Ahmadian, G., Ranjbar, G., Dehestani, A., Khalesi, N., Motallebi, E. and Vahed, M. (2010) *Bacillus pumilus* SG2 chitinases induced and regulated by chitin, show inhibitory activity against *Fusarium graminearum* and *Bipolaris sorokiniana*. *Phytoparasitica* 38, 141–147.

Schmid, G. (1965) Five and more years of observations on the proliferation virus of apples in the field. *Zaštita Bilja* 85–88, 285–289.

Schneider, B., Seemüller, E., Smart, C.D. and Kirkpatrick, B.C. (1995) Phylogenetic classification of plant pathogenic mycoplasma-like organisms or phytoplasmas. In: Razin, S. and Tully, J.G. (eds) *Molecular Microbial and Diagnostic Procedures in Mycoplasmology*. Academic Press, San Diego, California, pp. 369–380.

Shuman, S. (1994) Novel approach to molecular cloning and polynucleotide synthesis using vaccinia DNA topoisomerase. *The Journal of Biological Chemistry* 269, 32678–32684.

Sessitsch, A., Reiter, B., Pfeifer, U. and Wilhelm, E. (2002) Cultivation-independent population of bacterial endophytes in three potato varieties based on eubacterial and actinomycetes-specific PCR of 16S rRNA genes. *FEMS Microbiology Ecology* 39, 23–32.

Suárez-Moreno, Z.R., Caballero-Mellado, J., Coutinho, B.G., Mendonça-Previato, L., James, E.K. and Venturi, V. (2012) Common features of environmental and potentially beneficial plant-associated *Burkholderia*. *Microbial Ecology* 63, 249–266.

Sun, L., Qiu, F., Zhang, X., Dai, X., Dong, X. and Song, W. (2008) Endophytic bacterial diversity in rice (*Oryza sativa* L.) roots estimated by 16S rDNA sequences analysis. *Microbial Ecology* 55, 415–424.

Tamura, K., Dudley, J., Nei, M. and Kumar, S. (2007) MEGA4: molecular evolutionary genetics analysis (MEGA) software version 4.0. *Molecular Biology and Evolution* 24, 1596–1599.

Trivedi, P., Duan, Y. and Wang, N. (2010) Huanglongbing, a systemic disease restructures the bacterial community associated with citrus roots. *Applied and Environmental Microbiology* 76, 3427–3436.

Ulrich, K., Ulrich, A. and Ewald, D. (2008) Diversity of endophytic bacterial communities in poplar grown under field condition. *FEMS Microbiology Ecology* 63, 169–180.

Van Loon, C.L. (2007) Plant responses to plant growth-promoting rhizobacteria. *European Journal of Plant Pathology* 119, 243–254.

Van Luijk, A. (1938) Antagonism between various microorganisms and different species of the genus *Pythium*, parasitizing upon grasses and lucerne. *Mededelingen van het Phytophologisch Laboratorium Willie Cammelin Scholter* 14, 43–83.

Weisburg, W.G., Tully, J.G., Rose, D.L., Petzel, J.P., Oyaizu, H., Yang, D., Mandelco, L., Sechrest, I., Lawrence, T.G. and Van Etten, J. (1989) A phylogenetic analysis of the mycoplasmas: basis for their classification. *Journal of Bacteriology* 171, 6455–6467.

Wilson, D. (1995) Endophyte – the evolution of a term, and clarification of its use and definition. *Oikos* 73, 274–276.

Conclusions

As the reader has seen in this book, new biocontrol agents, endophytes, plant growth-promoting bacteria (PGPR) and fungi (PGPF), oomycetes and plant resistance inducers can be used to reduce phytopathogenic infections on grapevine. This strategy can be used under organic approaches or integrated pest management (IPM) systems.

Over the past few years, considerable information on the possibilities of using biocontrol agents or plant resistance inducers to fight a variety of grapevine diseases, thereby improving yield and productivity, has become available. Additionally, differences in the microflora between diseased and non-diseased plants have been partly revealed. In this book, the focus has been on the control of grey mould, trunk diseases, powdery and downy mildews, phytoplasma diseases and the occurrence of ochratoxinogenic fungi.

There are several examples of how we can reduce infections in vineyards that are caused by a variety of microorganisms, and new strategies to accomplish this can also be found in the book. Continuous research for effective biocontrol agents and plant resistance inducers, and studies into their mechanism of action, are, however, very important for the development of effective biocontrol agents that will allow sufficient disease management for specific grapevine diseases in viticulture. There are, though, not yet enough examples of effective biocontrol agents. We need more examples of new biocontrol agents and plant resistance inducers, and of their use to counteract phytopathogenic diseases of grapevine. There is a current need for a portfolio of different products for practical use. This will allow more efficient disease treatment. More research on the applications of each biocontrol agent or plant resistance inducer will also be of high importance. This should subsequently lead to their better application in disease management.

Additionally, the development of new types of biocontrol products as well as efficient formulations and delivery mechanisms are key to success in reducing grapevine diseases. Study of the effects that new biocontrol agents have in controlling grapevine diseases, or in stimulating plant defence reactions, is of special importance for improving fundamental knowledge on and the development of new products that would be allowed under organic or IPM systems. New sources of biocontrol agents and an understanding of the native microflora that can be responsible for reducing disease on plants is also of special interest. The use of new technologies to understand mechanisms of action as well as microbial ecology is also of importance.

Although it is an established strategy to use biocontrol agents and plant resistance inducers to reduce grapevine diseases, what is certain is that we will continue research on this topic. Various stakeholders, including farmers, industries, policy makers and scientists, can benefit from this. We do not only make science, we also drink and make wines. We also appreciate the moment of tasting and want to guarantee a new age for our products. Time has moved on from the era when a lot of pesticides were used, which, as we now realize, had dramatic effects on the ecosystem and on human health. Better products that do not harm human health or the environment should be guaranteed. We are now at the point where we can reduce the use of chemical phytosanitary products and develop a better and sustainable management of viticulture, for science and for us.

S. Compant and F. Mathieu

Index

Page numbers in **bold** type refer to figures and tables.

Acinetobacter lwoffii 18, 20
PTA-113 strain 64, 68
acremines 39, 100, 110, 135, **136**
Acremonium spp. 10
A. byssoides 39, 135
A. persicinum 101
A20 isolate 99, 100–101, 103–107, 110
distribution, as endophytes on Italian cultivars 101, **102**
active oxygen species *see* reactive oxygen species
Alternaria spp. 10
A. alternata 140, 183
aluminium chloride 3, 8
Ampelomyces quisqualis 10, 113–116, **114**, 120–121
antibiotics
acremines from *Acremonium* A20 100, 110
antifungal agents
from *B. amyloliquefaciens* AG1 185–186, 188, **188**
from *S. algeriensis* strain 87, 89–90
effective against downy mildew 130–139
identification and testing
assay by plate diffusion method 78, 80, **80**, 85, **88**
dual culture assay 152, 183, **186**
purification and characterization 78–79, 81, **81**, 140, 184
sandwich agar plates for volatile antibiotics 152, 154
produced by *Pantoea ananatis* 56–57
scope of *Bacillus* spp. antimicrobial activity 72, **72**
Aphanocladium album 10
Arabidopsis thaliana
defence signalling-pathway mutants 85, 89, **90**, 90–91
growth conditions for challenge testing 86–87
used in *Streptomyces* J-2 extract bioassay 77, 79, 82, **82**
Ascophyllum nodosum (seaweed) extract 8, 130
Aspergillus spp. 183
A. carbonarius as assay indicator 184, 185, **188**
A. ochraceus 205
section Nigri (black aspergilli) 191–192, 194, 196–197, 205
Aureobasidium pullulans
abundance in organic viticulture 40, 55, **55**
as biocontrol agent 14, 16, 40
effect on organic wine flavour 41
automated ribosomal intergenic spacer analysis (ARISA) 41–42, **57**

BABA (beta-aminobutyric acid) 6, 142–143
Bacillus spp. 20–21, 37
antimicrobial activity 72, **72**, 226
mechanisms 72–74, **73**
sensitivity of fungi to crude protein extract 185, **187**
B. amyloliquefaciens AG1 182, 183–184, 185, 188
B. subtilis PTA-271 64, 66, 67
phylogenetic analysis **74**, 74–75
bacteria
antibiotics production 56–57, 72, 76–77, **131**
associated with grapevines 53–56, 216
endophytes, diversity 36–39, **45**, 55, 182, 221
tested for plant resistance enhancement 64–65, 71
used as biocontrol agents 18–21, **19–20**
visualization by FISH techniques 93–96, **96**, 182–183, **184**
Balkans endemic nephropathy (BEN) 191
BCAs *see* biocontrol agents
beauvericin (antibiotic) 135
benzothiadazole (BHT) 143
beta-aminobutyric acid (BABA) 6, 142–143
biocontrol agents (BCAs)
laboratory versus field performance 22–23, 53, 113, 179
combined agent strategies 121
evaluation against canker/dieback 153–157
microbiome colonization studies 59
sources 231
culture collections 113–114
grapevine microhabitats 53–54, **54**, 151
harsh environments 84

biocontrol agents (BCAs) (*continued*)
types used in viticulture
bacteria 18–21, **19–20**
fungi 10–14, **11–12**, 39–40
oomycetes 22
yeast-like fungi 14–18, **15–16**
biocontrol mechanisms 13, 52, 76, 161
antibiosis 56–57, 72, 80–81, 130–139
endophytes activated by pathogen infection 101, 103, 110
hyperparasitism 113, 120–121
inhibition of pathogen protein kinase C 140–142
see also induced systemic resistance
black dead arm (BDA) disease 162, 165, **165**
BLAST sequence analysis 71, 219
comparison with reference databases 78, 218
bois noir (BN) disease
causal pathogen 210
incidence and control 211, 216
Bordeaux mixture 2–3
Botector® (*Botrytis* biocontrol product) 16, 40, 56
Botryosphaeriaceae family
causal agents of grapevine canker 150, 151
isolation of potential biocontrol agents 151–152, 154
pathogen–antagonist interactions
evaluation methods 152–154
field trials of protective agents 156–157, **157, 158**
observations and test results 154–156, **155, 156**
Botrytis cinerea (grey mould pathogen)
antagonistic microbes/bioproducts 56–59, 72, 81
agar plate antifungal assay 85, **88**, 183
tested with detached grapevine leaves 79, **82**
whole plant testing by root treatment 86
conidia used for challenge inoculation 65, 70, 86, 87
infection and disease in grapevines 52
microscopic appearance 72, **73**
necrotic spot symptoms on leaves 86, **88**
ISR, inducers in grapevine protection 64
Bacillus strains B27 and B29 73–74
Saccharothrix algeriensis NRRL B-24137 87–88, **88**, 90
see also grey mould
Bulleromyces albus 18
Burkholderia spp.
Burkholderia phytofirmans 21, 43
Burkholderia sp. antifungal compounds 37, 226

calcium 3
Candida spp. 16–17
canker diseases, symptoms and impacts 150
cell walls
effects of calcium 3
elicitin-like proteins of *Pythium oligandrum* 117, 122
fungal, degradation by antagonists 40–41, 72
Chaetomium spp. 10, 12
chelerythrine (PKC inhibitor) 141, 142
chemical pesticides
banned fungicides 160
for control treatments in antagonist testing 153, 155
limited effectiveness for pruning wound protection 150
preharvest, compared with bioelicitor treatment 193, 196–203
risks and costs of use 1–2, 52, 125, 192, 231–232
chitin and chitinase activity 6, 176, **177**
chitosans 3, 6–7, 176, 177, 192
Cladosporium spp. 153, 156, 183
clearing methods, for leaf microscopy 99, 183
Clonostachys rosea see Gliocladium roseum
colour evaluation, grape juice 195–196, 201–202, **203**
compost extracts (inducing resistance) 9–10
confocal laser scanning microscopy (CLSM) 43, 56, **59**, **103**, 183
copper sulfate 2–3
correspondence analysis (CA) 220, 223–225, **224**
costic acid 129, **130**
critical dilution assay, antifungal activity 184, 185–186, **187**
Cryptococcus spp. 17
cryptosporiopsin A (antibiotic) 134–135, **135**
Cryptosporiopsis sp. 133–135, **135**
culture collections, *Ampelomyces quisqualis* 113–114, **114**
culture-independent microbial analysis 35–36, 41–45, 216, 220
culture media *see* growth media
cuttings (vine) used in laboratory studies 118, 172, 175–176
cystospores (downy mildew) 126

denaturing gradient gel electrophoresis (DGGE) 41
2,4-diacetylphloroglucinol (DAPG) 132–133, **133**
dieback
causal pathogens 151, 210
disease prevalence, France 160
symptoms and preventive measures 150
see also Eutypa lata
diketopiperazines (DKPs) 140, **140**
Diplodia seriata (canker pathogen) **155, 156**

diseases
counteracted by beneficial bacteria 84–85
critical stages in cycles, for pathogenesis 125
economic costs of control 150
factors influencing expression 161
pressure level, and efficacy of control products 10, 22–23
prevalence in French vineyards 160
DNA
extraction from leaves 211, 217
extraction from wood samples 175
method of extraction from microbial isolates 77
tagging analysis 44
dodder, for phytoplasma transmission 211, 212
DOPE-FISH bacterial visualization tool 94, 95–96, **96**
downy mildew
control strategies 125–126, 139–140, 143–144
isolation of pathogen antagonists 99
symptom emergence on leaves 103, **104**, 104–105, **106**
reduced by *Acremonium* A20 105, 107, **107**
visual assessment for bioassay 128, **129**
see also Plasmopara viticola
dual culture assay for antagonists 152, 183, **186**

elicitin-like proteins, *Pythium oligandrum* 117, 122
Endodesmia calophylloides
plant extract bioactivity 130, **131**
source of antibiotic-producing endophyte 135
endophytes
bacteria 36–39, **45**, 55, 182, 225
definition 35, 215
fungi 39–41, 98, 133–136
isolation assays, extraction method 99, 151
molecular research techniques 35–36, 41–45, 96, 216
roles in plant health 35, 36, 53, 98
visualization by FISH/CLSM 184–185, **185**, 186–188
Epicoccum nigrum 12, 40
epiphytic microorganisms 54, **54**, 151, 185, **185**
epothilone A (antibiotic) 136
Erwinia herbicola see Pantoea agglomerans
Erysiphe necator (powdery mildew pathogen) 117, 121
strains used for biocontrol assays 118, 120, 122
esca decline of vines
disease prevalence, France 160
FISH observation of bacteria on/in leaves 184–185, **185**, 187–188
slow-developing form 167–168, **168**, 172
symptoms 161, 162, 165, **165**
virulence factors and cultivar tolerance 171
Esquive® WP (GTD biocontrol product)
application and evaluation methods 162, 165
bioactivity of control agent 161, 169
effects on vine plant health 166–168, **167**
efficacy against esca/BDA 165, **166**
French trial sites 162, **163**, **164**
essential oils, bioactivity
cinnamon oil 153
Cymbopogon citratus (lemon grass) 140
Viola odorata 8
ethylene signalling, role in induced resistance 89, 90–91
European Union (EU), directives/regulations 1, 191, 216
Eutypa lata (dieback pathogen)
infection and pathogenesis 161
pathogen–antagonist interactions
evaluation methods 152, 153–154
field testing results 156–157, **157**, **158**
potential biocontrol agents 151–152, 154
Ewingella americana 220

fenugreek extract (FEN 560) 192, 193
compared with standard fungicides 196–203
protection mechanisms 204, 205
flavescence dorée (FD) disease
causal pathogen 210
control of insect vector 216
effect on bacterial endophytes 37–38
fluorescence *in situ* hybridization (FISH) 42–43, **53**, 93–94, 182–183, **184**
bacteria in asymptomatic and esca leaves 184–185, **185**, 186–188
doubly-labelled probe (DOPE-FISH) method 94, 95–96, **96**
fluorochromes
DAPI stain for oospore nuclear DNA 101
labelling of FISH probes 94, 95
Fomitiporia mediterranea (FOMED) 183–184, **186**
fungi
associated with grapevines 56, **57**
endophytic communities 39, 40–41, 101, **102**
ochratoxinogenic, diversity study 192–194, **193**, **196**, 196–197
pathogens of major grapevine diseases 1, 52, 150, 161
as biocontrol agents 10–14, **11–12**, 39–40
variation between strains **115**, 115–116
extracts inducing resistance 9, 133–136, **134**
see also Botrytis cinerea; yeast-like fungi
Fusarium proliferatum 12–13, 135–136

gene expression bioassay (method) 79, **82**, **174**, 174–175
giant knotweed *(Reynoutria sachalinensis)* extract 8
Gliocladium roseum 13
Gram-positive bacteria
ancestral to phytoplasmas 215
antagonistic *Streptomyces* spp. 137–139, **138**, **139**
grape marc extract 9–10
grapevine yellows (GY) diseases
recovery, and endophytic bacteria 37, 216, 225–227, **227**
symptoms and causal agents 210
grapevines (*Vitis vinifera* L.)
berry and juice characteristics 192, 195–196
effect of farming method on microbial populations 54, **54**
endophytic communities 36, 44–45, **45**
Acremonium spp., distribution 101, **102**
habitat niches for microorganisms **53**, 53–54
major disease pathogens 1, 52, 125, 150, 210
used for experiments, growth conditions 85–86, 118, 162
grey mould
causal pathogen *(Botrytis cinerea)* 52
current control strategies 3, 70, 71
reduced by root treatment with bacteria 67, **68**, 90
see also Botrytis cinerea
growth media
bacterial
Bennett medium 78
ISP2 (International Streptomyces Project) medium 77, 85
LB (Luria-Bertani) medium 65
tryptic soy agar (TSA) 217
cell suspension culture (*Vitis* cells), Nitsch medium 65
experimental plants/plantlets
Gamborg agar medium 79
Murashige and Skoog (MS) medium 65, 86
sterilized soil-based media 85–86, 172
fungal
malt extract agars/broths 99, 100
potato dextrose agar (PDA) 70, 151
synthetic grape medium (SGM) 205
obligate biotrophs
phytoplasmas, culture difficulty 210
use of host tissue 127, 128, **128**
GTDs *see* trunk diseases of grapevines

Hanseniaspora uvarum 18
homogenous solution assays 127–128
HPLC analysis methods 78, 81, 194, 195
human health
benefits of red wine polyphenols 204
effects and sources of ochratoxin A (OTA) 191
pathogens present in grapevines 38

image analysis, grapes 195, **195**
indolocarbazole alkaloids 138–139, **139**
induced systemic resistance (ISR) 2, 64, 93
activation of defence gene expression 82, **82**, 176, **177**, 226
assay by challenge inoculation 65, 71, 86–87, 118
bacteria-induced, against *Botrytis cinerea* 65–68, 73–74, 87–91, **88**, **89**
mechanisms against *P. viticola* on grapevine 142–143
primed by natural products 126, 205
by *Pythium oligandrum* strains/products 117–118, 120–122
inorganic compounds, resistance inducers 2–3
integrated pest management (IPM) 231
inter-simple sequence repeat (ISSR) analysis 118–119, **119**
internal transcribed spacer (ITS) analysis 41, 55
Inula viscosa extract 9, 129, **130**
isocoumarin antibiotics 138, **139**

Jaccard similarity index 220, 223, **224**
jasmonic acid (JA) 2, 7
jasmonate pathway role in induced resistance 89, 90–91, 212
juice analysis, methods 193
ochratoxin A 194
quality evaluation (pH, colour, components) 195–196

khatmiamycin (antibiotic) 137–138, **138**

laminarin 7, 143, 176, 192
Lasiodiplodia theobromae (canker pathogen) 154, **155**, **156**, 183
leaf assays for biocontrol
detached leaves 79, **82**, 128
plated leaf discs 118, 120, 128, **129**
leaf grafting for phytoplasma transmission 212, **212**, **213**
length heterogeneity PCR (LH-PCR) 42, 216
used for endophyte diversity database 218–219, 220–221, **223**
lipoxygenase activity
assay method 71
elicitiation at model pruning wounds 176, **177**

induced by chitosans 6
stimulated by *Bacillus* spp. pretreatment **73**, 73–74
Lysobacter capsici 21, 132

malic acid content, grape juice 199–200, **200**
media (laboratory culture) *see* growth media
Melaleuca alternifolia (tea tree) extract 8
methyl jasmonate 7, 192
Metschnikowia spp. 17
microbe-assisted molecular patterns (MAMPs) 7, 64, 65, 67
microbial communities
characterization by molecular profiling 35–36, 41–45, 216, 218–220
ecological interactions 35, 36, 59
localization on/inside plants, FISH techniques **53**, 93–96, **96**
niches/habitats associated with grapevines 53–54, **54**
grape berry surface 192, **196**, 196–197
wood/vine trunk 151, 172
microscopy
confocal laser scanning (CLSM) technique 43, 56, **59**, **103**, 183
observations of fungal hyperparasitism 103, **104**, **105**, 107
quantitative bioassay of spore inhibitors 127–128, 152–153
scanning electron microscopy 195, **198**
study of grey mould infection on leaves 72, **73**
techniques for endophytic fungi 99, 101, 183, 184–185
mitochodrial inhibition by DAPG 133
molecular research techniques 35–36, 41–45, 172
see also sequencing, genetic
mycotoxins 191

natural extracts inducing resistance **4–5**, 8–10, 129–130
next generation sequencing (NGS) 43–44, 59
nodal model (grapevine cuttings)
applications 178–179
esca pathogen development in wood 177–178, **178**
induced resistance study 176, **177**
technique **173**, 173–174, 175–176

ochratoxin A (OTA)
content, in grape juice **196**, 198, 203–205
detection and quantification methods 194
effects on human health 191
sources of grape juice/wine contamination 191–192

oligandrin 117, 122
oligogalacturonides (OGAs) 7
oomycetes (peronosporomycetes)
as biocontrol agents 22, 117
pathogens 1, 125, 132
see also Plasmopara viticola; Pythium oligandrum
organic acids, grape juice 199–201, **200**, **201**, 204–205
organic chemical inducers 3–8, **4–5**, 176
organic farming 231
and *Aureobasidium pullulans* on grapes 40–41, 55, **55**
effect on microorganism abundance 54, **54**
OTA *see* ochratoxin A
oxidative burst *see* reactive oxygen species

Pantoea spp. 225
P. agglomerans 21, 42, 220
PTA-AF2 strain 64, 68
P. ananatis 56–57, **59**, 220
pathogens
infection responses of plants 2, 7
interactions with endophytes 36
Penicillium spp.
P. chrysogenum, extract 9, 143
P. verrucosum 183, 185, **187**
peronosporomycetes *see* oomycetes
peroxidase activity
assay method 71
stimulated by *Bacillus* spp. pretreatment **73**, 73–74
pH, grape juice 195, 198–199, **200**, 205
Phaeoacremonium aleophilum (esca pathogen, PAL) 161, 177, 183, **186**
Phaeomoniella chlamydospora (esca pathogen, PCH) 161, 183, **186**
infection and virulence 171–172
PCR quantification in vine wood 175, 177–178, **178**
source of reference strain 172
phloroglucinol derivatives 132–133, **133**
Phomopsis sp. 135, **136**
phylogenetic analysis
Ampelomyces quisqualis strains 114
characterization of *Bacillus* isolates **74**, 74–75
dendrograms constructed from LH-PCR data 219, **222**
phytoalexins
accumulation induced/primed by bacteria 65–67, **67**
detection and measurement methods 65
induced by chitosan treatment 6
phytoplasmas
artificial infection methods 212
detection by nested PCR 211, 217, **218**, 220

phytoplasmas (*continued*)
diseases caused, and treatment 210–211, 215–216
insect vectors 216
interactions with endophytes 37–38, 39, 55, 211–212, 225–226
disease recovery model 226–227, **227**
responses of plants to infection 225
Pichia spp. 17
P. membranifaciens 57–58
plant defence mechanisms 2, 7, 226
genetic control in *Arabidopsis thaliana* 77, 82
induced by natural products 142–143, 192
measurement of gene induction 172, 176, **177**
primed by beneficial bacteria 64, 66–67
plant extracts (inducing resistance) 8–9, 129–130
Plasmopara viticola (downy mildew pathogen)
antagonistic bacteria
Gram-negative 136, **137**
Streptomyces spp. (Gram-positive) 137–139, **138**, **139**
antagonistic fungi
antibiosis by endophytes 133–136, **134**
invasion and parasitism observations 100, 103, 107, **107**
isolation methods 99
biological characteristics 143
oospore development 100–101, 107–108, **108**, **109**
role of zoospores in disease cycle 125, 126
screening of inhibitory substances
detached leaf/leaf disc methods 128, **129**, 139–140
homogenous solution bioassay 127–128
sporangial germination 39, 100, 103–104, **105**, **106**
see also downy mildew
plate diffusion methods (antibiosis activity assay)
dual culture inoculation technique 152, 183, **186**
indirect, using crude protein extract 183–184, 185, **187**
Saccharothrix algeriensis and *Botrytis cinerea* 85, 87, **88**
Streptomyces J-2, and various pathogens 78, 80, **80**
Podosphaera spp. (powdery mildew pathogens) 114, **115**
polyphenols, grape juice 196, 203, 204, **204**
population dynamics
bacterial endophytes 37–38
tracked by quantitative PCR 93
potassium bicarbonate 3
potassium silicate 3
powdery mildew
causal pathogens 114, 117
evaluation of mycoparasite aggressiveness 114–116, **115**
factors affecting spread of disease 120–121
reduced by plant extracts 8
preharvest grape treatments 192
probes, oligonucleotide, for FISH analysis
design, and species specificity 95, 188
fluorochrome labelling 94, 95
universal probe EUB338 for bacteria 183
Propionibacterium acnes 38
protein kinase C (PKC)
inhibitors 140–141, **141**, **142**
role in flagellar motility 141–142
pruning wounds
chemical fungicide treatment, limitations 150
colonization by *Trichoderma* spp. 58, 161, 169
technique for nodal model testing **173**, 173–174
testing method for potential BCAs 153–154
Pseudomonas spp. 21
P. aeruginosa 65, 67, 68
P. fluorescens
DAPG (antibiotic) production 132
PTA-CT2 65, 67, 68
P. putida 65, 68
Pseudozyma flocculosa 17
pyrosequencing analysis 38, 43–44
Pythium spp. 22
P. oligandrum 117
culture filtrate/mycelium extract bioassays 121–122, **122**
genetic diversity of strains 118–119, **119**
oospore treatment and disease control **120**, 120–121, **121**

reactive oxygen species (ROS)
H_2O_2 oxidative burst induced by bacteria 65, **66**
detoxification by catalase 219, 221
produced as infection response 2, 216, 226
tolerance of vine microbes after disease recovery 37, 55
research laboratories 162, 169
resistance inducers (elicitors) 2–10, **4–5**, 52, 126, 176
see also induced systemic resistance
resveratrol
elicitation in grapevine by bacteria 65–67, **67**
induced by inorganic treatments 3
reverse transcriptase (RT-PCR) analysis 79
quantitative (RT-qPCR) 172, 175, 178–179
Reynoutria sachalinensis (giant knotweed) extract 8
rhamnolipids 7–8
rhizosphere
antifungal activities of bacteria 76
bacterial populations in old vineyards 55
colonization by microbes, FISH visualization 95
as source of beneficial bacteria 84–85

RNA extraction (methods) 79, 174
root treatment (grapevines)
inoculation with protection agents, methods 86, 94–95, 118
with potential BCAs, effect on disease
beneficial bacteria 67, **68**, 90
P. oligandrum oospores **120**, 120–121, **121**
ROS *see* reactive oxygen species

Saccharomyces spp. 18
S. cerevisiae extract 9
Saccharothrix algeriensis NRRL B-24137
potential for *Botrytis cinerea* biocontrol 84–85, 87–91, **88**
tracked on grapevine by DOPE-FISH 94–96, **96**
salicylic acid (SA) 2, 8, 226
Salvia officinalis (sage) extract 9, 130
sampling methods
black aspergilli diversity study 192–193, **193**
epigeal organs of grapevine 99
leaves 118, 217
grapes, for berry and juice analysis 193
wood, in nodal model **173**, 173–174
Schizosaccharomyces pombe 18
seaweed *(Ascophyllum nodosum)* extract 8, 130
sequencing, genetic
16S rDNA analysis by PCR, methods 71, 77–78
for characterization of fungal strains 118–119, **119**
leaf gene libraries, grapevine endophytes 42, 216, 217–218, 220
PCR quantification of pathogen development 175, 177–178
for study of defence induction in wood samples **174**, 174–175
tuf gene analysis, for phytoplasma infections 211–212
used for *Bacillus* spp. phylogenetic analysis **74**, 74–75
whole genome 143–144, 171
see also next generation sequencing
Serratia spp.
S. liquefaciens 21
S. marcescens, antibiotics 136, **137**
size analysis, grape berries 195, **195**, 197, **197**
skin thickness, grapes 195, 197–198, **198**, **199**, 204
sodium bicarbonate 3
soils
isolates with potential for biocontrol 71, 76–77, 84–85
suppressive potential in old vineyards 55
types related to *P. oligandrum* strain diversity 118, 119
Solidago canadensis extract 8
spore suspensions
germination assay 152–153
preparation methods 70, 86, 154, 183–184, 194
used for nodal model inoculation 173
see also zoospores
statistical analysis methods 79, 87, 114, 165, 219–220
graphic plots of results 221–225, **224**
staurosporine (antibiotic) 138–140, **139**, 141, 142
Stolbur disease (tomatoes) 212
Streptomyces spp. 21, 76
antibiotic secondary metabolites 137–139, **138**, **139**
S. violaceusniger YCED9 76
Streptomyces ANK302 138
Streptomyces ANK313 137
Streptomyces B5136 (marine strain) 139
Streptomyces J-2 77, 80, **80**, 81–82
subtilisins 188
sugars, grape juice 195, 201, **202**, 205
systemic acquired resistance (SAR) 2, 211

tartaric acid content, grape juice 199, 200–201, **201**, 204
thiamine (vitamin B_1) 142
tomatoes, Stolbur phytoplasma transmission 212
tomentosin 129, **130**
Trichoderma spp. 13–14
T. atroviride I-1237 161, 169
T. harzianum 14, 58, 130, 143
trunk diseases of grapevines (GTDs)
aetiology and research progress 171–172
fungal pathogens 81, 150, 161, 179
recent increase in prevalence 160
slow-developing forms of disease 167–168, **168**, 172
symptoms, severity assessment 162, 165, **165**
testing of antagonists
field evaluation 153–154, 156–157, **157**, **158**, 162–169
potential *Trichoderma* BCAs 13
wood block bioassay method 153, 154–156, **155**, **156**

Ulocladium spp. 14
U. oudemansii 6, 58–59

Verticillium spp.
V. dahliae 183, 185, **187**
V. lecanii 14
vineyard management
comparison of vines in organic and IPM systems 42, 56, **57**
economic sustainability 150

vineyard management (*continued*)
integration of plant protection methods 59–60, 231–232
long-term monoculture consequences 55
low environmental impact strategies 22–23, 142
preharvest treatments for OTA contamination 192, 193
ε-viniferin elicitation by bacteria 66–67, **67**
Viola odorata essential oil 8
viticulture *see* vineyard management
Vitis vinifera L. *see* grapevines
VitiSeal™ paste (wood treatment) 153, 155, **155**
volatile organic compounds (VOCs)
antifungal activity 152, 154, 205
produced by grapevine microorganisms 56, **58**

Wickerhamomyces anomalus 18
wild grapevine species 54, **54**, 125
wood (grapevine) diseases *see* trunk diseases of grapevines

xylem, colonization by pathogens 171

yeast-like fungi as biocontrol agents 14–18, **15–16**, 40
Yucca schidigera extract 9, 130

zoospores
downy mildew infection process 126
inhibition of sporogenesis 142
lysis and motility disruption 132, 133–135, 137–139
observations for bioassay of inhibitors 127–128